8-16-01

Learning to Manage Global Environmental Risks

Learning to Manage Global Environmental Risks

Volume 2

A Functional Analysis of Social Responses to
Climate Change, Ozone Depletion, and Acid Rain

The Social Learning Group

The MIT Press
Cambridge, Massachusetts
London, England

This book was set in Times by Interactive Composition Corporation, and printed and bound in the United States of America.

Printed on recycled paper.

Library of Congress Cataloging-in-Publication Data

Learning to manage global environmental risks / the Social Learning Group.
  p. cm.—(Politics, science, and the environment)
  Includes bibliographical references and index.
  Contents: v. 2. A functional analysis of social responses to climate change, ozone depletion, and acid rain
  ISBN 0-262-19445-7 (v. 2: hc: alk. paper)—ISBN 0-262-69239-2 (v. 2: pbk.: alk. paper)
  1. Environmental policy. 2. Environmental management. 3. Global environmental change. I. Social Learning Group. II. Series.

GE300 .L43   2001
363.7'05—dc21

To the John D. and Catherine T. MacArthur Foundation,
for early and continuing promotion of social
learning about the global environment.

Contents

VOLUME 2

III   STUDIES OF MANAGEMENT FUNCTIONS

15
Risk Assessment in the Management of Global Environmental Risks   7

Jill Jäger with Jeannine Cavender-Bares, Nancy M. Dickson, Adam Fenech, Edward A. Parson, Vassily Sokolov, Ferenc L. Tóth, Claire Waterton, Jeroen van der Sluijs, and Josee van Eijndhoven

16
Monitoring in the Management of Global Environmental Risks   31

Jill Jäger with Nancy M. Dickson, Adam Fenech, Peter M. Haas, Edward A. Parson, Vassily Sokolov, Ferenc L. Tóth, Jeroen van der Sluis, and Claire Waterton

17
Option Assessment in the Management of Global Environmental Risks   49

William C. Clark, Josee van Eijndhoven, and Nancy M. Dickson with Gerda Dinkelman, Peter M. Haas, Michael Huber, Angela Liberatore, Diana Liverman, Edward A. Parson, Miranda A. Schreurs, Heather Smith, Vassily Sokolov, Ferenc L. Tóth, and Brian Wynne

18
Goal and Strategy Formulation in the Management of Global Environmental Risks   87

Marc A. Levy, Jeannine Cavender-Bares, and William C. Clark with Gerda Dinkelman, Elena Nikitina, Ruud Pleune, and Heather Smith

I OVERVIEW

1
Managing Global Environmental Change: An Introduction to the Volume

William C. Clark, Jill Jäger, and Josee van Eijndhoven

2
Acid Rain, Ozone Depletion, and Climate Change: An Historical Overview

William C. Clark, Jill Jäger, Jeannine Cavender-Bares, and Nancy M. Dickson

II STUDIES OF ARENAS

3
Developing a Precautionary Approach: Global Environmental Risk Management in Germany

Jeannine Cavender-Bares and Jill Jäger with Renate Ell

4
Institutional Cultures and the Management of Global Environmental Risks in the United Kingdom

Brian Wynne and Peter Simmons with Claire Waterton, Peter Hughes, and Simon Shackley

Tables, Figures, and Boxes

Volume 1

Tables

Figures

Boxes

Volume 2

Tables

Figures

Boxes

Editors and Authors

(Superscripts refer to institutional affiliations listed on the right)

Editors

William C. Clark[11]
Jill Jäger[12,13,38,14]
Josee van Eijndhoven[21,36]
Nancy M. Dickson[11]

Authors

Jeannine Cavender-Bares[11,13]
William C. Clark[11]
Ellis Cowling[17]
Nancy M. Dickson[11]
Gerda Dinkelman[9,36]
Rodney Dobell[35]
Renate Ell[14]
Adam Fenech[8,5]
Alexandre Ginzburg[1]
Elena Goncharova[22]
Peter M. Haas[31]
Éva Hizsnyik[19,13,3]
Michael Huber[27,10]
Peter Hughes[28]
Jill Jäger[12,13,38,14]
Marc A. Levy[7,37,20]
Angela Liberatore[6,10]
Diana Liverman[24,18]
Justin Longo[35]
David McCabe[15,31]
Donald Munton[33,25]
Elena Nikitina[1,2]
Karen O'Brien[34,16,18]
Edward A. Parson[11]
Vladimir Pisarev[1,2]
Ruud Pleune[36]
Miranda A. Schreurs[30,15]
Simon Shackley[29,28]
Peter Simmons[28]
Heather Smith[33]
Vassily Sokolov[1,2]
Ferenc L. Tóth[19,13,4]
Jeroen van der Sluijs[36]
Josee van Eijndhoven[21,36]
Claire Waterton[28]
Cor Worrell[36]
Brian Wynne[28]

Institutional Affiliations of the Authors during the Project (1991–1999)

1. Academy of Sciences, Russia
2. Academy of Sciences, Union of Soviet Socialist Republics
3. Budapest Institute for Environmental Studies, Hungary
4. Budapest University of Economics, Hungary
5. Centre for Inland Waters, Canada
6. Commission of the European Communities, Belgium
7. Consortium for the International Earth Science Information Network, United States
8. Department of the Environment, Canada
9. Energy Research Center, The Netherlands
10. European University Institute, Italy
11. Harvard University, United States
12. International Human Dimensions Programme on Global Environmental Change, Germany
13. International Institute for Applied Systems Analysis, Austria
14. Jäger International, Germany
15. University of Michigan, United States
16. North American Commission for Environmental Cooperation, Canada
17. North Carolina State University, United States
18. Pennsylvania State University, United States
19. Potsdam Institute for Climate Impact Research, Germany
20. Princeton University, United States
21. Rathenau Institute, The Netherlands
22. Russian Information Agency
23. Stockholm Environment Institute, Sweden
24. University of Arizona, United States
25. University of British Columbia, Canada
26. University of Economics, Hungary
27. University of Hamburg, Germany
28. University of Lancaster, United Kingdom
29. University of Manchester, United Kingdom
30. University of Maryland, United States
31. University of Massachusetts at Amherst, United States
32. University of Michigan, United States
33. University of Northern British Columbia, Canada
34. University of Oslo, Norway
35. University of Victoria, Canada
36. Utrecht University, The Netherlands
37. Williams College, United States
38. Wuppertal Institute for Climate, Environment, and Energy Policy, Germany

Series Foreword

As our understanding of environmental threats deepens and broadens, it is increasingly clear that many environmental issues cannot be understood, analyzed, or acted on simply. The multifaceted relationships between human beings, social and political institutions, and the physical environment in which they are situated extend across disciplines as well as geopolitical confines and cannot be analyzed or resolved in isolation.

This series addresses the increasingly complex questions of how societies come to understand, confront, and cope with both the sources and the manifestations of present and potential environmental threats. Works in the series may focus on matters political, scientific, technical, social, or economic. What they share is their attention to the intertwined roles of politics, science, and technology in recognizing, framing, analyzing, and managing environment-related contemporary issues and their relevance to the increasingly difficult problems of identifying and forging environmentally sound public policy.

Peter M. Haas
Sheila Jasanoff
Gene Rochlin

Foreword

More than a hundred years have gone by since an awareness of human-induced changes of the environment emerged. Air pollution became a political issue when industrial development and uninhibited emissions of smoke and gases in the United Kingdom caused serious social problems. Legislation became necessary in order to protect people's health and living conditions. The first law to prevent damage was passed by the British Parliament in the 1870s. Nevertheless, the use of coal as the prime source of energy continued to increase, and pollution was by no means eliminated. The solution of industrialists was of course to increase the height of chimneys and to spread the pollution over greater distances. Studies of turbulence and mixing in the lowest layers of the atmosphere became a profession of its own of obvious practical importance. The environmental problem of local pollution had been recognized, and means for solving it were developed, but the full scope of the issue was not yet understood. It took until the early 1950s before regional problems were gradually brought into focus. The seriousness of the issue was recognized early in the Los Angeles area, where special meteorological and climatic features of the region were of central importance, as well as in the industrialized parts of Britain and Western Europe. Some may still remember the disastrous smog that hit the London area in December 1954. Still, progress in mitigating and preventing serious incidents of this kind was slow. Damaging emissions often could not be seen far away from the source area and, after all, the incidents were temporary.

Similarly, people were generally unaware of the risk of emissions from nuclear plants. The accident at Windscale in England in October 1957 changed the public attitude drastically. Radioactive iodine originating from the accident was discovered far from the source over the European continent. The invisible radioactive threat became a reality, and the safety problems for the nuclear industry a political issue. Still the regional pollution due to everyday human activities was in general not thought about much.

In the mid-1950s C. G. Rossby recognized the importance of the natural large-scale dispersion by the winds of key chemical trace components of the air. A network of observational sites was established in northwest Europe to determine the chemical composition of the air and precipitation. It was soon shown that the long-distance transport of nutrients was of basic ecological significance. Nitrogen compounds emitted from agricultural activities in Denmark were of some importance for the farmers in southern Sweden. Similarly, it was realized that boreal forests to some extent depend on the supply of airborne nutrients in addition to what nature provides by chemical weathering of the bedrock. This network of observations also provided the data that later were used to study the regional dispersion of sulfur pollution, which led to the discovery of the acidification of precipitation and fresh water systems. The conclusion was obvious: The atmosphere must not be treated as an unlimited wastebasket for human activities. Regional pollution emerged as an important issue for the future. This development was largely science-driven. The pollution could not be seen, and the dangers remained abstract to the general public for a long time. Matters changed when fishing in the lakes of Scandinavia deteriorated and when the forests in central Europe were damaged in obvious ways.

At about the same time, another global environmental issue emerged, again not visible to the general public, and its appearance on the political agenda was therefore completely science-driven: The ozone layer might be damaged because of the emissions of chlorofluorocarbons (CFCs). In this case there were not even data that could validate the theoretical deductions by scientists, but if true it meant a serious threat to life on earth. It was difficult to reach agreements on whether protective actions should be taken now or later. Some modest measures were agreed on, and negotiations for a convention on the protection of the ozone began. Progress was slow, and an agreement of a framework convention, however with no legally binding commitments, was not settled until 1985. The public awareness of the issue was slight. Merely a year later the ozone hole over the Antarctica was discovered. Scientists were able to explain the reason for its temporary appearance, and observations were now available that could validate the theory. Major reductions of the emissions of CFCs were called for, and an agreement was reached within a year. The issue still remained a mystery for most people, except in a few countries in the Southern Hemisphere, where there was a need for protective measures.

A possible global human-induced change of the environment—that is, the composition of atmosphere—already had been recognized implicitly by Svante Arrhenius in 1895, when he showed that changes of the concentration of atmospheric carbon dioxide might change the climate of the earth. The prime aim of his analysis was, however, to explain the last glaciation that had been discovered a few decades earlier, not to warn about human-induced climate change.

A possible human-induced climate change was not generally recognized outside the scientific community for many years. C. D. Keeling was the first to demonstrate clearly that the atmospheric concentration of carbon dioxide was increasing by about 0.5% per year and must have increased significantly (about 10%) above the preindustrial level. It remained, however, exclusively a scientific issue until the middle of the 1970s. At that time—that is, eighty years after Arrhenius's original analysis—Syukuro Manabe showed convincingly that Arrhenius was largely right. From then on, the threat of a possible human-induced global climate change, "global warming," gradually received more attention, first in scientific circles and then also in the political realm.

Global warming is undoubtedly the most complex environmental threat that we have been confronted with so far. Observations now provide considerable evidence that human-induced climatic change is occurring. Although more detailed scientific analyses of what happens now and may happen in the future are needed, the real challenge now is rather to try to understand better how countries and people will respond. How will different segments of society with different interests react? What will be the road from recognition of the issue to concerted action? How can controversies between developed and developing countries and between rich and poor be prevented? What key technical developments will be needed? Which institutions, nationally and internationally, will be required to manage their development without imposing unduly on people and the global market?

The present book does not provide answers to these questions. But it does contribute to a better understanding of the long-term development of efforts to manage interactions between society and the environment. The study looks in depth at the three issues discussed above—acid rain, stratospheric ozone depletion, and climatic change—and traces the evolution of efforts to deal with these issues over the period extending from the International Geophysical Year in 1957 to the United Nations Conference on Environment and Development in Rio in 1992. By taking this long-term perspective and by looking at developments in a range of countries as well as in international institutions, the study is able to illustrate the basis on which current efforts to respond to global environmental change can build. An important contribution of the study is the inclusion of a wide range of actors, rather than just focusing on scientists and legislators. The book itself is an important contribution to social learning about the management of global environmental risks.

Bert Bolin
Stockholm

Preface

This book emerged from the growing recognition during the late 1980s of the need for better understanding of how human societies might perceive, evaluate, and respond to global environmental change. As participants in a number of early attempts to articulate those needs and the research that would be necessary to meet them, we were particularly struck by three shortcomings of existing work. First, although experts, advocates, and political leaders in different parts of the world clearly encountered global environmental change in very different ways, most of our understanding reflected the perspectives of a very narrow range of countries and groups. Second, although society's response to global environmental change was clearly a long-term process unfolding over decades, most of our understanding focused on key discoveries and decisions and paid little attention to the historical connections among them. Third, although most debates on how to improve social response were replete with analogies and lesson drawing, there was little critical discussion of what might be appropriately learned from the experiences of other problems and places. The need to develop a long-term, comparative perspective on the evolution of social responses to global environmental risks—and of the role of learning in that evolution—therefore seemed evident.

The broad outline of the study reported here was developed by one of us, Bill Clark, with planning support from the Stockholm Environment Institute and the U.S. National Science Foundation. At an early stage, the project was taken under the wing of the Committee on the Human Dimensions of Global Environmental Change of the U.S. Social Science Research Council (SSRC). The project benefited substantially from the tough but constructive criticism of the remarkable group of scholars committed to promoting excellence in cross-disciplinary research that the SSRC had assembled. The development and implementation of the project as a truly international team effort, however, was made possible by the John D. and Catherine T. MacArthur Foundation. The Collaborative Studies Program that the Foundation began in 1990 was virtually unique at the time in providing opportunities for sustained multinational, interdisciplinary research teamwork on global environmental problems. We used the opportunity of Clark's stay as Jean Monnet Visiting Professor at the European University Institute in the spring of 1990 to bring together a team of coprincipal investigators for the purposes of developing a proposal to the MacArthur Foundation for a study on Social Learning in the Management of Global Environmental Risks. These individuals—the three of us plus Ida Koppen, Vassily Sokolov, and Brian Wynne—provided an initial core to the Social Learning Group that eventually grew into the collaborative network of scholars listed in the front of this volume. Shortly after notification by the MacArthur Foundation of favorable action on our proposal, the core group accepted an invitation by Dr. Peter de Janosi, director of the International Institute for Applied Systems Analysis (IIASA), to use the Institute's facilities for a retreat to develop detailed plans for implementing the project.

The basic design to emerge from the IIASA meeting was a project organized around teams recruited by the core group members. Each team ultimately consisted of senior scholars, junior faculty, and graduate students in various proportions. Each was selected to allow research on the history of social responses to global environmental risks in a specific arena—initially Germany, the Netherlands, the United Kingdom, the Soviet Union, the United States, the European Community, and the family of international institutions. In particular, we decided to have each arena study develop a comparative analysis of the response of that arena to the three risks of acid rain, stratospheric ozone depletion, and climate change. In addition, a series of management functions were identified that would be addressed in each arena history through a common research protocol. The results would then be synthesized into cross-cutting "function" chapters for the final report, authored by groups consisting of contributors from the arena teams.

The substantive aspects and rationale of this design are described in chapter 1. Procedurally, the project adopted an iterative strategy to promote design and implementation of a truly comparable and comparative approach to research. The core element in this strategy was an annual summer study that brought all project participants together for a week of intensive discussions in plenary and smaller drafting group sessions. For the first iteration of the strategy during the spring and early summer of 1991, the draft research protocol developed by the core group at the initial IIASA meeting was applied by each arena team to the case of climate change. Our first summer study, held in 1991 at Bad Bleiberg in Austria, reviewed these results, revised the research protocol, recommended improvements in project management and direction, and planned a series of smaller meetings on cross-cutting topics for the following year. In an effort to expand the range of countries studied by the project, scholars familiar with the response to global environmental issues in Japan, Hungary, and Mexico were invited as observers to Bad Bleiberg and, based on their contributions there, were subsequently asked to join the group as full partners. Subsequent summer studies were held in 1992 at Canada's Dunsmuir Lodge (focused on the ozone case), in 1993 at Germany's Wuppertal Institute (focused on the acid rain case), and in 1994 again at Dunsmuir Lodge (focused on cross-case comparisons and further work on the climate change case). Within the broad structure provided by the annual projectwide summer studies, smaller and shorter meetings were held periodically to advance work on particular cross-cutting themes and chapters. From the earliest (pre-Internet) days of the project, an active electronic network was used to bind the group together between meetings.

In parallel with our largely internal cycle of summer studies and smaller technical meetings, we engaged in a continuing program of outreach to bring the insights of a larger community of scholars and practitioners to bear on the project's evolution. As already noted, the SSRC organized an early critical review of the conceptual foundations of the study. A few outsiders were usually invited to our summer studies and provided valuable independent criticism—at least until they were coopted into the project as full-time participants. In May of 1993 we brought together at the European University Institute a small group of distinguished scholars who had written on various aspects of social learning—Emanuel Adler, Klaus Eder, and Sheila Jasanoff—and asked them to review the project's preliminary findings and to advise us on potential orientations and audiences for what has become the present book. A presentation to Canadian government officials after the 1994 summer study provided valuable feedback on some of our emerging conclusions about the practical implications of our study for risk management. The synthesis chapters were initially presented in draft form at the First Open Meeting of the Human Dimensions of Global Environmental Change Research Community at Duke University in 1995. An even broader audience was addressed through a panel presentation organized at the annual meeting of the American Association for the Advancement of Science in 1996.

Taken together, our joint activities at these internal and external project meetings forged the Social Learning Group—a multidisciplinary, multinational collaborative team that, though replete with differences of opinion, perspectives, and research styles, nonetheless developed a shared set of concepts, data, methods, and commitment to the larger project. More prosaically, the meetings produced continuing refinements of the research protocol and project design, began drafting and reviewing the cross-cutting studies of management functions, and added an arena study on Canada to the overall project. By the end of the 1994 summer study we had reached decisions on a table of contents for the project's final report, adopted length targets and editorial guidelines for the written material, negotiated responsibilities for completing those chapters, decided how credit would be allocated to contributors, and designated an editorial board consisting of the three of us plus Nancy Dickson to shepherd the final report to publication. Over the next two years, first drafts of most of the chapters included in this volume were completed, subjected to internal review, and revised extensively. The editors met again at IIASA in the summer of 1995 to review progress and—in light of the research results then emerging—to prepare a substantially revised outline of three synthesis chapters to pull results of the study together. This outline was circulated to other Group members for suggestions, with drafting of the revised synthesis chapters beginning in early 1996.

Any project as wideranging and multifaceted as that undertaken here faces enormous challenges of quality control. The basic housekeeping of fact checking, sequence verification, and secondary sources confirmation have been daunting—all the more so given the absence of consolidated archives for the relevant documents and the long historical time span and multiple languages involved in the study. Well aware from our initial research of the high proportion of elementary factual errors in the literature dealing with the history of global environmental change, the Group imposed on itself early on an especially rigorous program of peer review. For each of the core arena and function chapters, this has meant not only critical reading by a cross-section of project members and editors but also an external blind review by at least three external reviewers. These reviewers were selected by the editors in consultation with Group members for their familiarity with both the factual and conceptual aspects of the chapter. Care was taken that the reviewers selected for each chapter represented a wide range of national and disciplinary backgrounds. The reviews were blind in that only the editors, not the authors, were aware of the reviewers' identities. Beginning in the summer of 1996, most chapters had passed internal review and were ready for this external process. Reviewers were given a set of specific questions to answer and asked to reply in writing. They did—often at great length. The editorial board returned reviews to the chapter authors and monitored revisions to make sure that reviewers' concerns were addressed. Many of these revisions were relatively minor matters of fact, attribution, and emphasis. Several, however—especially in the function chapters—constituted major rewriting of the material. Revisions to reflect outside review were carried out through 1997. Final work on the synthesis chapters to incorporate those revisions was completed the following year. Along the way, results were critiqued by participants at the 1997 Bologna Summer School in Environmental Policy, participants in the 1998 meeting of the European Forum on Integrated Environmental Assessment, and members of a panel on Learning and Belief Change among Policy Elites held at the 1998 Annual Meeting of the American Political Science Association.

The final product that emerged from the Group project in the form of this book is a uniquely collaborative endeavor. The question of authorship therefore merits special comment. One of the toughest challenges in running any collaborative project is to balance the individual scholar's need for individual credit, the project's interest in getting collaborators to share their ideas and criticism with one another, and the practical requirement that someone have incentive and authority to focus and complete a multicontributor work. This challenge is particularly acute in projects such as this one that involve collaborators ranging from senior professors to junior graduate students. But to the extent that it cannot be satisfactorily resolved, collaborative research will remain the exception rather than the rule. This project took very seriously the need for collaboration and the challenges of devising appropriate incentives and credits to promote it.

We concluded, reasonably amicably, as follows. The chapters in this book grew out of multiple working group meetings by our arena teams and at our summer studies and other meetings. Each has drawn on the primary research memos and draft text of many individuals. In the end, however, one or a few individuals have taken responsibility for giving the chapter its present form. These individuals are listed as the lead author(s) on each chapter. In most cases, they share credit "with" a second group of authors who contributed to the conceptual content, but not the specific language, of the chapter. Finally, most chapters list in their first endnote a series of acknowledgments to others who contributed primary research or criticism on which the chapter is built but who did not shape the chapter in its present form. The resulting impression that many members of the Group contributed in multiple ways to multiple chapters is both intended and true. The book as a whole is presented and cataloged as a work of the corporate author the Social Learning Group in recognition of the collaborative character of not only the writing but the design and execution of the overall study. All of the contributors listed at the front of this volume are members of the Group and thus authors of the book. Finally, the designation of editors for the book as a whole reflects the Group's recognition of the extra effort needed from a few people in drawing the physical book together and ensuring consistency throughout.

A gratifyingly large number of people and institutions have supported this project in a variety of ways. Core funding, as already noted, was provided by the John D. and Catherine T. MacArthur Foundation. Other major supporters of the project included the U.S. National Science Foundation, Canada's University of Victoria, the Netherlands Energy Center, the IBM Foundation, the Canadian Atmospheric Environment Service, the German Ministry for Research and Technology, the European University Institute, the Stockholm Environment Institute, and Utrecht University. Additional support was provided by the German Research Society, the U.S. Social Science Research Council, the USSR (later Russian) Academy of Sciences, the U.S. Department of Energy, the U.S. National Institute for Global Environmental Change, the International Institute for Applied Systems Analysis, Germany's Wuppertal Institute for Climate, Environment, and Energy, the Mobil Foundation, the U.K. Economic and Social Research Council, the Fulbright Foundation, and the Hungarian National Scientific Research Fund. Finally, the home institutions of the Group members, listed at the front of this volume, contributed more than most of them know to making the project possible.

For their contributions as participants in the external reviewer process, the project is indebted to Robert Boardman, Harvey Brooks, James Bruce, Tom Brydges, Lynton Caldwell, Peter Chester, Ellis Cowling, Peter Fabian, Tibor Farago, Carlos Gay, Anver Ghazi, George Golitsyn, Len Good, Loren Graham, Hartmutt Grassl, Nigel Haigh, Maarten Hajer, Leen Hordijk, W.J. Kakebeeke, Yoichi Kaya, P.M. Kelly, Jeremy Leggett, Ronnie Lipschutz, Mike MacCracken, Gordon MacDonald, Margaret McKean, Erno Meszaros, Alan Miller, Mario Molina, William Moomaw, Tsuneyuki Morita, Friedemann Mueller, Stephen Mumme, Ted Munn, Hiroshi Ohta, Michael Oppenheimer, Tim O'Riordan, Ian Rowlands, Milton Russell, Roberto Sanchez, Peter Sand, Rolf Sartorius, Steve Schneider, Toni Schneider, Ian Simms, Udo Simonis, Rob Swart, Peter Thacher, Arild Underdal, Peter Usher, David Victor, Arpad von Lazar, Konrad von Moltke, Helmut Weidner, Gilbert White, Pieter Winsemius, George Zavarzin, and Charles Ziegler. Numerous other individuals contributed critical insights on individual chapters and are named in those chapters' acknowledgments.

The production of a volume involving dozens of authors, several languages, numerous time zones, and a variety of word processing programs is a task that no personnel officer would allow in a job description. Fortunately, this project has been supported by a cast of the least flappable and most efficient, resourceful, and downright nice people we have ever had the pleasure of working with: Kristen Eddy, Nora O'Neil, Bonnie Robinson, Rebecca Storo, and Ingrid Teply-Baubinder. This is their product, too, and we are immensely grateful for their support and good cheer through the toils of bringing it to fruition.

The Social Learning Group itself was a unique collection of scholars. This collaboration has had its own decade-long social history—one marked by trials and errors, hurrying and waiting, job changes and promotions, and unprecedented changes in the worlds inhabited by our Russian and Hungarian colleagues. Through it all, the Group's inventiveness, energy, commitment to understanding one another, and willingness to subsume individual agendas within a common endeavor defined for us a new standard for international, interdisciplinary collaboration. That many Group members did this while they completed dissertations and as their families grew at rates far exceeding those at which the study progressed is all the more testimony to the remarkable cast of characters the project entrained. Finally, we must single out for special mention our coeditor of this volume, Nancy Dickson. She joined the Group as project manager at its first summer study and, surviving that baptism under fire, has been its chief cat herder ever since. But she did far more than the complex and often thankless tasks of designing and maintaining our communications, organizing our meetings, pushing our schedules, and supervising the production of this book. She also emerged as a resourceful and accomplished researcher in her own right, as indicated by her coauthorship of several of the chapters in this volume.

For what they have accomplished, and for the colleagues and friends they have become, we are grateful to Nancy and all the members of the Group in more ways than we will ever be able to express.

William C. Clark
Jill Jäger
Josee van Eijndhoven

III
STUDIES OF MANAGEMENT FUNCTIONS

Introduction

Part III of the book focuses on risk management "functions." As indicated in chapter 1 (Managing Global Environmental Change: An Introduction to the Volume) the "functional" characterization of issue development is derived largely from the literatures of policy analysis and risk management. While remaining skeptical of the linear or sequential relations among stages of issue development and the performance of particular tasks that is assumed in much of the literature, we nonetheless found particularly useful the common functional categories adopted by works as different as Kates, Hohenemser, and Kasperson's (1985) studies of technological hazards and Kay and Jacobson's (1983) early work on international environmental policy. Kay and Jacobson argue, and we agree, that through its focus on *what* is done rather than *who* does it, this functional framework is particularly appropriate for long-term comparative studies in which comparability of actor groups and institutions might otherwise be problematical. Our functional framework for the description of issue development, somewhat modified from that of Kates and Kay and Jacobson, addresses the following management activities:

• Risk assessment: "What is the problem?"

• Monitoring: "What is happening?"

• Option assessment: "What could be done?"

• Goal and strategy formulation: "What should be done?"

• Implementation: "What is being done?"

• Evaluation: "How are we doing?"

A Taxonomic Framework

Our studies of these individual risk management functions used a common taxonomic framework to characterize and classify the content of discourse about global environmental issues and their management. Did scientists present end-to-end, "integrated" assessments of the issue, or did they concentrate on particular facets of the overall story? Did policy advocates focus on measures to address causes or effects? Did controversies range over all aspects of the issue, or were they more narrowly confined? As pointed out in chapter 1, the beginnings of a taxonomic framework that would allow classification of empirical evidence relevant to such questions was developed in the 1980s by scholars of technological-hazard analysis (Kates, Hohenemser, and Kasperson 1985) and environmental-impact assessment (Beanlands and Duinker 1983). Figure IIIA summarizes the taxonomy developed from those initial works for this project. This framework is used in most of the chapters of Part III.

Panel A of figure IIIA lists our categories for classifying discourse about environmental issues. Our use of these categories is intended to be purely descriptive; no ordering or priority in how or when society addressed them is assumed:

• *Demand for goods and services* Any environmental concern (such as energy) may be traced back to its origins in human demands for goods and services. Conversely, the environmental implications of particular social demands for goods and services may be explored;

• *Choice of technologies and practices* The implications for the environment of particular technologies or practices (such as coal versus natural gas fuels) may be discussed, with selection driven by interest in the technologies themselves, as a means for meeting basic demands, or in sources of pollutants of concern;

A. Issues	Demand for goods and services	Choice of technologies and practices	Flux of materials	Valued environmental properties	Exposure of people and things	Consequences to people and the things they value
B. Actions (Options)	Change demand	Change choice	Change flux	Change environment	Change exposure	Change consequences
C. Groups of actions (options) used in this study	Emissions		Environment		Impacts	
D. Other groups of actions used by actors documented in this study	Mitigation options				Adaptation options	
	Preventive		Offset		Adaptation	
E. Framing categories used in this study	Causes		Environment		Impacts	

Figure IIIA

A taxonomy of hazard management

• *Flux of materials* The release of certain materials (such as sulfur dioxide, chlorofluorocarbons, and carbon dioxide) to the environment may become the subject of attention—perhaps in their own right, perhaps as a possible threat to valued environmental properties, or perhaps as a possible consequence of certain development choices people make;

• *Valued environmental properties* Certain properties of the environment (such as global climate, stratospheric ozone, and precipitation acidity) are singled out by scientists, advocates, or political leaders as meriting concern;

• *Exposure of people and things* Discussions of global environmental change (such as coastal locales exposed to global sea-level changes) may highlight the exposure of specific local places to different sorts of stresses;

• *Consequences to people and the things they value* People may discuss possible impacts of global environmental change (such as crop loss and health implications) on themselves or on other things they value.

Panel B of figure IIIA lists our basic categories for classifying discourse about actions that might be undertaken in response to concern for environmental issues. It simply reflects the obvious but important fact that actions could in principle be undertaken within every one of the categories used to characterize the issue itself (Schelling 1983). We employed this symmetrical classification in our basic research protocol and analysis.

For this study it was important to distinguish what advocates of particular actions were actually talking about from how they were seeking to package their proposals. We therefore focused our descriptive taxonomy of action proposals on the same basic categories outlined above. When grouping was called for, we adopted the relatively neutral and descriptive terms shown in figure IIIA, panel C:

• *Emissions* This category captures measures (such as energy taxes and bans on CFC propellants) that would directly affect emissions of pollutants of interest through changing demand or changing the choice of technologies and practices.

• *Environment* This category captures measures (such as carbon sequestration though forest plantations and liming of acidified lakes) that would directly affect the amount of emissions remaining in the environment or would alter valued environmental properties.

• *Impacts* This category captures measures that alter the impact of changes in the environment on people and things they value. Such measures (such as shielding people from ultraviolet radiation and air conditioning places where people work) can work by changing exposure or changing vulnerability.

Panel D of figure IIIA shows how our descriptive taxonomy relates to various categories of actions used by the actors we studied in their discussions about global environmental problems. Finally, panel E introduces terminology employed in our analysis of issue framing and relates this to the other categories and underlying descriptive taxonomy.

Evaluating Functional Performance

Each chapter in Part III asks whether functional performance is getting better over time. To answer that question some evaluation criteria were required. As discussed in chapter 1, we use the metacriteria *adequacy, value, legitimacy,* and *effectiveness* in Part III for critical discussion on the question of what might be meant by "improvements" or "progress" in the performance of risk management functions.

Our working definitions of these metacriteria (Clark and Majone 1985) are as follows:

• *Adequacy* The role of criteria of adequacy is to permit the accumulation of certified facts, thus providing what historian Oscar Handlin has called the "grounds for peaceful discourse." Two potential uses of such criteria stand out as particularly relevant for efforts to link knowledge with action in the management of global environmental risks. The first is the simple posting of known pitfalls: methodological blunders and inappropriate use of data that immediately vitiate any assessment that fails to avoid them. The second is the channeling of disputes into well-defined categories where focused and informed discussion can be carried out.

• *Value* The role of criteria of value is to help channel inquiry into important areas where it has some prospect of making contributions that extend beyond the immediate gratification of those performing the inquiry. At one level, such criteria address such commonsense notions of worth or relevance. At another, somewhat deeper level, they include evaluations of feasibility, encompassing exhortations from a number of fields that temper inclinations to attack only the really important problems with due respect for "the art of the possible."

• *Legitimacy* In political contexts, legitimacy rests on questions of majority and minority and how to control the treatment of the latter by the former. In scientific contexts, it has been centrally bound with "the fair play of ideas," and how skeptical questioning of accepted interpretations, can be simultaneously encouraged yet kept from arbitrarily dismantling consensual understanding.

• *Effectiveness* The role of criteria of effectiveness is simply to evaluate whether knowledge- or action-based efforts undertaken to help resolve problems actually do so. Effectiveness is viewed not in terms of the creation of solutions but rather in terms of the ability of a given endeavor to shape the agenda or advance the state of the debate.

Linking Risk Management Functions

In part III the performance of the risk management functions is discussed in six separate chapters (chapters 15 to 20), each of which includes specific examples of performance, more general patterns and a discussion of whether performance is improving over time. Chapter 21 analyzes how linkages among the risk management functions have changed as the issues moved from the science agenda onto the policy agenda.

References

Beanlands, G.E., and P.N. Duinker. 1983. *An Ecological Framework for Environmental Impact Assessment in Canada.* Halifax: Institute for Environmental Studies, Dalhousie University.

Clark, William C., and Giandomenico Majone. 1985. The critical appraisal of scientific inquiries with policy implications. *Science, Technology, and Human Values* 10(3): 6–19.

Kates, Robert W., Christoph Hohenemser, and Jeanne X. Kasperson, eds. 1985. *Perilous Progress: Managing the Hazards of Technology.* Boulder: Westview Press.

Kay, D.A., and H.K. Jacobson. 1983. *Environmental Protection: The International Dimension.* Totowa, N. J.: Allanheld, Osmund.

Schelling, Thomas C. 1983. Climatic change: Implications for welfare and policy. In *Changing Climate*. National Research Council. Washington: National Academy Press.

15

Risk Assessment in the Management of Global Environmental Risks

Jill Jäger with Jeannine Cavender-Bares, Nancy M. Dickson,
Adam Fenech, Edward A. Parson, Vassily Sokolov, Ferenc L. Tóth,
Claire Waterton, Jeroen van der Sluijs, and Josee van Eijndhoven[1]

15.1 Introduction

A risk assessment provides information about the causes, possible consequences, likelihood, and timing of a particular risk. Risks by definition involve uncertainties, and especially for global environmental processes these uncertainties are so large that the usual features of risk assessment—namely, the calculation of probabilities of specific harm from particular activities, natural or manmade—are swamped by larger uncertainties and ignorance about key processes, interactions, and effects (Funtowicz and Ravetz 1992; Wynne 1992). Risk assessment as a formal quantitative tool originated in the analysis of the risks of failure of well-defined mechanical systems such as aircraft and industrial engineering plant. The endpoints of such analysis were defined accidents such as massive releases of toxic chemicals or radioactive isotopes into the environment.

By looking at risk assessments for global environmental issues made at different times in various countries, it is possible to trace differences and changes in understanding about the nature, causes, consequences, likelihood, and timing of the risk in question. These differences and changes bear implications for other risk management functions. For example, if a new risk assessment concludes that an environmental change will reach a certain magnitude within the next thirty years instead of the 100 years previously assumed, goal and strategy formulation could change. If a new risk assessment concludes that an important cause has been neglected, response options assessments might have to be modified to take account of this new information.

To trace changes or differences in risk assessments, a number of indicators are very useful. One indicator, which makes cross-arena and cross-actor group comparisons possible, is the time at which a certain way of characterizing the risk becomes established. For example, in the case of the risk of anthropogenic climate change, the time at which the risk became characterized as a greenhouse gas issue as opposed to a carbon dioxide (CO_2) issue is an important indicator. Equally important are estimates of the probability or timing of the risk. Risk assessments for the issue of stratospheric ozone

depletion, for example, show important changes over time in the assessments of the probability of the risk. In all cases, it is important to note the origins of innovations in characterizations and estimations so that it is possible to trace the innovations with their selection and diffusion across arenas and actor groups. Finally, it is important to look for evidence of confirmation of the risk assessment, either in studies using other methodologies, at international meetings, in peer-reviewed publications, and so on.

As discussed in the introduction to part III of this book and illustrated in figure IIIA of that introduction, risk assessments can be made for one or a number of elements within the *causal chain* of the risk in question. The first link in this causal chain is the demand for goods and services—for example, a warm house, a means of getting from Vancouver to Wuppertal, or a refrigerator. Although assessments of the demand for goods and services have been made, few major innovations in methodologies or input have been implemented over the past thirty years. The second link is the choice of technologies and practices. For example, the magnitude of the risk of climate change depends on whether the demand for energy is met by fossil or nonfossil fuel combustion. Similarly, the demand for refrigeration can be met by refrigerators using chlorofluorocarbons or their substitutes. Risk assessments have looked at the question of choice of technologies and practices in a variety of cases. For example, assessments of the rate of tropical deforestation influenced the debate on anthropogenic climate change from the mid-1970s onward, while assessments of the role of supersonic air transportation at the beginning of the 1970s led to dramatic improvements in the understanding of stratospheric chemistry.

The choice of technology or practices leads to a change in the fluxes of material in the environment. In the case of the three risks in this study, most of these fluxes are between the earth's surface and the atmosphere. Fossil fuel combustion and tropical deforestation lead to an increased flux of carbon dioxide into the atmosphere. Fossil fuel combustion can also lead to an increased flux of sulfur dioxide into the atmosphere depending on the choice of technology, and similarly the choice of propellant technology is linked with the flux of chlorofluorocarbons

(CFCs) into the atmosphere. Assessments of the changes of such fluxes have been central to the studies of the three global environmental risks.

The next link in the causal chain is valued environmental components. That is, as a result of changes in material fluxes other environmental components—such as the acidity of the precipitation, soils and groundwater, the global average surface temperature, the sea level, the amount of ultraviolet (UV) radiation at the earth's surface—are changed. These changes are generally the main cause for interest in the global environmental risk in question. In particular, assessments of the risk of global climate change have often focused on the global average surface temperature.

The changes in valued environmental components are linked to the consequences via differential exposure. There are latitudinal differences in exposure to both temperature and ultraviolet radiation changes, while the exposure to changes in acidity of the precipitation depend on the existing buffering capacity of the soil. This leads then to the last link in the causal chain: the consequences, such as changes of ecosystems, agricultural production, human mortality and morbidity, property values, water supply, and water quality.

This framework shows which aspects of the global environmental risks have to be assessed and how they are linked. Most risk assessments have not included all links but have concentrated on parts of the causal chain.

The rest of this chapter is divided into four parts. Section 15.2 looks at a selection of risk assessment stories and traces important innovations that were made in risk assessments and how they spread. These illustrate both the process of diffusion and significant links to other risk management functions. For example, changes in the assessment of how urgent the problem is led to new goals and strategies. Section 15.3 answers a number of more general questions on the observed patterns in risk assessment. An analysis of whether there is evidence that risk assessments improve with time and whether improvements lead to better management of the risk is presented in section 15.4. Finally, section 15.5 presents the main conclusions and discusses the major links to other risk management functions. The empirical material on which this chapter is based consists largely of major published risk assessments conducted by individuals and national and international bodies.

15.2   Risk Assessment Stories

15.2.1   Introduction
This section presents ten stories that illustrate various aspects related to the role of risk assessment in the management of global environmental risks. Each of the global environmental risks is taken as an example: three stories address the climate issue, then three stories involve stratospheric ozone depletion, and four involve acid rain. The stories share some common features. For each of the issues there is an example of how risk assessments affected the perceived urgency of a response: in the climate case by additionally considering gases other than carbon dioxide; in the ozone depletion case by including new knowledge on the lifetimes and potentials of ozone-depleting substances; and in the acid-rain case by considering the cumulative effects instead of individual effects. Similarly, for each of the cases there is a story about the assessment of impacts: on sea-level rise in the climate case, on human health in the ozone case, and on multiple stress effects in the case of acid deposition. These stories illustrate the linkages (or lack of them) between risk assessment and other risk management functions and the diffusion of some assessments from one arena to the other.

Lastly, some of the stories illustrate the nature of the risk assessment process itself. In the climate case, the assessments of the temperature increase due to a doubling of the carbon dioxide concentration arrived at the same range since 1979, despite major advances in climate modeling. In the ozone case, the impact of an early, major assessment on the subsequent management of the risk is examined. In the acid rain case, two stories are presented to illustrate in the first case the diffusion of a set of ideas less comprehensive than those held in previous years and in the second case the diffusion of the idea that acid pollutants can be transported over long distances.

15.2.2   The Introduction of Non-Carbon Dioxide Greenhouse Gases into the Climate Change Debate
For many years the risk of anthropogenic climate change was seen as a "CO$_2$ problem." After it was recognized that other greenhouse gases played an important role, the issue became more urgent. The change in risk assessment led to a new goal and strategy formulation after the Villach Conference in 1985.

For a long time, interest in the risk of the enhanced greenhouse effect focused on the increase of the atmospheric carbon dioxide concentration. Very little attention was devoted to the other greenhouse gases.

In its assessment of the climatic effects of human activities, the Study of Critical Environmental Problems (SCEP 1970) considered a wide range of gases but concluded that the gases do not have globally significant impacts except when they form particles. The Study of Man's Impact on Climate (SMIC 1971) basically reconfirmed the findings of SCEP.

In 1975 the World Meteorological Organization (WMO) report on climate and climate modeling concluded that constituents like N_2O, CH_4, etc. are present in such small concentration that their direct radiative effects would be negligible and that these trace gases could have only an indirect effect on the energy budget of the planet (through participating in the photochemistry of ozone or the production of particulate matter) (WMO 1975).

V. Ramanathan, at the U.S. National Center for Atmospheric Research (NCAR), proved the fallacy of this theory in 1975 when he tested CFCs for infrared absorption and found that on a molecule-for-molecule basis the CFCs were 10,000 times more effective than carbon dioxide in trapping heat (Ramanathan 1975). He speculated that this enhancement may lead to an appreciable increase in the global surface temperature (0.9°C), if the atmospheric concentrations of these compounds reach values in the order of 2 parts per billion. He based this statement on Lovelock's 1974 publication that showed that recent measurements indicated substantial concentrations of CCl_4 within the troposphere and the suggestion of the presence of other chlorocarbons within the atmosphere (Lovelock 1974). Ramanathan's model indicated a 2.0°C temperature increase in response to the doubling of CO_2 and an additional 1.6°C warming when changes in CFC-12 and CFC-13 are included (Ramanathan 1975).

In 1976, atmospheric physicists at Goddard Institute for Space Studies recognized that N_2O, CH_4, NH_3, HNO_3, C_2H_4, SO_2, CCl_2F_2, CCl_3F, CH_3Cl, CCl_4, H_2O, CO_2, and O_3 are greenhouse gases (Wang, Yung, Lacis, Mo, and Hansen 1976). Their analysis discussed the nature and climatic implications of possible changes in the concentrations of N_2O, CH_4, NH_3, and HNO_3 and focused on two major anthropogenic perturbations: stimulation of agriculture by chemical fertilizers and the combustion of fossil fuels. Wang et al.'s 1976 model predicted equal roles for CO_2 and trace gases in raising the average global surface temperature 0.8°C by 2010 and predicted a 2°C warming for a doubling of CO_2. They calculated that the warming due to increased CO_2 would be 48 percent of the total and that the warming due to the direct radiative effects of the trace gases other than ozone would be 26 percent of the total. The total amount of ozone would contribute the remaining 26 percent of the warming (Wang, Yung, Lacis, Mo, and Hansen 1976).

During the following years, these results were picked up and used by scientists in the United States and other countries. In 1978 at an international workshop held at the International Institute for Applied Systems Analysis in Austria, Hermann Flohn from Germany used the U.S. results to compute a *combined greenhouse effect* that showed that by adding the effects of the other gases, the timing of a temperature change equivalent to that caused by a doubling of the CO_2 concentration was significantly advanced (Flohn 1978). However, at the end of the 1970s and beginning of the 1980s there was still a tendency to refer to the "CO_2 problem" and ignore the proven capacity of other greenhouse gases to change the earth's radiation balance.

In 1982 there was international scientific recognition of the importance of trace gases. A meeting of experts on Potential Climatic Effects of Ozone and Other Minor Trace Gases was held September 13 to 17, 1982, in Boulder, Colorado (WMO 1982). The WMO panel of experts concluded that, remaining uncertainties not withstanding, the available studies strongly suggested that the combined climatic effects of potential future alterations in minor trace gases can be as large as those estimated due to a CO_2 increase.

Important international acknowledgment of the effect of non-CO_2 greenhouse gases came at the UNEP, WMO, and ICSU International Conference on the Assessment of the Role of Carbon Dioxide and of Other Greenhouse Gases in Climate Variations and Associated Impacts held in Villach in 1985 (UNEP, WMO, and ICSU 1986). The Conference concluded that "The role of greenhouse gases other than CO_2 in changing the climate is already about as important as that of CO_2. If present trends continue, the combined concentrations of atmospheric CO_2 and other greenhouse gases would be radiatively equivalent to a doubling of CO_2 from preindustrial levels possibly as early as the 2030s" (2). This risk assessment led to the recommendation for both further research and policy development in response to climatic change.

The fact that other trace gases exert a greenhouse effect *and* have an increasing atmospheric concentration was scientifically established in the United States in the mid-1970s. In the United States the greenhouse effect of CFCs was also discussed in political debate at the end of the 1970s. Although individual scientists in other countries picked up the information and developed it further by the end of the 1970s, the information influenced the international debate on anthropogenic climatic change only in the mid-1980s, ten years after it was picked up in scientific circles. The initial preparations of the Villach Conference held in 1985 were concerned with CO_2. After the other greenhouse gases were included, there was a general realization that the problem was more acute than previously assumed, and a call was made to examine policy responses.

15.2.3 The Carbon Dioxide Doubling Temperature

All major climate change risk assessments have included an estimate of the global temperature change that would occur if the atmospheric carbon dioxide concentration were to double from its preindustrial concentration. The range adopted by the U.S. National Academy of Sciences (NAS) in 1979 has dominated the assessments.

Swedish chemist Svante Arrhenius made the first assessment of the effect of increasing carbon dioxide concentration on global temperature in 1896. He was concerned that the rapidly increasing rate of fossil fuel use in Europe would result in increasing concentrations of CO_2 that would alter the thermal balance of the atmosphere. He estimated that if the atmospheric concentration of CO_2 doubled, the surface of the planet would warm by about 5°C at the equator (annual mean) rising to about 6°C at 70°N and 60°S (Arrhenius 1896).

An often-cited estimate of the effect of CO_2 was made by the British scientist Callendar (1938), and this estimate stimulated the German scientist Hermann Flohn to begin looking at the impact of human activities on the climate.

After controversies about the role of carbon dioxide compared with water vapor, Gilbert Plass resuscitated Callendar's work. Plass was keenly aware of the estimates of Arrhenius and Callendar when he conducted this research. In 1956 Plass made laboratory measurements of the absorption in the CO_2 band and calculated the radiation flux in the atmosphere with the aid of a computer. None of these approximations had been attempted before. With this new information, Plass calculated that the surface temperature must rise 3.6°C if the CO_2 concentration is doubled (Plass 1956).

In the early 1960s a German scientist estimated that a doubling of the CO_2 concentration would lead to a temperature increase of 9.6°C (Möller 1963). However, Möller's model assumed that relative humidity remained constant, which added a strong positive feedback.

The differences in the early CO_2 doubling estimates can be attributed to the different assumptions made and the different feedback processes that were taken into account.

More recently, the climatic effects of CO_2 were studied using climate models. Several types of models have been developed, ranging from one-dimensional globally averaged energy balance models (EBMs) and radiative-convective models (RCMs) to comprehensive three-dimensional general circulation models (GCMs). In the last twenty years, numerous studies were conducted using GCMs. These studies have been conducted in the United States (at four institutions), the United Kingdom, Germany, Russia, Canada, Australia, and Japan.

In 1979 a group of scientists met under the auspices of the U.S. National Academy of Sciences (NAS) and concluded that a doubling of the atmospheric CO_2 concentration would lead to a globally averaged surface temperature increase of 3° plus or minus 1.5°C. This range was used in subsequent years in a large number of national and international assessments of the risk of global climatic change.

After 1979 there were major changes in the climate models (GCMs) used in climate change risk assessments, especially by including ocean features, cloud modeling, and longer-term computations. Despite all of these changes, the range presented for the CO_2-doubling temperature did not change.[2] The major scientific review of the Intergovernmental Panel on Climate Change (IPCC) (Houghton et al. 1990) concluded that the range was 1.5° to 4.5°C. While dissenting views were expressed by individual scientists (e.g., Idso 1981), the scientific consensus continued to affirm the range found by the NAS in 1979.

In the 1990 IPCC assessment (Houghton, Jenkins, and Ephraums 1990), the model results compiled there showed results ranging from 1.9° to 5.2°C for a doubling of the CO_2 concentration. A selection of these model results was made by taking only "recent" results. However, as van der Sluijs, van Eijndhoven, Wynne, and Shackey (1998) have shown, this selection process was inconsistent. The IPCC (Houghton, Jenkins, and Ephraums 1990) concludes that "On the basis of evidence from the more recent modeling studies . . . it appears that the equilibrium change in globally averaged surface temperature due to doubling CO_2 is between 1.9° and 4.4°C. The model results do not provide any compelling reason to alter the previously accepted range 1.5° to 4.5°C" (Houghton, Jenkins, and Ephraums 1990, 138). Similarly, the 1992 supplement to the IPCC Working Group I report concluded that there was no compelling new evidence to warrant changing the equilibrium sensitivity to doubled CO_2 from the range of 1.5° to 4.5°C as given by IPCC 1990.

Van der Sluijs, van Eijndhoven, Wynne, and Shackley (1998) concluded, however, that those making the assessments persisted with the same range of 1.5° to 4.5°C by narrowing "the domain of types of uncertainty" and by screening out GCM-results that lay outside of the previously accepted estimate. Van der Sluijs et al. suggest that the experts were not necessarily engaged in these processes consciously but were responding to contingencies that were broader than just scientific considerations. They conclude that the remarkable stability of the climate-sensitivity range may be significant as a way of holding together the fragile process of building a global policy community.

15.2.4 Sea-Level Rise as a Result of the Melting of the West Antarctic

Concern about the instability of the West Antarctic Ice Sheet led scientists to predict that a large rise of sea level could occur within the next 100 years. After the mid-1980s the assessments concluded that the risk of an imminent catastrophic sea-level rise was very low. The change in risk assessment influenced the assessment of response options and strategy formulation with regard to sea defenses.

In the debate on sea-level rise as a result of global warming there was a closure in the mid-1980s of the question of the likelihood and timing of disintegration of the West Antarctic Ice Sheet (WAIS). Associated with this innovation, a problem shift occurred from "A 5 to 8 meter sea-level rise in the coming century cannot be excluded" (from 1978 until the mid-1980s) toward "We don't have to worry about dramatic sea-level rise in the coming century" (from the late 1980s onward). This shift in assessment had a direct impact on the coastal-defense policy of some coastal countries.

In the assessments of the risk of sea-level rise as a result of anthropogenic climate change, three factors play a role: natural trends, thermal expansion of sea water, and ice-sheet dynamics. Ice-sheet dynamics constitutes the most problematic factor in the assessments of future sea level because it harbors the largest uncertainties. It is estimated that if all of the ice on earth melted, then the worldwide average sea level would rise 80 meters.

The research into the behavior of ice sheets was originally part of a scientific discussion on the causes of sea-level changes in the recent geological past. The debate on the stability of the West Antarctic Ice Sheet (WAIS) was initiated by John H. Mercer, researcher at the Institute of Polar Studies at Ohio State University. Mercer studied the glacial geology of the WAIS for years (e.g., Mercer 1968b). He investigated the stability of glaciers and especially the reasons for the very dramatic changes during the last glaciation (Whillans 1993). In the late 1960s Mercer (1968a) presented a hypothesis that sought to explain interglacial high sea levels by the deglaciation of West Antarctica.

In 1978, Mercer linked up the stability of the WAIS with anthropogenic climate change. He suggested in an article in *Nature* that "If the global consumption of fossil fuels continues to grow at its present rate, atmospheric CO_2 content will double in about 50 years. Climate models suggest that the resultant greenhouse-warming effect will be greatly magnified in high latitudes. The computed temperature rise at lat. 80°S could start rapid deglaciation of West Antarctica, leading to a 5 m rise in sea level" (Mercer 1978, 321). He stated that "deglaciation of West Antarctica would probably be the first disastrous result of continued fossil fuel consumption. . . . If so, major dislocations in coastal cities, and submergence of low-lying areas such as much of Florida and the Netherlands, lies ahead" (Mercer 1978, 321).

Mercer's theory gave rise to public concern and to a scientific debate on the stability of the WAIS.[3]

In the late 1970s Schneider and Chen (1980) calculated the potential impacts of a hypothetical 5 to 7 meter rise of sea level that some scientists believed could occur within the next few decades as a result of WAIS disintegration.

In the Netherlands the Carbon Dioxide Committee of the Health Council (Gezondheidsraad) recommended in its first report on the CO_2 problem (1983) that high priority be given to research into the dynamics of ice sheets because of their exceptional importance for the Netherlands (Gezondheidsraad 1983).

By the mid-1980s, studies concluded that the WAIS is more stable than Mercer had suggested. In the United States, scientists showed in the early 1980s that the disintegration would take 200 to 500 years. These figures were also quoted in the 1984 report of Environment Canada. In the Netherlands, as a consequence of the priority given to ice-sheet research in the first recommendation of the Health Council, an international scientific workshop on the dynamics of the WAIS was organized in May 1985 by the Institute of Meteorology and Oceanography Utrecht (IMOU) (van der Veen and Oerlemans 1987). The workshop led to the conclusion that disintegration of the West Antarctic Ice Sheet was not likely to occur in the coming century. Van der Veen (1986) also concluded that the collapse and deglaciation of the West Antarctic Ice Sheet is not very likely. A similar conclusion was reached by the second report of the Carbon Dioxide Committee of the Health Council (1986).

Since that time this assessment has remained more or less unchanged. As the IPCC report concluded in 1990: "Within the next century, it is not likely that there will be a major outflow of ice from West Antarctica due directly to global warming," and "there is no firm evidence to suggest that the Antarctic ice sheet in general, or the West Antarctic ice sheet in particular, have contributed either positively or negatively to past sea-level rise. On the whole, the sensitivity of Antarctica to climate change is such that a future warming should lead to increased accumulation and thus a negative contribution to sea-level change" (Houghton, Jenkins, and Ephraums 1990, 274).

15.2.5 Unanticipated Effects of Risk Assessments: The CIAP

The Climate Impact Assessment Program (CIAP) was a large-scale risk assessment that had, with the benefit of hindsight, significant unanticipated effects for conducting atmospheric environmental risk assessments in the United States and for the role of the scientific community within the Executive Branch.

CIAP was authorized by Congress in 1971 to assess (1) the potential impact of fuel emissions from a large fleet of high-flying supersonic transport (SSTs) on stratospheric ozone concentration and (2) the hypothesized effects of the resulting ozone depletion on the incidence of skin cancer as well as on climate. It was funded for three years at a total cost of about $20 million, with an unofficially estimated $40 million of contributed research (Dotto and Schiff 1978). Administered by the U.S. Department of Transportation, the study involved hundreds of scientists (some estimates running as high as 1000) in its deliberations and associated congressional testimony and drew members from a number of countries, including Australia, Belgium, Canada, Germany, Italy, Japan, Poland, and the USSR. It found that a small fleet of SSTs would not constitute a serious threat to stratospheric ozone.

It concluded with a lengthy (six-volume) appraisal of the potential effect of SST traffic on stratospheric ozone, which was generally well regarded, although the report had no influence on the immediate target for which it had been designed to assess. Congress voted not to continue funding a commercial U.S. SST well before CIAP's final report was published in November 1975 (Glantz, Robinson, and Krenz 1982).

Moreover, although the substantive report went through a peer-review process, the reception of this main report was complicated by the release of an executive summary that had not been peer-reviewed. The executive summary was released two months prior to the body of the final report to meet the congressionally imposed deadline. The executive summary that was rushed out by the project managers without peer review (and later was included in the *Report of Findings*) offered a far less daunting representation of the risks to stratospheric ozone and human health from SSTs than was determined in the actual body of the report.

CIAP mobilized researchers in stratospheric science, an area of atmospheric sciences that had been relatively neglected. A multinational community of atmospheric scientists emerged that bridged the gaps between various atmospheric-science-related disciplines. CIAP helped to foster the personal and institutional links between international scientists that would prove instrumental in subsequent efforts to identify and respond to subsequent ozone and climate change risks. CIAP had the effect of contributing to the social-capital formation for performing subsequent global risk assessments in the United States, the U.S. NAS Climate Impact Committee, France's Comité sur les Conséquences des Vols Stratospheriques (COVOC), and the United Kingdom's Committee on the Meteorological Effects of Supersonic Aircraft (COMESA).

Interdisciplinary communication was helped by the fact that the study structure cut across disciplinary lines between dynamicists and chemists within the atmospheric sciences (Glantz, Robinson, and Krenz 1982). The risk assessment was less effective at integrating the involvement of biological and economic assessments in the atmospheric appraisals.[4]

CIAP also had the inadvertent effect of limiting scientific access to the White House on all technical issues. President Nixon, angered by the absence of scientific support for the SST, which he favored, responded to CIAP by eliminating the Office of Science and Technology Policy, which had been responsible for advising the President on scientific and technological matters.[5]

15.2.6 Appreciation of the Urgency of the Stratospheric Ozone-Depletion Problem

International institutions helped to catalyze convergence between elements of national risk assessments and helped to diffuse the assessments to other countries. The growing consensus on lifetimes and ozone-depletion potential (ODP) led to accelerated goal and strategy formation (the Montreal Protocol).

An acceptance of the full urgency of dealing with the threat of stratospheric ozone depletion hinged on an appraisal of the lifetime and effects (or reactivity) of CFCs. If persisting for only a short period of time in the atmosphere (as was originally believed) or with an uncertain chemical reactive effect in the stratosphere, it was impossible to accurately assess the extent and timing of the risk of CFCs to stratospheric ozone integrity. Assessments of these two elements converged via the mechanism of deliberate international institutional efforts to foster consensus among the scientists most active in determining and modeling these questions.

Individual assessments of the lifetimes of CFCs, and hence their potential contribution to collective risks, varied widely. CFCs were originally viewed as having long lifetimes because of their inert nature. Lovelock (1971) used CFCs as atmospheric tracers and assumed a lifespan of one year. Molina and Rowland (1974) gave ranges of forty to 150 years in their original calculations for both CFC-11 and CFC-12, but the United States–based Chemical Manufacturers' Association (CMA), which included members from both Europe and the United States, sponsored extensive research on the environmental effects of CFC use and suggested in 1976 that the lifetimes for CFC-11 and CFC-12 were likely to be under thirty years and possibly less than ten. Combined with this disagreement among individual assessments about the atmospheric lifetimes of CFCs was a disagreement about the effects of the CFCs while they were in the stratosphere.

Few countries appear to have engaged in sustained efforts to establish CFC lifetimes. Through the late 1970s and early 1980s debates over risk came less and less to be framed in terms of atmospheric lifetimes, as estimates of the true atmospheric lifetimes of the basic CFCs gradually converged through collective research programs sponsored by the United Nations Environment Programme (UNEP) and WMO. The WMO (1985) Assessment established a lifespan of 75 years for CFC-11 and 110 years for CFC-12. UNEP's Würzburg meeting in April 1987 came up with estimates of 65 years for CFC-11 and 130 years for CFC-12, and by the beginning of the 1990s accepted values were 55 years for CFC-11 and 116 years for CFC-12 (WMO 1991). Canada's estimates of lifetimes and effects were drawn from the WMO consensus, and lifetime figures usually cited in Germany usually fell well within the range endorsed internationally (although the direction of influence is unclear).

Controversy persisted over the magnitude of ozone loss that CFCs were likely to cause, but this came increasingly to reflect different estimates of emissions and their rate of growth, of the contribution of other ozone-depleting gases, and of complications to the relevant chemistry introduced by increasing anthropogenic emissions of other gases (e.g., WMO 1985). National estimates of CFC atmospheric effects varied, and many countries did not attempt such estimates. UNEP catalyzed consensual estimates of the chemical reactive effects of individual CFCs (and thus made possible a comprehensive assessment of the environmental consequences of CFC emissions) at the April 1987 Würzburg meeting by bringing together key modelers to coordinate their models and resolve disagreements. Participants were G. Brasseur (Belgium), I. Isaksen (Norway), G. Jenkins (United Kingdom), M. Ko (United States), D. Sze (United States), R. Watson (United States), D. Wuebbles (United States), and P. Usher (UNEP). They adopted the concept of the ozone-depleting potential (ODP) originally developed in 1981 by D. Wuebbles of Lawrence Livermore Laboratory as a rough yardstick to subsume information about different compounds' atmospheric lifetimes, reactivity, and altitude of dissociation. This provided a single metric for comparing total ozone-depleting impact. Assessments applying the ODP demonstrated the need for rapidly cutting CFC emissions by at least 50 percent in the major industrial economies, while also demonstrating that there would be little environmental consequence if developing countries were granted ten-year grace periods for their reductions. These assessments, and the legitimacy accorded them by the UNEP

auspices, conferred increased urgency on the ozone issue and helped to accelerate agreement on the Montreal Protocol. Chemical models, rather than physical ones, came to prevail in governmental risk assessments in the United Kingdom after the circulation of the Würzburg consensus by UNEP. By 1989 the ODP had been superseded by scientific consensus about the chemical models of stratospheric-ozone depletion, and risk assessments based on the chlorine-loading potential (CLP) replaced the ODP assessments in UNEP's ozone assessment report.

International institutions helped to catalyze convergence between elements of national risk assessments and helped to diffuse the assessments to other countries, which had not conducted their own assessments. The effects of growing consensus on lifetimes and ODP led to accelerated goal and strategy formation (the Montreal Protocol).

15.2.7 The Impacts of Stratospheric Ozone Depletion on Human Health

Relatively little attention was focused on assessments of the human health impacts of stratospheric ozone depletion, although this is the main reason for concern about the issue.

Few studies of the human health effects of ozone depletion were conducted. Early U.S. estimates of cancer effects from increased UV exposure were accepted early on and were widely applied in other countries.

That basal- and squamous-cell skin cancers increase with UV exposure had been known since the 1940s. The first association of projected ozone loss with skin cancer increases was offered by J.E. McDonald in congressional testimony in 1971. McDonald estimated that 1 percent ozone loss would lead to a 6 percent increase in the incidence of cancers in humans and an additional 7000 annual cases of cancer in the United States (U.S. Congress 1971). At a scientific meeting a few weeks later, McDonald slightly revised his estimates to 6000 new cancers per year. T.M. Donahue, chair of atmospheric and ocean sciences at the University of Michigan, suggested in congressional testimony in 1974 that a 1 percent ozone loss contributes to a 2 percent increase in UV radiation and a 2 percent cancer increase.

Most subsequent risk assessment of health impacts of ozone depletion continued to use the "multiplier" formulation articulated by Donahue and CIAP—2 to 1. The report of the U.S. Committee on Inadvertent Modification of the Stratosphere (IMOS 1975) used the 2 to 1 ratio for UV increase to ozone loss. In Germany, a 1.5 to 1 ratio for UV increase to ozone loss was used. A 1976 report of the U.K. Department of the Environment (DOE), while noting some of the shortcomings behind such a figure, cited the prior IMOS report and adopted a "generally accepted" 2 to 1 ratio. The same ratio appeared in the U.S. NAS reports of 1975, 1978, and 1984. The Coordinating Committee on the Ozone Layer (CCOL) also used the same ratio at the end of the 1970s. The same figure appeared in U.S. Environmental Protection Agency (EPA) (1987) and the 1988 report of the German Enquete Commission. The Enquete Commission estimated the ozone-cancer multiplier to be 4 to 6 percent (basal and squamous cell carcinoma, respectively), and the U.S. EPA estimated it to be 3 to 6 percent. The German scientists had no access to reliable, age-correlated data, so they based their assessments on random samples and studies made in other countries, including an extrapolation of estimates from the Netherlands (German Bundestag 1989).

Subsequent risk estimates for nonmelanoma skin cancer spanned a substantial range. The 1989 Dutch Environmental Plan estimated an ozone-cancer multiplier of 4 to 6; the 1989 UNEP Impacts Assessment Panel estimated 3; and the 1991 Panel estimated 2.6, giving 300,000 more cancers per year worldwide.

The relationship between UV and the more dangerous melanoma skin cancers remained problematic. While a 1987 U.S. EPA report suggested an ozone-melanoma multiplier of 1 to 1.5, it was felt that this value was too high and too simplistic. Melanoma appears to be associated with exposure but shows no correlation with occupational exposure to UV radiation. This and other evidence led to the suggestion that intense episodes of exposure, particularly blistering sunburns in childhood, are much more strongly implicated than cumulative exposure (UBA 1989; Glass and Hoover 1989). The Japanese generally dismissed concerns about increased skin cancer as a problem limited to fair-skinned peoples.

Beyond skin cancers, the set of other health effects attributed to ozone depletion and resulting UV exposure changed over time. In the United States, the 1975 IMOS report listed noncancer health effects as sunburn, eye damage, and excessive vitamin D synthesis. Some assessments mentioned eye cancer and macular degeneration, but it became clear that since most incident UV is absorbed by the lens, the most significant form of eye damage from UV is cataract formation (Taylor et al. 1988). Cataracts were mentioned in U.S. EPA reports in 1986, 1987, and 1990 and were the only form of eye damage that assessments after the mid-1980s sought to quantify. The 1989 Netherlands Environmental Plan gave an ozone-cataract multiplier of 0.2 to 0.6, while the 1989 UNEP Impacts Assessment Panel gave 0.6. The 1991 UNEP Impacts Panel without explanation increased the multiplier to 2.6, yielding the estimate that a 10 percent

ozone loss would lead to 1.6 to 1.75 million additional cataracts per year worldwide.

The USSR emphasized studies of immune-system suppression, while these concerns received more limited attention in the West, and the possible links remained controversial. Increasing evidence of immune suppression from UV exposure was mentioned in the Netherlands Environmental Plan and in both the 1989 and 1991 UNEP Impact Assessment Panel reports.

In general, little attention was given to improving numerical estimates of the health impacts of stratospheric ozone depletion. The examples cited in this section suggest that there has been considerable "borrowing of numbers."

15.2.8 The Introduction of Non–Sulfur Dioxide Gases in the Acidification Debate

The dominance of sulfur dioxide (SO_2) in risk assessments concerned with acidification in the 1970s illustrates a widespread diffusion of a direct-effect hypothesis brought into international prominence by Swedish scientists. We observe the diffusion of a set of ideas that were less comprehensive than those held in previous years and yet, nevertheless, were widely taken up in a response to what was perceived as a new and urgent international issue.

Risk assessments looking at sulfur dioxide as the main pollutant in the acidification debate certainly predominated in the period of activity and research beginning in the early 1970s. However, well before this time, accompanying SO_2 were often considerations of nitrogen oxides (NO_x) and other air pollutants. Hermann and Gorham's (1957) work in Canada identified SO_2 and nitrogen acidity in rainfall. Work carried out in Germany cited both SO_2 and hydrogen fluoride (HF) as presenting risks to forest health (Wentzel 1967). British scientists had carried out work on determining ammonia, chlorine, and sulfuric acid in rain at Rothamstead Research Station as early as 1854, and in the mid-1970s they were well aware of the shortfalls of monitoring only SO_2 for risk assessment purposes. In the United States, Likens, Bormann, and Johnson (1972) considered NO_x along with SO_2 in the discussion of major sources of acidifying pollutants.

However, despite the recognition of a relatively catholic range of pollutants contributing to acidification, many major risk assessment programs beginning in the 1970s tended to focus on SO_2, with NO_x very much a subsidiary, when it was considered at all. Examples include the OECD (later the Cooperative European Monitoring and Evaluation Programme) (EMEP) monitoring and risk assessment program in Europe that focused exclusively on SO_2 in the 1970s (Dietrich 1992), the dominance of SO_2 in risk assessments in the United States and Canada, emphasis on sulfur as a pollutant in German risk assessments in 1972, 1975, and 1980 published in *Bild der Wissenschaft,* and the exclusion of pollutants other than SO_2 in the international Regional Acidification Information and Simulation (RAINS) model (Alcamo, Shaw, and Hordijk 1990) until 1986, when NO_x was included.

The dominance of SO_2 in risk assessments concerned with acidification in the 1970s illustrates a widespread diffusion of a direct-effect hypothesis brought into international prominence by Swedish scientists in the United Nations in 1971 and at the Stockholm Conference in 1972. Particular aspects of the Swedish theory narrowed the focus down to sulfur, stripping away (for a good part of the 1970s at least) the relevance of the inclusion of non-SO_2 pollutants. Thus we see the selection and diffusion of a set of ideas that were *less* comprehensive than those held in previous years and yet, nevertheless, were widely taken up in a response to what was perceived as a new and urgent international issue. Factors important in the bias toward the role of sulfur in acidifying processes therefore include the following considerations:

• During the late 1960s and early 1970s it was thought that processes of acidification were linked to the *long-distance transport* of acidifying substances;

• Sulfur was thought to be important in this transport because the main source of emissions was thought to be from power stations and because sulfur dioxide may have a long residence time in air masses;

• The Swedish case study presented to the United Nations in 1971 and to the U.N. Stockholm Conference in 1972 featured the long-range transport of sulfur dioxide and subsequent international negotiations mimicked this focus;

• In some countries (such as the United Kingdom) large-scale monitoring of air pollution had largely focused on sulfur pollution associated with health effects. There was insufficient data on the status of other pollutants in the acidifying process.

The inclusion of non-SO_2 pollutants in risk assessments appears to have increased in the early 1980s.

By looking at the sources of the risk assessments, it is clear that in some instances they were carried out by institutions with vested interests in diverting research toward pollutants that they were either not involved with or had no means of control over. For example, a report from the German Hard Coal Association (1982) identified SO_2,

NO_x, and HCl as main acidifiers but cited the emissions of sulfur from *natural sources* before emissions from power plants. The assessment that NO_x is a main acidifier was tied to the later statement that the majority of NO_x emissions were from traffic sources. The *naturalization* of acidification in rhetorical terms and in terms of risk assessment research has also been observed in other countries—for example, in the United Kingdom in the case of the Central Electricity Generating Board (CEGB).

Risk assessments carried out by the U.K. CEGB in the 1970s and 1980s likewise stressed the importance of oxidants in the transport of gaseous pollutants, thus establishing a nonlinear ratio between emissions reduction and deposition over Europe.

The country studies also indicate that while certain actors may at a certain point in time have considered a very comprehensive range of pollutants (e.g., Wentzel and Zundel 1984 identify SO_2, HF, HCl, NO_x, photooxidants, and heavy metals as potential acidifiers), others at around the same time in the same country continued to focus on a narrower range (e.g., Griesshammer 1983 focused on SO_2 and NO_x). In the early 1970s a reversal of this trend was seen, when the range of acidifiers identified was narrowed down to a focus almost exclusively on sulfur by the influential governmental and industrial institutions involved in risk assessments at that time, although in universities and research institutions, research taking a wider range of potential pollutants (such as NO_x and ozone) into account continued to be carried out. This points to an institutional and political shaping of the issues. International pressure and the requirements to "get something done" quickly may have been factors in this particular narrowing of the subjects considered appropriate for assessments in the early 1970s.

15.2.9 The Urgency of the Acid Deposition Problem Reflected in the Assessments of Cumulative Effects
Establishing the urgency of the risk of acid deposition hinged on the recognition that the lifetime of the risk in the environment (cumulative effects) was longer than originally anticipated. Different actors used differing interpretations of the assessment of cumulative effects to justify widely varying goal and strategy formulations.

Cumulative effects came into focus as a contribution to the risk of acidification as research itself gathered cumulative results over time. As with the inclusion of non-SO_2 pollutants and the multiple-stress hypotheses, methods to assess cumulative effects and cumulative effects theories became more widespread in the 1980s through parallel developments in the United States, the United Kingdom, Canada, and Germany. Major conferences on the risks of acidification were held in 1975, 1980, 1983, and 1988. National experiences were exchanged and compared.

The cumulative effects of air pollution on soils were acknowledged in the 1980s by scientists from the United States, the United Kingdom, Canada, and Germany. In the United States in the 1960s it had been thought appropriate to think of pollutant exposure in terms of air concentrations and/or annual amounts of deposition. It was later realized that cumulative (multidecade) exposures are also important and that it is necessary to understand that soils, for example, have a finite "sulfate absorption capacity" or "assimilative" capacity for acidifying substances and other airborne pollutant chemicals. In Germany the overloading of the buffer system of the soil was considered as a causal factor of forest dieback (Ulrich, Mayer, and Khanna 1979).

In Britain the notion that SO_2 could accumulate in the soil for hundreds of years was initially put forward as an argument by the CEGB not to reduce sulfur emissions, suggesting that costly reductions in emissions would still not result in a corresponding drop in acid runoff from soils contributing to the acidification of fresh waters. The CEGB position was reversed after a period of research and a 1986 visit to Scandinavia by CEGB's chair, Lord Marshall, and his scientific advisor. They became convinced that lakes were threatened, although they remained skeptical about the vulnerability of other ecosystems. The accumulation of sulfur in soils became considered part of the logic in reducing emissions as soon as economically possible and the installation of scrubbers on British power plants. Canadian concern over cumulative effects of sulfur in soils was related to the potential effects on agricultural crops.

The implications of the acceptance of a cumulative effect hypothesis have remained subject to actors' broader policy orientations. In most countries the acceptance of cumulative effects led to increased concern and a stronger sense of urgency in pursuing goals and strategies to mitigate the risk. Actors concerned about not acting on pollution of soils via industrial emissions applied the cumulative-effects hypothesis in a self-serving way to argue that they were not responsible for observed acidification and that the sources of current forest damage lie in historical emissions. In Germany the Hard Coal Association used the risk assessment to dismiss the urgency of action. In the United Kingdom the CEGB continued to use it in a rhetorical sense long after the results of its own paleolimnological research suggested that acidity in lakes was related to more recent increases in emissions and that, although applying the concept in its acid rain risk assessments, it had virtually no lasting or significant effects on policy. European nongovernmental organizations (NGOs) used the cumulative-effects findings to press their governments for more stringent and more rapid cuts in SO_2 emissions.

15.2.10 The Transportation of Acid Pollutants

The idea that acid pollutants could be transported over long distances and, in particular, from country to country originated in Scandinavia. This led to the development of a monitoring network that played a major role in the management of the risk.

Discussion of the transport of pollutants has been an important part of the acid rain debate. The phenomena of transporting substances over long distances through the atmosphere had long been recognized in dust storms in Europe from the Sahara desert, volcanic eruptions, forest-fire effects detected hundreds miles away from the source, and radioactive fallout from nuclear testing. Therefore, from the beginning of the acid rain debate many scientists considered distant sources of air pollution. For example, in the USSR the issue emerged from and was closely related to the studies of the long-range transport and deposition of nuclear particles that were being carried out at the end of the 1960s and the beginning of the 1970s (Izrael and Petrov 1970; Karol 1972).

Scandinavian Origins The sampling network was set up first in Sweden to investigate the fertilization of crops by nutrients from the atmosphere.

An extensive network of sampling stations was established in Scandinavian countries in the early 1950s and operated over the following years. Accumulated data showed that they were regularly receiving highly acidic rain. Svante Oden (Oden 1967, 1968) from Sweden was the first to recognize the transportation of sulfur emissions from Britain and Central Europe to Scandinavia.

The problem was recognized more widely after the United Nations Conference on the Human Environment (UNCHE) in Stockholm in 1972, where the results of Swedish investigations were reported. In the Netherlands the first model of long-distance transport of air pollution was developed in 1978. In Great Britain a special group to review the evidence of long-range dispersion of air pollutants in response to the Scandinavian claims was established as early as 1976 by the Department of the Environment in collaboration with the National Environment Research Council.

North American Experience The Scandinavian data were widely used by North American scientists to show the significance of the transportation problem, first of all by E. Cowling of North Carolina State University, H. Harvey from the University of Toronto, and G. Likens from

Cornell University. For example, Likens and Bormann reported in 1974 that SO_2 could be transported more than 1000 kilometers before being deposited. In Canada, the first study using a methodology similar to that of Svante Oden appeared in 1976 (Summers and Whelpdale 1976).

In the mid-1970s the transport of emissions was recognized as a regional problem in the United States (Likens and Bormann 1974). Likens was the first to show that industrial development and tall stacks were contributing to the expansion of zones of acid rain in the United States (Likens, Wright, Galloway, and Butler 1979).

Later, especially in the first half of the 1980s, regional transportation was the focus of scientific and political communities of the United States that were concerned about Midwest sources polluting the Northeast and other regions.

The U.S. National Research Council (NRC) reported in 1983 on an empirical analysis of pollutant transport up to 1000 km (U.S. NRC 1983). The first modeling efforts by the National Acid Precipitation Assessment Program (NAPAP) in 1990 showed that about half of the total sulfur deposition from major source regions in the eastern United States occurs within 500 to 700 km of the source.

International Community Response In the international context the issue first appeared on the agenda of the U.N. Economic Commission for Europe, which established a Working Party on Air Pollution Problems in 1969 and a Special Group on Long-Range Transmission of Air Pollution in 1978. In 1971 the Council of Europe adopted a resolution that governments grant residents of border regions of adjacent nations the same protection they grant their own inhabitants. Just before the Stockholm Conference in April 1972 the OECD inaugurated a Cooperative Technical Programme to Measure the Long-Range Transport of Air Pollutants.

In 1977 the first findings of the OECD measurement program were published and offered the first independent international verification of Scandinavian charges that imported sulfur pollution was primarily responsible for the acidification of lakes. In particular, it was shown that the United Kingdom was the largest exporter of sulfur to Norway (14 percent) and Sweden (20 percent).

According to estimates and results of modeling of transboundary fluxes of pollutants in the USSR by the end of 1970s, the annual import of SO_2 across the western boundary was five to ten times higher than its export (Problemi 1981). Later it was specified that only about 30 percent of sulfur and 60 percent of nitrous compounds were being deposited on the European part (*Nacionalnyi* 1992, 158).

Similar efforts were made in other countries. For example, W. Kuttler from Bochum University in Germany estimated in 1982 that 28 percent of all European sulfur emissions were from the European part of the Soviet Union. Deposition models developed by MIT scientists suggested that the United States contributed approximately three times as much sulfur to Canada as Canada did to the United States (Fay, Golomb, and Kumar 1985).

The discussions on transboundary air pollution in Europe led to the establishment of a monitoring network, particularly the European Monitoring and Evaluation Programme (EMEP) since 1978 and regular publications in different countries of data reflecting sources and recipients of acid pollution. (This is discussed further in chapter 16 on monitoring.)

15.2.11 The Introduction of the Concept of Multiple Stress in the Acid Deposition Debate

In the 1980s multiple stress was increasingly included in risk assessments for the issue of acidification. The diffusion of the multiple-stress hypothesis has differed according to different actors and different scientific cultures.

A consideration of multiple stress associated with acidification looks at the interactive processes and effects of acidification on soils, water, and forests. The consideration of multiple stress became increasingly widespread in the 1980s across most countries considered here. This coincided partly with a perhaps more general trend toward complicating the reductionist focus held in risk assessments in the 1970s on the (predominant) role of sulfur as a pollutant in acidification. However, although in general a widespread trend can be seen, diffusion of the multiple-stress hypothesis has differed according to different actors and different scientific cultures.

The stresses introduced into the debate during the 1970s included heavy-metal contamination and microbiological processes in the soils. By the 1980s it was beginning to be recognized that nitrogen deposition in soils may have not only a beneficial fertilizing effect on plant growth but also a deleterious effect in the case of nitrogen-saturated soils. This was found to be occurring in some forest soils of the United States by the 1980s.

A report by the German Forestry Association (1984) discussed the synergistic effects of SO_2 and ozone contributing to forest dieback. In addition, the role of vehicle emissions as NO_x producers that are precursors of ozone was analyzed. In the United Kingdom, experiments looking at the influence of sulfur on plant growth soon turned to consider the effects of NO_2 and ozone in the late 1970s, considering also the synergistic effects these gases could have. Similarly, in the United States attention to the risks posed by SO_2 to plant growth (in particular, agricultural crops) was superseded in the 1980s by a recognition of the impact of tropospheric ozone on crops. Such multiple-stress hypotheses run parallel to the increase in risk assessments made into non-SO_2 substances. Typically, plant scientists looking at the problem of acidification were led to theories along the lines of multiple-stress hypotheses, due to the evidently complex nature of the organisms studied at a detailed level.

Scientists in the United States, the United Kingdom, Germany, and Canada and at the international level came to realize the multivalent nature of forest decline by the 1980s.

Research in Britain and in the United States found that exposure to ozone and acid-cloud water increased tree susceptibility to winter frosts. Furthermore, the process was found to be extremely spatially and temporally variable. Research conducted by the CEGB concluded that climatic extremes could interact with SO_2, ozone, acid mist, or fungal pathogens to produce different symptoms in different areas.

An interesting aspect of multiple-stress hypotheses was the different ways that such findings were interpreted by different scientific communities. For example, in the United Kingdom, while tree surveys carried out by the Friends of the Earth considered various factors involved in air-pollution effects on tree decline as evidence that air pollution was occurring in interaction with a host of other factors, the government-owned Forestry Commission stated that the multicausality of the evidence suggested that "damage" due to air pollution could not be concluded. This is an interesting example of the way in which the softness of the science (no specific damage could be attributed to any one "cause") can be interpreted as "sound" or "unsound" science by different actors. Multiple-stress hypotheses are sites where "learning" can diverge dramatically according to cultural scientific, political, or economic factors.

15.3 Patterns across Time

15.3.1 Introduction
This section looks at a number of patterns in risk assessment for global environmental risks. It examines the patterns over time for each of the issues, the differences in the risk assessments made by various actors, and the differences between arenas. In doing so, the section is only to a small extent dependent on the material presented in section 15.2. It relies much more on the material used in the individual arena studies and on broader discussions of the background material during the course of the study.

15.3.2 Across Issues
In all three cases (acid rain, stratospheric ozone depletion, and climate change) clear patterns emerge in the

development of risk assessments over time. These developments are evaluated here with reference to three criteria:

• The *focus and extent* of the risk assessment: Which elements of the causal chain were considered? Were they treated separately or in an integrated fashion?

• The *detail* of the risk assessment: Did it result in one globally averaged number or a detailed regional analysis? Were the results of the assessment based on empirical analysis or a one-, two-, or multidimensional theoretical model?

• The *pluralism of participation* in the risk assessment: Was the assessment produced by a single individual, a team, or an intergovernmental committee, for example?

In the case of climate change risk assessments, there was a pattern over time toward covering more elements of the causal chain, but very few assessments considered all elements in one analysis. Only with a complex integrated global model with numerous submodels would it be possible to analyze the emissions of all greenhouse gases, the resulting changes in atmospheric concentrations, the climatic changes, and the consequent socioeconomic impacts. Most assessments concentrated on the link between concentrations and climate change, and toward the end of the time period more attempts were made to examine the links between climate change and socioeconomic impacts.

The detail of the climate change risk assessments clearly increased over time. While early assessments tended to characterize the risk in terms of one globally averaged number (for example, if the carbon dioxide concentration doubles, the globally averaged temperature of the earth's surface will increase by x degrees Centigrade), many recent assessments considered all important greenhouse gases and began to look at the consequences for several climatic variables, sometimes on a regional scale.

The methodological basis of climate change risk assessments shifted from the use of simple climate models and empirical analysis to the use of complex global climate models that considered processes in the atmosphere and the oceans. The latter formed the basis of the major risk assessments, although simpler models are still used (and useful) for some sensitivity analyses.

Participation in major climate change risk assessments clearly expanded over the period studied here. Before climate change arrived on the political agenda, risk assessments were carried out in general by individual experts or small teams. As will be seen in the other cases, the major risk assessments since the end of the 1980s (the IPCC assessment) were produced by an intergovernmental panel (relying, however, on the work of individual national and international experts and teams), and this assessment also became the basis of subsequent national assessments of the risk of climate change. In terms of the number of actor groups participating in climate change risk assessments, the scientific nature of the assessments (including the need in many cases for very large computers and volumes of data) precluded widescale participation of other actor groups in the initial analysis, although other actor groups were active in the wider use or communication of the results of the assessments.

These basic trends are generally a result of scientific developments in the field as a whole. The availability of increasingly powerful computers led to the development of complex global models. A dense scientific network supported the rapid diffusion of new knowledge and methodologies.

Patterns over time in the assessments of the risk of stratospheric ozone depletion were similar to those for climate change. The scientific advances in the understanding of atmospheric chemistry led to the development of increasingly complex models used to predict the amount, timing, and location of stratospheric ozone depletion. Assessments generally did not consider all elements of the causal chain but concentrated on the middle portion (concentration changes leading to ozone depletion), while the impacts on human health and ecosystems were studied separately. Indeed, there appeared to be conscious efforts in the latter part of the time period considered here to decouple the assessments of ozone depletion from those of health and ecosystem effects. In contrast, early assessments such as the CIAP assessment looked at all elements of the causal chain. As in the case of climate change, participation in the ozone-depletion risk assessments was generally limited to the scientific experts, with a trend toward wider international participation over time.

Assessments of the risk of acid rain often considered the majority of elements of the causal chain, but the detail within each element changed over time. Thus, early risk assessments tended to consider only the emissions of SO_2, while later assessments also included NO_x and other acidifying compounds. The impacts of acidification were often not included in the assessments, although the specific impacts of interest varied, with early studies in some arenas concentrating on the impacts on lakes and freshwater and assessments in the 1980s looking more at the impacts on forests after the interest in forest dieback in Germany stimulated widespread attention to this issue. Due to the nature of the problem, risk assessments were for regions, but there was a trend toward taking assessments for one region and using a modified model to make

an assessment for another region (for example, using the RAINS model for Europe in Southeast Asia). There was also a trend toward the preparation of comprehensive, "policy-oriented" risk assessments for acid rain as illustrated by the National Acid Precipitation Assessment Program (NAPAP) of the United States and the RAINS project at the International Institute for Applied Systems Analysis (IIASA).

15.3.3 Similarities and Differences between Issues

When comparing the risk assessments made for the three cases, there are two possibilities. The first is that the risk assessments show major differences due to the varying natures of the problems being analyzed or differing extents of scientific development. It is also possible that experience with the risk assessment of one problem is reflected in assessments of another problem.

There are in fact few differences between the risk assessments of each case. In all three cases, the assessments are based on complex models of the atmosphere or climate coupled sometimes with other models to assess the impacts. Over time in all cases the risk assessments incorporated scientific developments, especially progress in the modeling of the atmosphere and climate. In the cases of climate change and ozone depletion, the assessment of impacts was made less consistently than in the case of acid rain. This is presumably because the assessment of impacts of global risks like climate change and ozone depletion require regionalized input and accurate information on regional scales was not always available. Thus the impact assessments took the form of sensitivity analyses rather than risk assessments in the case of climate change.

There is evidence that assessments of one risk led to changes in the assessments of another risk in terms of physical and chemical connections, methodological innovations, and institutional change. An example of the physical and chemical connection is provided by the Canadian case study, in which it is shown that modeling of the stratospheric ozone depletion led to changes in the assessment of the risk of climate change.

The emergence of the concept of critical loads in acid rain risk assessments[6] is a major methodological change that is also found in the assessments of climate change and ozone depletion. In the climate change risk assessments, almost the same concept—perhaps in this case a critical level—emerged in the form of the ecological standards developed at the Villach and Bellagio workshops in 1987, stating that there exists some maximum rate of temperature change that ecosystems can take (0.1 degree Centigrade per decade with an absolute maximum of 2°C in total). The study by Florentin Krause (for the

European Commission) played an important role in increasing the awareness of this standard in the Netherlands. In the stratospheric ozone case example, a global cap for chlorine loading was discussed to limit the peak in stratospheric chlorine—the magnitude of the peak being the potential for the most ozone depletion to be experienced.

In the United States and in the Netherlands, there is evidence for institutional change in the way the research programs were structured over cases. This was especially so in the case of research programs for climate change copying those for acid rain. In organizing the risk assessment of acid rain, the United States experimented with a relatively new institutional approach to dealing with interdisciplinary environmental problems—namely, the interagency task force approach of the National Acid Precipitation Assessment Program (NAPAP), a ten-year research program started in 1980. When NAPAP began, both the climate change and the stratospheric ozone issues were being dealt with through the traditional "lead-agency" approach, with the Department of Energy and the National Aeronautical Space Administration, respectively, playing dominant roles. U.S. government research and assessment efforts dealing with climate change and, to a lesser extent, ozone depletion were later being organized more or less on the NAPAP model, with a formal coordinating body in the President's Office of Science and Technology Policy and an operations office now being established in Washington. In the Netherlands, the National Research Program on Climate Change was almost a copy of the one on acid rain; not only were the approach and institutional settings copied, but even the budget was equal.

15.3.4 Differences between Actors

For the actor groups studied here, there appear to be significant differences in the kinds of risk assessment made and of the risk communication. For example, experts and scientists communicated their methods and results via well-established channels such as professional journals and conferences, whereas environmental NGOs built up their own networks and publications for the exchange of information. Industry actors generally had different motives than other actors had for making risk assessments, and this influenced both the kind of risk assessment that was made and the ways in which communication occurred.

Experts made the majority of the risk assessments in all three cases. This is an obvious result of the high level of scientific expertise as well as computer resources, data, and models required for global environmental risk assessment. As the issues became politicized, there was a tendency to have more government-sponsored risk assessments. However, these government risk assessments were performed by the expert community in most cases.

NGOs were increasingly active in communicating the results of risk assessments as well as testing the assumptions and validity of them. For example, risk communication on the issue of climate change was supported by the international Climate Action Network.

NGOs also facilitated the "importation" of risk assessment methodology for specific issues. For example, in 1984, Friends of the Earth (FoE) (Scotland) in the United Kingdom "imported" a German forest researcher who had been studying the phenomenon of tree damage in Germany to conduct a survey of damage to British forests. Criteria of tree damage for quantitative risk assessment were defined using the same criteria as German researchers were using at this time. The context in which this transfer took place is important: it arose in a situation in the United Kingdom in which British NGOs did not trust the official status of risk assessment with respect to forests by the agency responsible for forests in the United Kingdom—the state-owned Forestry Commission. Epistomological commitments, interest-based commitments, and cultural needs and constraints were important elements involved in the way that these two institutions (differently) framed the issue, leading to different conceptions of how to frame the risk assessment issue and thus leading to distrust and the NGO's need to consult sources perceived to adopt a similar framing of the issue.

For the issue of global warming, the U.K. Coal Research Board served as an NGO-industry hybrid, facilitating seminars in which speakers from the other countries were invited to contest the risk assessments predominantly accepted at the time.

NGOs both used existing risk assessments *and* conducted their own risk assessments. The extent to which NGOs conducted their own assessments was often a function of influential constraints (for example, lack of resources) and of tacit skepticism about the prevailing official scientific risk assessments being used for any one framing of an issue at any one time. The existence of international NGO fora reduced the need for national NGOs to conduct risk assessments: this proved important in the climate change issue. A common practice among NGOs seemed to be to adopt officially accepted risk assessments, to deconstruct the assumptions built into such assessments, and then to reinterpret the assessment using alternative assumptions based on different epistomological, cultural, and scientific commitments. A prime example of this is the work of Jeremy Leggett of Greenpeace, who derived a very different interpretation of climate-warming risk from essentially the same scientific knowledge base as the IPCC, by surveying many of the scientists involved as to what they understood about the likely sign of uncertainties in feedbacks omitted from the existing models.

At the beginning of ozone campaigns in the United Kingdom, environmental NGOs such as Greenpeace and FoE used existing risk assessments that relied on National Aeronautics and Space Administration (NASA), UNEP, and U.K. Stratospheric Ozone Research Group (SORG) science. However, they put their own interpretation on the assessments—for example, in stressing the commitment to ozone damage already in the "natural pipeline" from past uses and emissions. Such interpretations were then tested with scientists as established and more powerful scientific mouthpieces (for example, with scientists in SORG). NGOs thus occupied a unique niche in creating an NGO-science relationship in which they received scientists' informed blessing for more acute interpretations of the science than public policy allowed.

After 1990, Greenpeace U.K. conducted its own risk assessments into hydrochlorofluorocarbons (HCFC) and hydrofluorocarbons (HFC) alternatives to CFCs. The uncertainties inherent in their assessments led Greenpeace to christen these "speculative risk assessments." Acknowledgment of uncertainty in this and other issues (NGO tree-damage surveys conducted in the United Kingdom) had the effect of breaking open the hard-science mold in which other relevant scientific institutions had cast the results of risk assessments into the same issues. Thus NGOs were able to ally sound scientific comment (recognition of uncertainty) and the tactics necessary to enter and influence risk assessment debates. This is seen in the instance of Jeremy Leggett's critique of risk assessment in the preparatory working meeting of the IPCC in 1989: emphasis on the implicit uncertainties in the science produced a conclusion that the risks were considerably larger than those that had been quantified through the assessment until then.

An example of NGOs using different assumptions within an existing risk assessment model and subsequently rerunning the model is seen in the case of Greenpeace U.K.'s adoption of the chemical industry's methodology for assessing the ozone-depletion effect of HCFCs. While the industry assumed a 200- to 500-year timescale for calculating ozone-depletion potential, Greenpeace used twenty to thirty years. HCFCs were calculated to have an ozone-depleting effect of five times that of the industry's estimate using this shorter timescale.

Examples from NGO assessment of the tree damage in the United Kingdom and in Germany illustrate the importance of different cultures within science. Whereas in tree damage surveys in the United Kingdom the Forestry Commission excluded old and exposed trees on the basis that they were more likely to suffer damage, the NGOs

(Friends of the Earth and Greenpeace) and later the Commission of the European Community (CEC) survey methodology included such trees. The British NGO scientists were trained ecologists and rested the case for the inclusion of old and exposed trees on the concept of the multiple-stress hypothesis. On the other hand, the search for conclusive causal evidence on the part of the Forestry Commission (based on a more reductionist science) required "noisy" factors to be screened.

Industry involvement in risk assessment is complex and in some instances puzzling. The general observation (if we restrict risk assessment to substantial, supported, public declarations regarding the character and severity of a risk) is that most risk assessment was done by governments, scientific bodies, and international organizations and that some was done by nongovernmental organizations. Industry did not do much risk assessment on the three global environmental risks studied here during the time period covered by this study.

Of the risk assessment that industry did, much might be called "defensive" risk assessment. Once a risk was identified and publicized by scientists, governments, or nongovernmental organizations and appeared to represent a threat of some costly regulatory action, there were many instances of industry groups that responded with statements, which in a simple sense were risk assessments, defending themselves against the perceived regulatory threat by contending that the severity or the imminence of the risk was overstated.

These defensive risk assessments had several defining characteristics. They tended to be late, following an existing debate on a risk, perhaps as part of a coordinated counterattack. They typically did not try to compete with expert or government risk assessments on the level of scientific argument. They normally involved the investment of rather modest resources; did not generate substantial reports; were undertaken to be presented in public fora rather than to further the organization's own understanding; were presented by executives, lawyers, and public affairs departments rather than scientists; and relied on a set of simple legalistic forms of argument.

These simple arguments in defensive risk assessment can be applied to almost any environmental risk. First, the effect is not proven, or alternatively, there is insufficient evidence of it to warrant action. Second, the effect is not new. Third, the effect exists, but human activities contributing to it are small compared to natural sources, or alternatively, the human-induced change is small compared to natural fluctuations. Or finally, the effect may be real and important, but it is insufficiently understood, and more research and monitoring must be undertaken before knowing whether and what to do about it. Examples of such defensive risk assessment are numerous on all three of the issues (for example, the British Aerosol Manufacturers, Global Climate Coalition, DuPont, and ICI all issued such statements).

A variant of the "defensive risk assessment" approach occurred when industry groups provided support and fora for prominent scientific dissenters with a prevailing risk assessment that the industry body deemed threatening. Examples include the DuPont-sponsored tour of the United States by U.K. scientist Richard Scorer in 1976 and the World Coal Council's presentation of prominent climate skeptic Richard Lindzen in a high-profile series of lectures in London in October 1991, where he criticized the IPCC report.

A second variant of "defensive" risk assessment occurred when a group representing one industry highlighted the importance of a risk for which the likely regulatory response would favor that industry. The most prominent examples were representatives of the nuclear-power industry who publicized the risk of climate change and presented nuclear power as a solution, as shown in the United Kingdom chapter.

There were two major classes of exceptions to the foregoing argument. First, in a few instances firms or industry groups were confident of their scientific competence, actively engaged in assessment of potential risks, and engaged the scientific risk assessment debate on their own terms.

Early industry research in environmental impacts of CFCs provides an example. Initiated at DuPont's urging, the Chemical Manufacturers' Association (CMA) sponsored a research program that began before Molina and Rowland and focused on the potential smog risks that CFCs posed in the troposphere. Following publication of the CFC-ozone theory, this program spent a couple of years supporting research clearly intended to seek out every possible chance to vindicate CFCs but thereafter funded high-quality independent research both in industry and universities. Much of the work sponsored, including atmospheric modeling by scientists within DuPont, indicated grounds for increased concern about the risk posed by CFCs.

Acid rain research sponsored by the CEGB in the United Kingdom provides a mixed case. This program, initiated in 1973, included research on atmospheric chemistry, catchments, soils, and tree health. While the work was independent and of high scientific quality, the timing of the particular topics that were supported suggests that the motivation for this work was at least partly defensive. In 1980 the research program turned to emphasizing ecosystem effects, which both provided grounds for delay during the lengthy study and suggested that intervention at the point of impact was more cost-effective than reducing emissions. A program of paleolimnology conducted in Scotland from 1981 served in part to test the hypothesis that acidification was a natural phenomenon.

Both the German Hard Coal Association and Shell in the Netherlands did risk assessment on climate. Also, the International Energy Agency has a Coal Research Institute in London, which did research on both climate and acid rain.

Why do industry groups do risk assessment of this kind? One proposition is that they are groups whose identity is strongly bound up with their image of themselves as scientists. They believe that whatever the regulatory environment they will be able to flourish by technical prowess. Industry scientists sometimes serve on government-sponsored risk assessment panels. A representative of the CMA program was on CCOL; M. McFarland of Dupont was on the Ozone Trends Panel.

The second and most interesting exception to the pattern of defensive risk assessment occurred when an industry group identified an environmental risk as representing a substantial business risk intrinsically, rather than merely through the accompanying threat of costly regulation. In this case, an effective long-run view of their own operations required strategic assessment of the extent and character of the environmental risk.

Of particular interest in the discussion of industry is the interest of the insurance industry in the management of global environmental risks. Travelers Insurance of Hartford established an excellent climate-research center at its headquarters in the 1950s. The enlightened president of Travelers became acquainted with Tom Malone, then at the MIT Meteorology Department, and invited him to establish the Travelers Research Center. Malone brought Robert White on board.

In the United Kingdom Commercial Union funded work done by the Climatic Research Unit in the 1970s. At the end of the period studied here, the Reinsurance companies became increasingly interested in the problems of global environmental change. In April 1993, Swiss Reinsurance Co. and Munich Reinsurance Co. published a report that attributed the huge increase in insured-catastrophe losses in 1992 to $27.1 billion, principally for natural catastrophes ($22.5 billion). The report also noted a large increase in natural catastrophe losses over the period 1982 to 1992 ($52 billion total losses in the decade) and warned insurers to expect stronger storms and larger claims in the future as a result of climate change (*Business Insurance,* April 5 and 19, 1993).

15.3.5 Differences between Arenas

There is substantial evidence to support the view that risk assessments made in one country can be influenced by risk assessments in another country. The one country that stands out as the source of most of the changes is the United States of America. This is not surprising given the large amounts of time, effort, and money devoted to atmospheric science in the United States. Other countries, when developing an indigenous risk assessment, often first reviewed the work of U.S. scientific institutions (especially reports from the U.S. Environmental Protection Agency, National Academy of Sciences, or National Research Council). For example, climate risk assessments in a number of European arenas and in Canada cited the reports of the U.S. NAS (1979) and U.S. NRC (1983).

It is interesting that the six major general circulation models (GCMs) used until the close of the period studied here for temperature-change and other climate change predictions originated from U.S. institutions—the Geophysical Fluid Dynamics Laboratory (GFDL) at Princeton University, the Goddard Institute of Space Studies (GISS), and the National Center for Atmospheric Research (NCAR). The subsequent additions to the basket of world GCMs were based on the fundamental principles established by the originals, with slight modifications to advance the modeling science.

Substantial amounts of resources, both money and time, are committed to the production of GCMs. For example, the Canadian Climate Centre's model was developed over a period of thirteen years and required the purchasing of a supercomputer every five years at about $40 million per computer. Approvals to begin developing a GCM and to maintain continued resources over a number of years were perhaps one reason for the consistent range of temperatures for a doubling of the atmospheric carbon dioxide concentration reported by several GCMs (see section 15.2). The United Kingdom Meteorological Office (UKMO) experienced what happens when results from a country's large investment in a GCM differ from the results of other groups. An ensuing media controversy erupted, especially coming parallel to the work of the Intergovernmental Panel on Climate Change (IPCC), which reported a higher temperature range.

Something can be said for the geographical proximity of countries in assisting experts learning from one another. Canadian experts learned mostly from their close neighbor, the United States. The close-proximity factor applies in the acid rain issue in Hungary as well. Observed dieback in the stands of oak trees in Hungary was immediately attributed to acid rain because of similar conclusions in the countries bordering Hungary. As it turns out, there were some interesting differences that were not picked up, including the fact that only the oak trees were dying.

An interesting note: scientists from some non-English-speaking countries often made illustrious careers by "learning" from the English-speaking scientific community and translating the "learning" as their own work. This has been cited in the arena study of the former Soviet Union but probably is becoming less possible with the proliferation of English-speaking scientists worldwide.

15.4 Is There Evidence That Risk Assessments "Improve" with Time?

As discussed in the introduction to part III, progress on both risk assessment and monitoring is assessed relative to three criteria: adequacy, effectiveness, and legitimacy. Adequacy is a measure of the quality of inputs to the study, on narrow professional standards. Was the study done right, with state-of-the-art methods? Effectiveness is an output measure of the study's impact on the policy debate. Was it effectively communicated, and did it change, stabilize, or frame the debate? Did others cite it, and did it lead to a change in views or in the terms used to frame the debate? Legitimacy is an input measure of the extent of scope, consensus, and review represented by the study. Did it frame the questions broadly enough and include a sufficiently representative community of scholars? Did it receive an appropriate level of peer review? Legitimacy can also have a political component, in that it can describe the extent of inclusion of affected interests in the study and the degree of public consultation and review it receives.

15.4.1 Acid Deposition

Acid rain risk assessments became more adequate over time but still were deficient in many areas. Acid rain risk assessments became more and more technically sound as the instruments they used became more efficient and effective. Many resources were placed into developing computerized numerical models used to determine the source and receptor relationships of acid rain-causing pollutant emissions. This was especially important for the Canadian-U.S. case of transboundary air pollution. These models had state-of-the art technological capabilities but were less effective in terms of ability to integrate. It was the group at the International Institute for Applied Systems Analysis (IIASA) that introduced the computerized integrative assessment model (RAINS) (Alcamo, Shaw, and Hordijk 1990). This model provided relatively instantaneous responses about the effects of proposed reductions in acid rain-causing pollutants on several receptors including socioeconomic effects. It can also be argued that acid rain risk assessments became more and more based on scientific results and monitoring data provided by instruments of a higher technological precision than before. As well as being more precise, monitoring data used in the acid rain risk assessments increasingly came from a more intensive and extensive array of sites (see chapter 16 on monitoring).

While the instruments used in assessments were usually state of the art, sometimes the most up-to-date science was not always addressed. Two important examples are *total acidic deposition* and *ammonia.* Both Canada and the United States focused their assessments on the expected rates of acid rain deposition entirely on *wet sulfate deposition,* while all European countries looked at the *total sulfate deposition,* which included both wet and dry sulfates. It is also striking that gaseous ammonia and ammonium-ion deposition were considered major additions to acid depositions for most European countries, while Canada and the United States provided just brief mention in their national assessments.[7]

The comprehensiveness of acid rain risk assessments increased over time, but it varied from country to country. Assessments moved from a single-cause, single-effect (sulfur dioxide affecting the fish population) examination to a multiple-cause, multiple-effect approach.

In terms of effects, countries focused their assessments traditionally on a single receptor. For example, Canada and the United States focused on aquatic effects, Germany on forests, the Netherlands on the natural terrestrial flora and fauna, Japan on human health, and the United Kingdom on soil acidification.

The move toward multireceptor examination required an evolution of the assessment approach from a single discipline (aquatic biology) to a multiple discipline (aquatic biology plus forestry plus agriculture plus engineering plus . . .).

More often than not the actual publications of final acid rain risk assessments were relatively ineffective in leading to policy changes but nevertheless were useful in the overall framing of the debate. For example, by the time that the United States' National Acid Precipitation Assessment Program (NAPAP) was published in 1990, the policy (the Clean Air Act Amendments of 1990) was already in the making. Even the key findings of the NAPAP as cited by the U.S. EPA administrator—that the United States had an acid-deposition problem rather than a crisis—was politically impotent. In Canada, similar circumstances existed whereby the assessment was published after the main policy decisions had been made. An independent Canadian assessment was effective, however, in providing the ammunition for federal-provincial emission reduction targets and in bilateral discussions with the United States.

NAPAP was able, however, to affect how North Americans came to view forest damage and the potential human health effects due to acid rain. In other countries, scientific research on the effects of acid rain on forests and human health were threatening to heighten the public concern for further actions on acid rain–causing pollutants. Research under the auspices of NAPAP concluded that the observed effects on forestry, agriculture, and human health were in fact caused by tropospheric ozone. This assessment not only deflected the acid rain focus from two emotionally charged issues (forests and human health) in both Canada and the United States, but it also created a new issue.

NAPAP set the tone for policy-relevant assessments by organizing the final assessment using a policy-relevant question-and-answer format about the effects of concern—whether they are related to acid deposition, sensitivity to change, and possible control scenarios. The Canadian assessments of 1985 and 1990 followed a similar question-and-answer structure.

One of the most policy-relevant notions to evolve from acid rain risk assessments was the concept of critical loads. The critical-load concept was felt to be a more effective and scientific means for assessing the expected effects and comparing the benefits and costs of various types of emissions controls. All countries proposed some form of critical-load concept in their assessments except the United States, which ignored the use of critical loads for acid rain in their NAPAP assessment. The critical-load concept is used extensively both in scientific discussions and in formulating environmental policies and goals. The concept itself has evolved from discussion of a wet sulfate effect on aquatics (Canada) to a more complete total acidification effect on various receptors (such as in the Netherlands).

There is a range of legitimacy of acid rain risk assessments. It appears that cooperation, flow of information, communications, and scientist access to decision making are traded off when more legitimacy is required.

The United States was perhaps most successful in encouraging a broader range of scope and including a range of ideas and interests through its interagency task-force approach of NAPAP. The United States experimented with NAPAP as a new institutional approach to dealing with interdisciplinary environmental problems. When NAPAP began, both the global warming and the stratospheric-ozone issues were being dealt with through the traditional "lead-agency" approach with the Department of Energy and NASA, respectively, playing dominant roles. It is questionable whether NAPAP was a success in these respects, as no formal evaluations were made of the relative merits of the interagency group as ways to manage effective research programs. Subsequently the U.S. government's research and assessment efforts dealing with climate change and ozone depletion were organized more or less on the NAPAP model. This ensured a wide scope of investigation and the involvement of a wide range of organizations, but many individual agencies did not appear to like the interagency arrangement.

In other countries, such as the Netherlands and Canada, assessments were prepared by a relatively small group of persons within or with close contact to the government. A high degree of consensus existed in these societies between the scientists, the government, and the general public, thus foregoing the need for a large legitimizing exercise. This consensus provided the scientists with a relatively strong influence on decision making. The relatively smaller science and policy communities also made it more necessary for scientists involved in research also to be involved in policy development. And finally, the closer relationship between scientists and policy makers allowed a freer flow of information and communication between the two groups.

On the basis of the above criteria, risk assessments of the acid rain problem were improving over time. Many countries recognized that complex environmental problems such as acid rain required a multidisciplinary scientific approach and an integrated and policy-focused research and assessment program for their solution. Acid rain risk assessments evolved from

• Single receptor (aquatic) to multiple receptor,

• Single cause (wet sulfate) to multiple cause (total acidifying contributions),

• Single discipline to multidisciplinary or interdisciplinary,

• Single scientists to single agency to multiagency, and

• Single issue (acid rain) to multi-issue (such as tropospheric ozone).

Assessments were also successful in reframing debates. While acid rain assessments were formulated to ask policy relevant questions, the timeliness of some of the responses can be questioned.

15.4.2 Stratospheric Ozone Depletion

In broad outline, the trend in major ozone risk assessments over time was that their adequacy was always high and did not change and that they consistently improved in effectiveness. The progression on legitimacy was more complex, for while later assessments gained consensus among increasingly international groups of scientists and gained increasing levels of formal joint sponsorship by international organizations, there was a progressive narrowing of the questions addressed to correspond more closely with those on which strong simple consensus is possible.

All major ozone assessments, from CIAP to the Scientific Assessment Panels under the Montreal Protocol,

were high in adequacy. All recruited eminent scientists to participate and applied state-of-the-art methods, which, of course, advanced greatly over time.

CIAP suffered from three major problems, not all as a result of the actions taken by those doing the study. Though the study had great influence in the long term, it had no opportunity to be of high value to immediate decisions, since the study was established after the crucial SST decision had been made. In seeking to frame the study question with appropriate breadth by undertaking a complete, end-to-end assessment, the study had to include areas where assessment methods were so weak and so little information was available that the work was vulnerable to severe attack. On legitimacy, the score is mixed; the monographs received broad scientific participation and review, but the report was mainly known through its Executive Summary, which was dashed off with little consultation in a great hurry and was widely charged with distorting the conclusions of the main assessment.

The major NRC assessments through the 1970s undertook a gradual retreat from the most highly problematic impacts end of the issue. They thus initiated a gradual decline in legitimacy, interpreted in terms of appropriate framing of the questions.

The 1985 WMO/NASA assessment (the Blue Books) was the first assessment to be directed at a specific international decision-making process, granting an opportunity for a large improvement in effectiveness that the authors partially declined by not writing an Executive Summary (the report has a science summary but no executive summary). This assessment, which was cited widely and changed the international policy debate completely, was extremely effective. For the criterion of legitimacy, the report continued the paradoxical trend of narrowing the disciplinary breadth (impacts were entirely excluded)—stopping at estimates of ozone depletion rather than considering effects on surface UV or impacts—while garnering consensus of an extremely large international set of scientists and the sponsorship of two European, two international, and three U.S. organizations.

These trends continued through the Ozone Trends Panel (OTP) and the Scientific Assessment Panels under the Montreal Protocol. The OTP represented an advance in adequacy, identifying a serious error in the satellite record and both identifying and correcting long-standing errors in the ground-based record. Effectiveness was also high, as the report answered a crucial question that had been identified by many major actors as the criterion for action. The report was identified as responsible for CFC phaseout decisions and endorsements by several major producers and national environmental agencies.

Effectiveness continued to advance with the Protocol Assessment Panels, which were established with authorization by the international negotiating body and charged with answering a specific set of questions on the negotiation body's decision timetable. These Panels continued the progressive narrowing of the question assessed, though, retreating from total-column ozone depletion to projections of future trends in stratospheric-chlorine concentration.

15.4.3 Climate Change

As in the other two cases, the adequacy of major climate change assessments was high, through the recruitment of the best scientists and the use of state-of-the-art methods. Of course, the adequacy improved over time, as a result, for example, of major advances in climate modeling.

Risk assessments on climate change were generally effective in stimulating changes or reframing of the debate. By adding the greenhouse gases other than carbon dioxide, the debate was reframed from being a concern about a problem that could happen sometime in the next century to being a concern about a problem that could be serious within the next fifty years. The international assessment made at the Villach Conference in 1985 was effective in stimulating the development of policy responses, first through the Advisory Group on Greenhouse Gases and then through the establishment of the IPCC. The IPCC report of 1990, particularly the assessment made by Working Group I, was effective in stimulating the negotiation of the Framework Convention on Climate Change. On the national level, there is also evidence of effective assessments. For instance, a report by the German Physical Society in 1986 received wide public attention and raised awareness of the issue, so that the ground was prepared for the politicization.

As for the issues of acid deposition and ozone depletion, the changes in legitimacy of climate risk assessment over time were complex. There was a lively debate about the legitimacy of the IPCC reports of 1990. Mostly, scientific consensus was sought in the development of climate risk assessments. The scope of the assessments tended to broaden over time, with more recent assessments attempting to include consideration of socioeconomic impacts of climate change. However, as this broadening took place, there were also losses of legitimacy. For example, while Working Group I of IPCC achieved scientific consensus with a broad participation of scientists and structured scientific peer review process, the topics addressed by Working Group III of IPCC in its pre-1990 structure were felt to be "political issues." Little consensus could be reached because those who prepared the final assessment were essentially delegates of national governments and not independent experts.

An important question mark on the legitimacy of climate risk assessments has arisen over the participation of developing countries in the debate. Scientists from developed countries dominated the preparation of climate risk assessments, largely because the computer models and data necessary for the work were available in these countries. However, experts from developing countries (e.g., Agarwal and Narain 1992) have expressed doubts about assumptions made in some assessments about levels of emissions or socioeconomic impacts in developing countries. Emissions from tropical deforestation or rice growing or the impacts of sea-level rise on small island states or countries like Bangladesh are examples. The Framework Convention on Climate Change thus institutionalized a mechanism for ensuring that developed countries assist developing countries in their assessments of sources and sinks of greenhouse gases.

Does an improvement in climate change risk assessments lead to better management of the risk? Looking at developments at the international level over the past ten years one might be tempted to answer this question with a yes. The Scientific Committee on Problems of the Environment (SCOPE) assessment in 1985 and the IPCC 1990 assessment show that there were improvements in adequacy (state-of-the-art methods) and effectiveness and the improvements led to the negotiation of a Framework Convention on Climate Change. Even at the national level there was evidence that the improvements in risk assessments led to the formulation of national emission-reduction strategies. However, the skeptics would point out that since further steps in the management of the risk (especially formulation and implementation of policy) have not been generally evident, and since most present goals are not long term, there is little evidence that management of the risk as a whole has improved. Serious discussion of the legitimacy of the IPCC 1990 assessment in terms of participation of a wider group of actors shows that while there have been improvements, they were perhaps not large enough to provide a basis for better management. That is, if the adequacy, effectiveness, and legitimacy had improved even more, firmer goals would have been set and implemented nationally and internationally.

15.5 Main Findings and Links to Other Functions

Looking at major assessments of the magnitude and timing of three global environmental risks over a period of twenty years or so gives clear indications of factors that can contribute to "successful" risk assessments. Major risk assessments that contributed to the further process of risk management have been multinational. The management of the climate change and ozone risks, in particular, were furthered by the use of international assessments, which were used to legitimize national assessments. However, it has been noted that problems arise if the representation of countries in the international groups, particularly the distribution of representatives of the developed and developing worlds, is not balanced.

Carefully managed, multidisciplinary risk assessments also contributed to more successful management of the risks. Multiagency design and avoidance of assessments that require inputs from one another are important elements of assessment planning. Furthermore, careful preparation of peer-reviewed executive summaries of major assessments increase the probability that the assessment plays a continued role in the management of the risk.

The assessments are made by experts within the various societal groups. Industry, NGOs, and the administrative sectors do produce risk assessments but for obvious reasons the scientists are the largest group involved in the assessment of global environmental risks. As the complexity of the assessments has increased over time, the fear has been expressed that there are too few scientists to cope with the demands for regularly repeated, multinational, multidisciplinary assessments.

As noted in the introduction, risk assessment has to make framing assumptions, in particular on the surrounding possibly interacting variables that are taken as fixed or nonexistent and the ranges over which given relationships are assumed to hold and about possible causal agents. For the three global environmental risks studied here, the risk assessments tend to concentrate on the middle of the causal chain—that is, on the emissions, concentrations, and environmental impacts. The human needs and wants that give rise to the emissions and the socioeconomic impacts of changes generally receive less attention. Particularly in the climate case, where large-scale climate models form the basis of many of the risk assessments, attention has been paid to the framing assumptions. In the early years the atmospheric models assumed a constant ocean surface temperature. In later studies a range of ocean models were coupled to the atmospheric models, so that the modeling of the interactions between the atmosphere and the oceans became more realistic. However, one of these framing assumptions has been used by dissenters from the general consensus to discredit the results and conclusions of assessments. For example, it has been argued that climate models overestimate the global warming due to the enhanced greenhouse effect because they wrongly model the formation of clouds, especially in the tropics. Over the time period studied here, the scientific assessments of global environmental risks have given increasing attention to the complexity of and interactions between these issues. Models have become important, if not the central, tools with which the issues are studied.

Within the risk management process, risk assessment is variously linked to other management functions. Rarely is risk assessment automatically linked with response assessment. In the climate change case there are some clear links between risk assessment and response assessment, in particular with regard to assessments of the timing and magnitude of sea-level rise, which led to consideration of response. But on the other hand, the assessment that about one-half of the increased greenhouse effect is the result of gases other than carbon dioxide was not reflected in immediate reassessment of response options.

Strong and obvious links exist between the functions of monitoring and risk assessment. In some cases the monitoring began before the magnitude and timing of the risk were evaluated (the motivations for such monitoring are discussed in more detail in chapter 16). In other cases, risk assessment led to new or renewed monitoring. Last but not least, risk assessment can be followed directly by goal and strategy formulation, as illustrated best by the assessments made at the Villach Conference in 1985 and at the Toronto Conference in 1988.

Appendix 15A.   Acronyms

CCOL	Coordinating Committee on the Ozone Layer
CEC	Commission of the European Community
CEGB	Central Electricity Generating Board (U.K.)
CFC	chlorofluorocarbon
CIAP	Climate Impact Assessment Program
CLP	chlorine-loading potential
CMA	Chemical Manufacturers' Association
CO$_2$	carbon dioxide
COMESA	Committee on the Meteorological Effects of Supersonic Aircraft (U.K.)
COVOC	Comité sur les Consequences des Vols Stratospheriques (France)
DOE	Department of the Environment (U.K.)
EBM	energy balance model
EMEP	European Monitoring and Evaluation Programme

EPA	Environmental Protection Agency (U.S.)		**SMIC**	Study of Man's Impact on Climate
FoE	Friends of the Earth		**SO$_2$**	sulfur dioxide
GCM	general circulation model		**SORG**	Stratospheric Ozone Research Group (U.K.)
GFDL	Geophysical Fluid Dynamics Laboratory (U.S.)		**SST**	supersonic transport
GISS	Goddard Institute of Space Studies (U.S.)		**UBA**	Umweltbundesamt (German Federal Environment Protection Agency)
HCFC	hydrochlorofluorocarbons		**UKMO**	U.K. Meteorological Office
HFC	hydrofluorocarbons		**UNEP**	United Nations Environment Programme
ICSU	International Council of Scientific Unions		**UNCHE**	United Nations Conference on the Human Environment
IIASA	International Institute for Applied Systems Analysis		**UV**	ultraviolet
IMOS	Inadvertent Modification of the Stratosphere		**WAIS**	West Antarctic Ice Sheet
IMOU	Institute of Meteorology and Oceanography Utrecht		**WMO**	World Meteorological Organization
IPCC	Intergovernmental Panel on Climate Change			
LRTAP	(Convention on) Long-Range Transboundary Air Pollution			
MIT	Massachusetts Institute of Technology (U.S.)			
NAPAP	National Acid Precipitation Assessment Program (U.S.)			
NAS	National Academy of Sciences (U.S.)			
NASA	National Aeronautics and Space Administration (U.S.)			
NCAR	National Center for Atmospheric Research (U.S.)			
NGO	nongovernmental organization			
NO$_x$	nitrogen oxides			
NRC	National Research Council (U.S.)			
ODP	ozone depletion potential			
OECD	Organization for Economic Cooperation and Development			
OSTP	Office of Science and Technology Policy (U.S.)			
OTP	Ozone Trends Panel			
RAINS	Regional Acidification Information and Simulation (model)			
RCM	radiative-convective model			
SCEP	Study of Critical Environmental Problems			
SCOPE	Scientific Committee on Problems of the Environment			

Notes

1. Many people contributed to the development of this chapter, in particular through research presented and discussed at the summer studies, as described in the preface, and two smaller meetings on the topic of risk assessment. Fen Hampson, Peter Hughes, Angela Liberatore, and Brian Wynne participated in some of those meetings, and their contributions are gratefully acknowledged.

2. However, van der Sluijs, van Eijndhoven, Wynne, and Shackley (1998) have noted that the interpretation of this range has varied over time. The 1979 NAS assessment, for example, included a margin for error based on an expert assessment of model uncertainties. The Villach 1985 estimate did not include such an assessment. Thus although the ranges quoted by the studies are the same, they differ significantly in their meaning.

3. For instance, in the Netherlands, a 5 meter sea-level rise was seen as a serious threat. Large parts of the country have been created by reclamation from the sea, and almost half consists of land 5 meters below sea level.

4. The biospheric and economic component studies were less successful in achieving interdisciplinary coordination and in producing research results. Both studies took place simultaneously during the last two years of CIAP. Hence, the economists were unable to receive the biologists' research results on the magnitude of the effect of UV radiation on crops until late in their research schedule.

5. A weaker OSTP was restored some years later.

6. Critical loads are discussed in more detail in chapter 18 on goal and strategy formation.

7. This is not because gaseous ammonia or ammonium-ion are not potential environmental problems for North America.

References

Agarwal, A., and S. Narain. 1992. *Global Warming in an Unequal World.* New Delhi: Centre for Science and Environment.

Ailing sugar bushes on the mend, study finds; Forestry study finds acid rain not culprit; Harsh weather took toll on roots and trunks. 1992. *Globe and Mail* (Canada), October 10.

Alcamo, J., R.W. Shaw, and L. Hordijk, eds. 1990. *The RAINS Model of Acidification: Science and Strategies in Europe.* Dordrecht: Kluwer.

Arrhenius, S., 1896. On the influence of carbonic acid in the air upon the temperature of the ground. *Philosophical Magazine* 41: 237.

Bolin, B., B.R. Döös, J. Jäger, and R.A. Warrick, eds. 1986. *The Greenhouse Effect, Climatic Change, and the Ecosystems.* SCOPE 29. Chichester: Wiley.

Callendar, G.S. 1938. The artificial production of carbon dioxide and its influence on temperature. *Quarterly Journal of the Royal Meteorological Society* 64: 223.

Dietrich, W.F. 1992. Monitoring of long-range transport of air pollutants and the policy debate on acid deposition. Contribution number I-19, Social Learning in the Management of Global Environmental Risks Project. Belfer Center for Science and International Affairs, Harvard University.

Dotto, L., and H. Schiff. 1978. *The Ozone War.* New York: Doubleday.

Fay, J., D. Golomb, and S. Kumar. 1985. Source apportionment of wet sulfate deposition in Eastern North America. Energy Laboratory Report No. MIT-EL85-001, Cambridge, Mass.

Flohn, Hermann. 1978. Estimates of a combined greenhouse effect as background for a climate scenario during global warming. In Jill Williams, ed., *Carbon Dioxide, Climate, and Society.* Proceedings of an IIASA Workshop cosponsored by WMO, UNEP, and SCOPE. February 21–24, 1978. Oxford: Pergamon Press.

Funtowicz, S., and J. Ravetz. 1992. Three types of risk assessment and the emergence of post-normal science. In S. Krimsky and D. Golding, eds., *Social Theories of Risk.* London: Praeger.

German Bundestag. 1989. *Protecting the Earth's Atmosphere.* An International Challenge. Interim Report of the Study Commission of the Eleventh German Bundestag on Preventive Measures to Protect the Earth's Atmosphere. Bonn, Germany: Deutscher Bundestag, Referat Öffentlichkeitsarbeit.

German Forestry Association. 1984. *Waldsterben. Argumente zur Diskussion.* Bonn: Deutscher Forstverein.

German Hard Coal Association. 1982. *Saurer Regen und Forstschäden.* Essen, Germany: Gesamtverband der deutschen Steinkohleindustrie.

Gezondheidsraad. 1983. *Deeladvies inzake CO_2 problematiek.* The Hague: Gezondheidsraad.

————. 1986. *Advies inzake CO_2 problematiek.* The Hague: Gezondheidsraad.

Glantz, M.H., J. Robinson, and M.E. Krenz. 1982. Climate-related impact studies: A review of past experience. In W.C. Clark, ed., *Carbon Dioxide Review, 1982.* New York: Oxford University Press.

Glass, A.G., and R.N. Hoover. 1989. The emerging epidemic of melanoma and squamous cell skin cancer. *JAMA* 262(15): 2097–2100.

Grießhammer, R. 1983. *Letzte Chance für den Wald.* Freiburg, Germany: Dreisam Verlag.

Herman, F.A., and E. Gorham. 1957. Total mineral material, acidity, sulphur, and nitrogen in rain and snow at Kentville Nova Scotia. *Tellus* 9(2): 180–183.

Houghton, J.T., G.J. Jenkins, and J.T. Ephraums, eds. 1990. *Climate Change: The IPCC Scientific Assessment.* Report prepared for the Intergovernmental Panel on Climate Change by Working Group I. Cambridge: Cambridge University Press.

Idso, S.B. 1981. Carbon dioxide: An alternative view. *New Scientist* 92: 444.

Inadvertent Modification of the Stratosphere Report (IMOS). 1975. *Fluorocarbons and the Environment.* Report of Federal Task Force on Inadvertent Modification of the Stratosphere. Washington: U.S. Government Printing Office.

Izrael, Y., and V. Petrov. 1970. Diffuzia i rasprostanenie radioaktivnyh productov jadernyh vzryvov na bolshie rasstojania. In *Atomnye vzryvi v mirnych celia.* Moscow: Gidrometeoizdat.

Karol, I. 1972. *Radioaktivnye izotopy i globalnyi perenos v atmosfere.* Leningrad: Gidrometeoizdat.

Likens, G.E., and F.H. Bormann. 1974. Acid rain: A serious regional environmental problem. *Science* 184: 1176–1179.

Likens, G.E., F.H. Bormann, and N.M. Johnson. 1972. Acid rain. *Environment* 14: 33–40.

Likens, G.E., R.F. Wright, J.N. Galloway, and T.J. Butler. 1979. Acid rain. *Scientific American* 241(4): 43–51.

Lovelock, J.E. 1971. Atmospheric fluorine compounds as indicators of air movements. *Nature* 230: 237.

Lovelock, J.E., and L. Margulis. 1974. Homeostatic tendencies of the earth's atmosphere. *Origins of Life* 5(1): 93–103.

————. 1978. West Antarctic Ice Sheet and CO_2 greenhouse effect: A threat of disaster. *Nature* 271: 321–325.

Mercer, J.H. 1968a. Antarctic ice and Sangamon sea level. International Association of Scientific Hydrology, Commission of Snow and Ice, General Assembly of Bern, Publication No. 79, 217–225.

————. 1968b. Glacial geology of the Reedy Glacier Area, Antarctica. *Geological Society of America Bulletin* 79: 471–486.

————. 1978. West Antarctic Ice Sheet and CO_2 greenhouse effect: A threat of disaster. *Nature* 271: 321–325.

Molina, M.J., and F.S. Rowland. 1974. Stratospheric sink for chlorofluoromethanes: Chlorine atom catalyzed destruction of ozone. *Nature* 249: 810–812.

Möller, F. 1963. On the influence of changes in CO_2 concentration in air on the radiative balance of the earth's surface and on the climate. *Journal of Geophysical Research* 68: 3877–3886.

Nacionalnyi Doklad SSSR k konferencii OOON 1992 goda po okruizhauishei srede i razvitiu. 1992. Moscow. Moscow: Minprirody.

National Acid Precipitation Assessment Program (NAPAP). 1990a. *Acidic Deposition: State of Science and Technology.* Vol. 1, *Emissions, Atmospheric Processes, and Deposition.* Washington: Superintendent of Documents.

————. 1990b. *Acidic Deposition: State of Science and Technology.* Vol. 2, *Aquatic Processes and Effects.* Washington: Superintendent of Documents.

————. 1990c. *Acidic Deposition: State of Science and Technology.* Vol. 3, *Terrestrial, Materials, Health, and Visibility Effects.* Washington: Superintendent of Documents.

————. 1990d. *Acidic Deposition: State of Science and Technology.* Vol. 4, *Control Technologies, Future Emissions, and Effects Valuation.* Washington: Superintendent of Documents.

Oden, S. 1967. Nederbordens Forsurning *Dagens Nyheter,* October 24, 4.

————. 1968. The acidification of air and precipitation and its consequences in the natural environment. In *Ecology Committee Bulletin* No. 1, Swedish National Science Research Council, Stockholm, Sweden; translation by Consultant Ltd., Arlington, Va.

Plass, G. 1956. Effect of carbon dioxide variations on climate. *Tellus* 8: 140.

Problemi ecologicheskogo monitoringa I modelirovanya ecosystem. 1981. (4: 235–250). Leningrad: Gidrometeoizdat.

Ramanathan, V. 1975. Greenhouse effect due to chlorofluorocarbons: Climatic implications. *Science* 190: 50–52.

Schneider, S.H., and R.S. Chen. 1980. Carbon dioxide flooding: Physical factors and climatic impact. *Annual Review of Energy* 5: 197–140.

Study of Critical Environmental Problems (SCEP). 1970. *Man's Impact on Global Environment.* Cambridge: MIT Press.

Study of Man's Impact on Climate (SMIC). 1971. *Inadvertent Climate Modification.* Cambridge: MIT Press.

Summers, P., and D.M. Whelpdale. 1976. Acid precipitation in Canada. *Water Air and Soil Pollution* 6: 447.

Taylor, H.R., et al. 1988. Effect of ultraviolet radiation on cataract formation. *NEJM* 319(22): 1429–1433.

UBA. 1989. *Verzicht aus Verantwortung: Maßnahmen zur Rettung der Ozonschicht.* Umweltbundesamt Bericht 7/89. Berlin: Verlag Erich Schmidt.

Ulrich, B., R. Mayer, and P.K. Khanna. 1979. *Deposition von Luftverunreinigungen und ihre Auswirkungen in Waldökosystemen im Solling.* Frankfurt am Main: Sauerländer.

UNEP, WMO, and ICSU. 1986. *Report of the International Conference on the Assessment of the Role of Carbon Dioxide and of Other Greenhouse Gases in Climate Variations and Associated Impacts.* Villach, Austria, October 9–15, 1985, WMO No. 661. Geneva: WMO.

U.S. Congress. House Committee on Appropriations. Subcommittee on the Department of Transportation and Related Agencies Appropriations. 1971. *Civil Supersonic Aircraft Development,* 92nd Congress, 1st Session.

U.S. Environmental Protection Agency (EPA). 1987. *Unfinished Business: A Comparative Assessment of Environmental Problems.* EPA/230/2-87/025a. Washington: EPA.

U.S. National Academy of Sciences (NAS). 1979. CO_2 *and Climate: A Scientific Assessment.* Washington: National Academic Press.

U.S. National Research Council (NRC). 1978. *Nitrates: An Environmental Assessment.* Washington: National Academy of Sciences.

————. 1983. *Acid Deposition: Atmospheric Processes in Eastern North America.* Washington: National Academy Press.

van der Sluijs, J., J. van Eijndhoven, B. Wynne, and S. Shackley. 1998. Anchoring devices in science for policy: The case of consensus around climate sensitivity. *Social Studies of Science* 28: 291–323.

van der Veen, C.J. 1986. Ice sheets, atmospheric CO_2, and sea level. Ph.D. thesis, IMOU, Utrecht University, Netherlands.

van der Veen, C.J., and J. Oerlemans. 1987. *Dynamics of the West Antarctic Ice Sheet.* Proceedings of a workshop held in Utrecht, May 6–8, 1985. Dordrecht: Reidel.

Wang, W.C., Y.L. Yung, A.A. Lacis, T. Mo, and J.E. Hansen. 1976. Greenhouse effects due to man-made perturbations in trade gases. *Science* 194: 685–690.

Wentzel, K.F. 1967. Die Belastung der Forstwirtschaft durch Immissionen und ihre technischen, waldbaulichen, raumplanerischen und rechtlichen Folgerungen. Veröffentlichungen der CEA Heft 35, Brugg, Switzerland.

Wentzel, K.F., and R. Zundel. 1984. *Hilfe für den Wald.* Niederhausen, Germany: Falken Verlag.

Whillans, I.M. 1993. Personal communication. Email message to Jeroen van der Sluijs, June 16.

World Meteorological Organization (WMO). 1975. *The Physical Basis of Climate and Climate Modelling.* GARP Publication Ser. 16. Geneva: WMO.

————. 1982. *Potential Climatic Effects of Ozone and Other Minor Trace Gases.* Report No. 14. Geneva: WMO.

————. 1985. *Atmospheric Ozone.* Global Ozone Research and Monitoring Project Report No. 16. Geneva: WMO.

————. 1991. *Scientific Assessment of Ozone Depletion: 1991.* Geneva: WMO.

Wynne, B. 1992. Carving out science (and politics) in the regulatory jungle. *Social Studies of Science* 22: 745–758.

16
Monitoring in the Management of Global Environmental Risks

Jill Jäger with Nancy M. Dickson, Adam Fenech, Peter M. Haas,
Edward A. Parson, Vassily Sokolov, Ferenc L. Tóth,
Jeroen van der Sluijs, and Claire Waterton[1]

16.1 Introduction

At the beginning of the period studied in this chapter monitoring of the environment was defined as the process of repetitive observing, for defined purposes, of one or more elements or indicators of the environment according to pre-arranged schedules in space and time, and using comparable methodologies for environmental sensing and data collections (Munn 1973).

Assessment is directly related to monitoring. In addition to what is called risk assessment in this study, which among other things examines the data collected in monitoring systems, there are two other meanings in the context of monitoring: (1) quality control and (2) examination of the efficiency of networks, including optimization of space and time densities of observations.

Munn (1973) discussed, in the context of setting up the Global Environment Monitoring System, the historical development of the scientific principles of environmental monitoring. The beginnings of international atmospheric monitoring networks were in the middle of the nineteenth century, after the invention of the telegraph.[2] Over the following hundred years the networks expanded, and there was increasing international standardization of observing procedures. In addition to atmospheric monitoring, networks were established in many countries in the last century or earlier to keep records of such things as the dates of the blossoming of fruit trees, spring breakup of ice, migration of birds, and so on.

Data on periodic biological phenomena (also known as phenological information networks) were maintained for decades, and indeed around the time of the Stockholm Conference in 1972 there was an upsurge of interest in the use of phenological events as indicators of anthropogenic environmental changes.

As Munn (1973) pointed out, a number of important issues need to be resolved in the design of monitoring programs, including the objectives for monitoring, the size of the network, and the need for series long enough to determine trends. Often it is necessary to undertake pilot studies with a time limitation on the data gathering to determine the feasibility and usefulness of the measurements.

The principles of monitoring were also discussed by the International Atomic Energy Agency (IAEA 1965), which distinguished between two objectives:[3] (1) to define the state of the environment and thus to provide a basis for predicting its future state and (2) to determine whether there is a present risk to human health and welfare.

Munn (1973) pointed out that a monitoring system must not be tied too closely to a particular model of the environment, since society's understanding of the environment or the intended use of the data may change over time. This is illustrated later in this chapter.

In reviewing the progress made in environmental monitoring between 1972 and 1992, Tolba et al. (1992) pointed out that during this period there was an enormous growth in the demand for information on the state of the environment from the public, scientists, resource managers, politicians, and many other actors. This demand stimulated both international and national initiatives. The management of the increasing flow of environmental data also became an important issue, and developments in computer technology (such as the storage of large amounts of data on compact disks) over the twenty-year period were extremely important in solving some of the problems.

Tolba et al. (1992) described a number of monitoring systems of relevance to the management of global environmental risks. These include

• The Global Observing System (GOS), which is coordinated by the World Meteorological Organization (WMO);

• The Global Atmosphere Watch (GAW), which monitors the chemical composition of the atmosphere and precipitation and which originated conceptually in the 1950s, when first systematic measurements of carbon dioxide concentration began (GAW started in 1988, the forerunner Background Air-Pollution Monitoring System (BAPMoN) began in the 1960s, and ozone monitoring began in the 1950s);

• The Global Environmental Monitoring System (GEMS), established by the United Nations Environment Programme (UNEP) in the early 1970s for monitoring atmosphere and climate, environmental pollutants, and renewable resources;

• The Long-Term Ecological Research (LTER) programs.

The other activity during the 1972 to 1992 period noted by Tolba et al. (1992) was considerable research into optimization of monitoring systems. Tolba et al. specifically mentioned the acid-deposition issue in Europe and North America in this regard, where optimization was used to reduce expected high costs.

Monitoring plays an important, if often neglected, role in dealing with the risk of global environmental change. As is shown in this chapter, the interactions between monitoring and other risk management functions are well defined and important. In the case of the climate change issue, for example, the results of monitoring of temperature at the earth's surface and the atmospheric carbon dioxide concentration were used repeatedly in assessments of the magnitude of the risk of climate change. The monitoring of the ozone concentration in the stratosphere above Antarctica, motivated by scientific interest and not by concern about the possible impacts of chlorofluorocarbons, led to major changes in the management of the issue of ozone depletion after the discovery of the Antarctic "ozone hole." Similarly, in the case of the issue of acid rain, the monitoring of acidification of Scandinavian lakes and of forest damage in Germany were important in the framing of the issue, risk assessment, and option assessment.

This chapter focuses mainly on the monitoring of natural systems and some related factors. Monitoring of human responses to global environmental risks is rarely considered here. To trace patterns in time, it is important to note the point in time when particular innovations in monitoring systems were put in place as well as the origin of each innovation. Another important indicator for changes in risk management is the time at which particular types of monitoring data are first used (such as collected, reported, or averaged) by an actor.

Section 16.2 of this chapter presents stories on monitoring of various aspects of the three global environmental risks considered in this study. Section 16.3 takes a more general look at patterns in and motivations for monitoring and concludes by looking for evidence that monitoring systems have improved over time. Technical adequacy of monitoring of global environmental risks has been consistently high over the time period considered in this study. Effectiveness and value of monitoring systems have improved. In particular, monitoring has been very effective in framing and reframing the debates on each of the environmental risks studied here. Section 16.4 draws from the stories and analyses in the rest of the chapter to discuss problems and pitfalls encountered in monitoring of global environmental risks.

16.2 Stories

16.2.1 Introduction

This section includes a set of stories that illustrate the development of monitoring systems, the use of monitoring data, and some of the encountered difficulties. Most of the stories show clearly that information from monitoring systems influenced the debate on the global environmental risks studied here. In the climate change case, the monitoring of the atmospheric carbon dioxide concentration was essential to demonstrate that the concentration was increasing from year to year. The monitoring of gas bubbles in ice, a very sophisticated kind of observation developed during the time period examined in this study, showed that changes in the concentration of greenhouse gases and related changes of the earth's surface temperature had also occurred in the past. This evidence of "the greenhouse effect" was important in the climate debate, especially in convincing people to take the issue seriously.

Similarly, in the cases of stratospheric ozone depletion and acid rain, monitoring influenced the debate, most particularly with the discoveries of forest dieback and the Antarctic ozone hole, discussed later in this section.

Some stories illustrate the development of important monitoring systems (such as EMEP, atmospheric carbon dioxide concentration, and integrated monitoring) and the role that individual or groups of scientific entrepreneurs have played. Four stories discuss various aspects of monitoring related to stratospheric ozone depletion and illustrate some of the difficulties and pitfalls that can arise.

16.2.2 Gas in Ice Bubbles

In the case of climatic change, monitoring of greenhouse gas concentrations in the ice of Antarctica and Greenland showed that greenhouse gas concentrations and the earth's temperature were correlated over many thousands of years of the earth's history. This was an important stimulus in some countries for setting goals to reduce greenhouse gas emissions.

The ability to recover long ice cores from Greenland and Antarctica, measure the concentrations of greenhouse gases in the different layers of the ice, and thus produce data on concentrations in the past has played an important role in the debate on the increased greenhouse effect. In particular, the results from the Vostok core in Antarctica, which showed concentrations of CO_2 and methane (CH_4) and temperature over the last 160,000 years, were important in showing a clear relationship between greenhouse gas concentrations and the global surface temperature.

Langway and Oeschger (1989) provided a summary of the findings of ice-core drilling projects. They pointed out that ice-core drilling into polar glacier ice for scientific purposes was first proposed on a major scale by H. Bader in 1954 (Bader 1958). The first pre-International Geophysical Year (IGY) ice-core drilling operations were conducted at Site 2 in northwest Greenland in 1956 and 1957. These ice cores reached depths of 305 and 411 meters, respectively. Langway and Oeschger (1989) concluded that surprisingly few ice cores have been obtained from various Greenland and Antarctic locations over the past thirty years. Only five ice cores had been recovered from depths of over 1000 m. Langway and Oeschger (1989, 2) described the significance of this research as follows: "These ice cores were studied by small numbers of individuals and international teams who mostly worked together in an exciting atmosphere of discovery and new ideas. The mutually collaborative field operations and laboratory studies cultivated scientific progress and opened up a new field in earth science which provided fresh information to modern thinking about earth processes."

Since the start of the first intermediate ice-core drilling project in Greenland, many new field methods were developed for handling, processing, sampling, and study of ice cores. The scientific results from the shallow, intermediate, and deep ice cores recovered through the past thirty years have been widely published. These studies demonstrated that ice sheets are unique natural depositional environments with an abundance of paleoenvironmental data.

Hecht et al. (1989) pointed out that much of the ice core research between 1971 and 1981 focused on two basic problems: obtaining a reliable chronology and deriving accurate estimates of climate variables. In the 1980s new findings in ice cores greatly expanded their use in the study of the earth's history. In particular studies were able to document the changing levels of carbon dioxide and methane in the atmosphere. In less than twenty years a wealth of environmental information was extracted from only five long ice cores.

Detailed long-term environmental records were derived from drilling performed on the high plateau of Antarctica. Three such series reached back at least into the last glaciation. The first was obtained at Byrd Station in West Antarctica, where the drilling reached bedrock in 1968. The ice-flow conditions there meant that the derived chronology became uncertain beyond about 20,000 years before present. Twelve years later at DOME C in East Antarctica a deep ice core was drilled, and the derived chronology reached back into the last ice age. At Vostok Station, East Antarctica, a first series of drilling

reached a depth of 950 m in 1974. At the same site a new deep hole was drilled in 1980 and reached 2083 m in 1982. The depth of penetration was later extended to 2200 m. For the first time an ice core extended as far back as about 160,000 years, completely through the last glacial-interglacial cycle.

Initial studies of the air trapped in Greenland and Antarctic ice cores revealed that the atmospheric carbon dioxide content could have been around 200 parts per million volume (ppmv) during the last glacial maximum compared with an average close to 270 ppmv during the postglacial Holocene period (Delmas, Ascencio, and Legrand 1980; Neftel et al. 1982). These studies were later confirmed by Antarctic Siple data showing a marked increase in CO_2 due to anthropogenic disturbance. The CO_2 record was extended over the last 160,000 years on the basis of the Vostok core (Barnola, Raynaud, Korotkevich, and Lorius 1987). The Vostok data showed a high correlation[4] between the CO_2 concentration and the temperature recorded by isotope measures in the ice cores.

Since the publication of the Vostok data major drilling projects have continued, and projects have also looked at other greenhouse gases (such as methane), but it was the Vostok data and in particular the high correlation between the CO_2 concentration and the temperature that led to a shift in the perspectives on the risk of climatic change. As it was possible to show clearly that the CO_2 concentration and the earth's temperature were correlated over long periods of the earth's history, decision makers—such as the European Commission (EC) and the German Enquete Commission (Enquete-Kommission, EK)—began to take the issue of global warming more seriously.

16.2.3   Measuring the Atmospheric Concentration of Carbon Dioxide

The original motivation for monitoring the atmospheric carbon dioxide concentration was unrelated to the risk of climate change. The curiosity and determination of one person were central in the creation of a set of monitoring data, subsequently used worldwide in assessment of the risk of climate change.

Jonathan Wiener (1990) in his book *The Next One Hundred Years* recounted in considerable detail the history of attempts to measure and record the concentration of carbon dioxide in the atmosphere. While early attempts were important especially in scientific circles, the efforts of one man, C. D. Keeling, are acknowledged as being central to the demonstration that the CO_2 concentration has been rising in recent decades.

Keeling did not begin to monitor the atmospheric concentration of CO_2 to show that it was increasing. He began to prove the validity of a hypothesis of Harrison Brown, the leader of the geochemistry program at the California Institute of Technology, where Keeling went to work after receiving his degree in chemistry.[5] To do this, he needed to develop a device for measuring carbon dioxide in air or water in small quantities (parts per million). It took about a year to develop this instrument, and Keeling began measuring the concentration of carbon dioxide in spring 1955. He was soon able to show by regular four-hourly outdoor measurements that the atmospheric concentration varies diurnally, decreasing when photosynthesis occurs and increasing when there is no daylight.

In 1955 Keeling also found that the concentration in the atmosphere as a whole was about 315 parts per million. His work was picked up by the scientists preparing for the International Geophysical Year (IGY). He transferred to Scripps Institution of Oceanography to work with Roger Revelle, who directed the institute and was one of the planners of the IGY. After the development of a new monitoring instrument, measurements began on Mauna Loa, Hawaii, in March 1958. Mauna Loa was chosen as a site because it is above the trade-wind inversion and far from industrial sources of CO_2, so the record would not be influenced by local sources and sinks. The first year's data showed a smooth annual cycle with lower values in the summer and higher values in the winter. In the second year, Keeling found that the atmospheric CO_2 concentration was one part per million higher than in 1958.

After observing for ten years that the concentration was increasing, Keeling began to look at the data on the production of fossil fuels and calculated the release of carbon dioxide into the atmosphere as a result of fossil fuel combustion and cement production in the world as a whole. He was able to show that in every year from 1959 to 1972 the emissions increased by an average of 4 percent per year.

Keeling's pioneering work in measuring the atmospheric CO_2 concentration at Mauna Loa and also within the IGY at the South Pole were the beginning of precise atmospheric CO_2 measurements. The Mauna Loa record is recognized as an extremely valuable time series. After the initial IGY monitoring efforts, other agencies and organizations implemented programs to monitor the atmospheric CO_2 concentration. Two of the larger programs are the BAPMoN of the World Meteorological Organization[6] and the Climate Monitoring and Diagnostics Laboratory (CMDL) of the U.S. National Oceanic and Atmospheric Administration.

The Mauna Loa atmospheric CO_2 concentration measurements became widely known and were used worldwide to illustrate the increase in concentration. However, as noted above, the measurements were not begun to show that increase. Furthermore, the determination of one person, C. D. Keeling, to develop instruments and to continuously sample the atmosphere played an important role in the development of CO_2 monitoring.

16.2.4 Monitoring of CFC Production and Emissions

To assess the risk of stratospheric ozone depletion, accurate estimates of the emissions of chlorofluorocarbons (CFCs) into the atmosphere are required. As a first step CFC production figures have to be analyzed. The availability of these data is very variable from country to country as is the time at which reporting began. This story illustrates the difficulties in assembling production statistics.

U.S. monitoring of the production and sales of CFCs began in 1959 with the inclusion of CFC-11, CFC-12, and HCFC-22 in the International Tariff Commission's (ITC) *Synthetic Organic Chemicals: Production and Sales—1958.* Carbon tetrachloride production and sales data have been reported since 1922 in this annual publication, which dates back to 1917. Data on methyl chloroform were included since 1967. All of the information came from questionnaires submitted to domestic producers and was not independently verified (ITC 1991). However, the ITC reports were not in response to any notion of a threat to the ozone layer; they were a compilation of trade statistics. Monitoring of production and sales of CFCs in response to the issues of stratospheric ozone depletion began in 1976, initiated by the Fluorocarbon Program Panel of the Chemical Manufacturers Association (CMA), an international industry group based in Washington, D.C. In 1990 the panel ceased its activity, and the Alternative Fluorocarbon Environmental Acceptability Study (AFEAS), also based in Washington, D.C., began issuing the reports.

The Canadian government began monitoring CFC production following the 1977 announcement that Canada would phase out certain nonessential uses of CFC aerosols. A 1975 Act granted the Environment Minister power to demand quarterly CFC production figures. This authority was first exercised in March 1977, calling for production figures for CFC-11 and CFC-12 from each producer and requesting production figures for the past six years. From 1980 onward, similar reports were required on production of all halocarbons.

Following the signing and ratification of the Montreal Protocol, the Canadian government introduced regulations requiring *all* CFC-producing companies to report every three months on the production and import of ozone-depleting substances. While there were no provisions in the regulation that required independent audits of the data to be undertaken, informal audits were conducted periodically. For example, a 1989 audit of the INCENDIX company of Montreal led to a fine of $3,000 for improper reporting of imported ozone-depleting substances.

In Germany there was considerable uncertainty about the production figures for CFCs, largely because the companies producing them did not want to publicize the figures for reasons of competitiveness. In the 1970s reports often relied on the production statistics compiled by the CMA.

In 1988 the Enquete Commission discussed the difficulties of obtaining reliable production statistics in its interim report. It pointed out that Hoechst and Kali-Chemie, the two CFC manufacturers based in the Federal Republic of Germany, did not disclose any figures on either their CFC production or imports or exports. Figures on consumption volumes were available only for subsectors. For the reference year 1986, the Enquete Commission obtained the information that 26,000 tons of CFCs were used in sprays, but the amount used in refrigerators had to be extrapolated. The consumption volumes in all the other major CFC applications (foamed plastics, solvents, and cleaners) had to be estimated on the basis of the production processes used and on the basis of data supplied by users.

Based on information supplied by industry, the Federal Environment Agency (UBA) estimated in mid-1987 that the CFC-consumption in Germany was around 60,000 tons per year. This figure was subsequently corrected to between 90,000 and 100,000 tons per year, since the amount used for solvents (in particular CFC-113 for degreasing and drying purposes) had been considerably underestimated. In April 1988 the Trade Association of the Chemical Industry (VCI) gave a figure of 75,000 tons for the consumption of fully halogenated CFCs in 1986 on the basis of information from Hoechst and Kali-Chemie. The CFC Working Group of the European producers pointed out that the actual consumption volume was 10 percent above or below the 75,000 tons. Therefore, the Enquete Commission concluded that it was not unrealistic to assume that consumption in 1987—and presumably also in 1988—amounted to at least 100,000 tons.

There was similar confusion about the production figures. In January 1988 the Federal Environment Minister estimated that German CFC production was 64,000 tons. The Öko-Institute used a combination of known data about inputs and the technically achievable outputs and calculated that the production amount must lie between 125,000 and 145,000 tons. Subsequently, the Environment Minister and the industry revised their estimates to 100,000 to 106,000 tons. The Enquete Commission

and the Federal Environment Agency estimated about 110,000 tons.

Nongovernmental organizations (NGOs) in Germany suggested that the refusal of industry to provide production statistics resulted in the following problems:

- The past emissions were not well known,
- The model calculations could not be verified, and
- Risk assessment and response development were handicapped.

In 1989 the UBA report also pointed out the difficulties associated with obtaining CFC-production figures. The report concluded that it was impossible to obtain an exact figure for CFC production in 1986, which meant that it was impossible to make a public verification of the compliance with national and international reduction goals.

In 1990 the situation had not changed. The Enquete Commission pointed out that there were no publicly verifiable data on national production and consumption volumes of CFCs, halons, and other ozone-depleting gases (ODGs). The German manufacturers still had not submitted any exact, verifiable, and broken-down figures on production volumes in 1986, the baseline year of the Montreal Protocol and the corresponding decision of the German parliament.[7]

In Russia the CFC-production data were not widely published and were available only for internal interdepartmental use. Before the Vienna Convention entered into force, the USSR had refused to provide CFC production data to the international community. One of the reasons for this was the involvement of the Ministry of Defense in ozone research, which resulted in substantial secrecy in this field.

The above points show that there were large differences in the availability of CFC production statistics and in the time that these statistics were made public. In the United States the statistics were reported early, but not because of concern about the stratospheric ozone layer. In Canada, the government made an early requirement that the CFC-production statistics should be regularly reported. In Germany there were difficulties in obtaining detailed, reliable production data. In European Community countries in general the production data were reported to an accountancy office to protect industry. In Russia the data were also apparently not publicly available and verifiable. Clear discrepancies arose between reported production and use volumes in individual countries. These studies were made at different times (before 1985 in the United States through the Rand Corp., in 1989 in Germany through the Federal Environment Agency, and in 1991 by Greenpeace Germany). Very few actors appear to have acknowledged that the *emissions* of CFCs are not the same as the annual production figures, since the CFCs are stored for variable amount of time in refrigerants and foams.

16.2.5 Discovery of the Antarctic Ozone Hole

The discovery of stratospheric ozone depletion in the Antarctic through monitoring systems that were not set up to look for such depletion had a major impact on other risk management activities. It led to renewed interest in risk assessment as scientists sought to explain the observed depletion and goal and strategy formulation as the image of the ozone hole became widely disseminated and people concluded that something must be done. Increased response assessment and monitoring of ozone concentrations were also a result. The story of the "discovery" shows how scientific curiosity and principles play a role in the management of global environmental risks.

Thinning of the ozone over Antarctica was first noticed by J. Farman of the British Antarctic Survey (BAS) in the late 1970s. In 1982 scientists in the United Kingdom, Japan, and the United States noticed decreased levels of ozone. Each group of scientists dealt with the discovery independently. While Japan's Chubachi Shigeru was the first to report his findings publicly in 1983, British scientist Farman received international recognition for the discovery of the hole.

Ozone measurements have been made by BAS from October to March at the Halley Bay station since 1957 using the Dobson spectrophotometer. Farman first became worried about the ozone loss in 1982 when a spring loss of ozone of approximately 20 percent was recorded, although Farman recalled that the data suggested "something interesting" as early as 1978 and 1979. Before sharing the observations with the wider scientific community, he decided to wait until the results of a more sophisticated instrument became available. He began using the new meter in 1982, and by 1984 the readings confirmed the ozone changes indicated by the old Dobson meter. This internal corroboration convinced Farman to go public with his findings. Farman chose not to seek external corroboration with his colleagues, either in Japan or the United States, claiming that the other programs had been gathering data for only relatively short time periods (which was not true for the United States or Japan) and that the differences in positioning of the two monitoring stations rendered a direct comparison very problematic.

The single biggest early contribution that Japanese researchers made to international studies of stratospheric ozone depletion was the work of Chubachi Shigeru and his colleague, Kajiwara Ryoichi. Chubachi, a specialist on the stratospheric ozone layer in the Meteorological Agency's Aerological Observatory, was a member of a winter research team sent by the Meteorology Agency to the Showa base in Antarctica. The Showa base, Japan's most distant meteorological observation outpost, was established in 1957 on East Ongul Island, about 5 kilometers from the Antarctic mainland. Stratospheric ozone had been measured at the base since 1966. Ozone measurements were made based on changes in the amount of ultraviolet radiation absorbed by the stratospheric ozone layer. Chubachi, however, set out to devise a new method that would make it possible to continue to obtain data on stratospheric ozone by moonlight. To do this, Chubachi incorporated a highly sensitive device into a Dobson spectrophotometer.[8]

On the coldest day on record at the base ($-45°C$) and the first day of sunlight observation in four months, September 4, 1982, Chubachi's data showed that the amount of stratospheric ozone was significantly less than the previous day's measurement. It was also extremely low when compared with data from the previous sixteen years of measurements. For the next month and a half, the data continued to show a decrease in the ozone layer that was considerably larger than estimates based on previous measurements would predict. Not only was the depletion rate high, but ozone depletion was also recorded at a height of 20 kilometers. At the time, scientists were talking about ozone depletion due to CFCs above a height of 40 kilometers but thought that the ozone-depleting chemical reactions would not occur at the height of 20 kilometers.

There was some suspicion that the low readings were due to some kind of problem with the new Dobson device. On returning to Japan in March 1983, therefore, Chubachi had the device tested. It was working properly, so Chubachi decided to report his findings. He first reported them in the spring of 1983 at the Ministry of Education's Polar Research Institute. He then traveled to Greece the following year and reported his findings at the 1984 Quadrennial Ozone Symposium (Chubachi 1984).[9] There were only seven other Japanese at this meeting of 300 scientists. Chubachi's findings received little reaction. Later that year, however, there was a conference on middle levels of the atmosphere in Kyoto. This is where atmospheric scientist Susan Solomon, who was later to play a role in the discovery of the ozone hole, heard about Chubachi's findings. Eight months after Chubachi's paper was presented in Greece, Farman presented his findings of an ozone hole over Antarctica to the world (Farman, Gardiner, and Shanklin 1985). His findings were similar to those of the Japanese Showa team, but whereas Chubachi's data showed a decrease in ozone levels only for a two-year period, Farman's data, beginning in 1957, showed a longer-term decline. Solomon, who had doubts about Farman's theory but needed supporting data, turned to Chubachi's paper. Thus, a Japanese scientist's discovery of the ozone hole became an integral part of subsequent international developments and the formation of an international consensus that stratospheric ozone depletion was a reality.

Farman and his colleagues submitted their finding in a paper to *Nature* at the end of 1984. They described the recorded depletion in ozone of over 30 percent and the correlation between CFC concentration and ozone depletion.

Why did the United States miss the ozone hole? The United States had been monitoring stratospheric ozone levels since 1957. The common belief is that the *Nimbus 7* satellite was programmed to reject the unusually abnormal data. However, Richard McPeters, head of the U.S. National Aeronautics and Space Administration (NASA) Ozone Processing Team, said that the abnormal Total Ozone Mapping Spectrometer (TOMS) data from 1983 above Antarctica were "flagged" rather than "thrown out." McPeters stated that in July 1984 a comparison was made with the Dobson station at the South Pole. The TOMS data did not match the Dobson data at all, so the satellite data were then considered anomalous; only later did NASA scientists realize that the Dobson station was in error. When they realized this, they submitted a paper for the 1985 meeting of the International Association of Meteorology and Atmospheric Physics (IAMAP), but Farman's article was published first (Pukelsheim 1990). This erroneous acceptance of the South Pole Dobson data would explain why neither U.S. ground-based systems nor U.S. satellite systems supported the existence of an ozone hole; the two systems performed a reverse "check" on each other.

16.2.6 Ozone Trends Panel

The Ozone Trends Panel (OTP) report demonstrates three important points about monitoring—that data series must be accompanied by interpretive information; that the menial tasks of reliable data gathering, storage, and calibration are crucially important; and that an authoritative statement of an apparently simple observation can influence the policy debate.

In early 1986, soon after the report of large seasonal ozone losses over Antarctica, a controversy arose over whether a substantial global-scale decline in total column ozone had also occurred. One interpretation of data archived from two instruments launched in October 1978 on the *Nimbus 7* satellite—the Solar Backscatter Ultraviolet (SBUV) instrument and the Total Ozone Mapping Spectrometer (TOMS)—seemed to show a loss in total ozone of several percentage points in the Northern

Hemisphere and up to 18 percent in the Southern between 1978 and 1986. Ozone losses appeared particularly severe in the upper stratosphere, more than 20 percent at middle latitudes.

Presentation of these observations to Congress in early 1986 resulted in substantial controversy, both because of the size of the apparent losses and because the Dobson network of ground-based instruments measuring total ozone appeared to show a global decline since 1979, but one much smaller than reported from the satellite observations. It was known that deposition of films on optical surfaces in the satellite instruments was degrading their performance, but it was not known whether the models being used to correct for this degradation in the archiving of data were adequate.

NASA managers responded to the controversy by establishing in October 1986 a panel of senior scientists to review all monitoring data available on global ozone trends. The Ozone Trends Panel included about 100 scientists from ten countries. Following the example of the highly successful 1985 international ozone assessment, the project (led by NASA), was jointly sponsored by NASA, the U.S. National Oceanic and Atmospheric Administration (NOAA), and the U.S. Food and Agriculture Administration (FAA) in the United States and by UNEP and WMO.

The Panel worked for sixteen months, conducting a comprehensive reanalysis of observations available from both ground- and satellite-based instruments. On reexamining the archived data from SBUV and TOMS, the Panel concluded that estimated ozone loss was highly sensitive to the model used to describe instrument degradation and that, since the model used to produce the data archived up to 1986 probably understated the effect of degradation, the archived observations probably overstated ozone loss. Other models generated estimated ozone trends consistent with a much smaller downward trend.

The correction of satellite-based observations yielded the first well-established observation of a global downward trend in ozone. After removing other, cyclic effects, the Panel concluded that the Dobson record showed a decline between 1969 and 1986 of 1.7 to 3 percent in annual average total ozone over northern temperate latitudes, with a somewhat larger decline, 2.3 to 6.2 percent, during the months December to March. These losses were roughly two to three times larger than those predicted by current models.

The Panel released the results in its Executive Summary at a press conference in Washington on March 15, 1988. The Panel also reported results from recent Antarctic investigations and stated that the weight of evidence strongly indicated that "man-made Chlorine species" were primarily responsible (NASA 1988).

The chair of DuPont had stated only weeks earlier in a letter to three U.S. senators that scientific evidence did not yet justify a change in their policy to continue producing CFC. The OTP Press Conference prompted a rapid reevaluation of the policy. DuPont announced one week later that it would cease producing CFCs by 1999 (Shabecoff 1988).

Many international sources identify the release of the OTP report as a key event in identifying the seriousness of ozone depletion. Its numbers were cited by the German Enquete Commission and the Dutch National Environmental Plan.

The OTP report demonstrates three important points about monitoring. First, the OTP examination of the TOMS and SBUV record reveals the contingent character of monitoring data—that no data series means very much without some body of accompanying interpretive information. The observations reported in 1986, seeming to show large ozone losses, were dependent on the model used to correct for diffuser-plate degradation, which had been applied to the data before archiving. Other degradation models applied to the same primary data stream could indicate losses even more severe or essentially no loss. Second, the OTP reexamination of the Dobson record reveals the crucial importance of the most menial, low-level tasks of reliable data gathering, data storage, and instrument calibration. Scientists had been looking at the Dobson record for decades without finding a trend, but the tedious, methodical work of cleaning up the basic data at each station revealed one. Third, the example reveals how an authoritative statement of an apparently simple observation, sanctioned by a sufficiently broad set of experts (internationally and including industry representation) can decisively advance debate over policy action.

16.2.7 Monitoring of Ultraviolet-Radiation Exposure

This story gives an example of an innovation in a monitoring system in response to an observed global environmental risk. Although the example is described for only one country, the innovation spread rapidly in the early 1990s, as both public awareness of the issue of ozone depletion and the incidence of skin cancer increased.

Monitoring can be undertaken for a variety of purposes and audiences, and it is common to imagine the primary audiences as scientists and policy makers. An innovation in monitoring in the 1990s, though, made the public the primary audience and sought to provide members of the public with the information that they needed to manage their own risk to UV exposure.

In Canada, after public and political scares that the Northern Hemisphere was primed for severe ozone depletion, Environment Canada established two information programs to make monitoring data available to the general public—the Ozone Watch and the UV Index.

The Ozone Watch, established in March 1992, was designed to report how current two-week average ozone levels over Canada compared to average levels from 1960 to 1980 at the same time of year. Developers of the program in part intended it to mute the intensity of public concern, since they believed measurements would show that ozone fluctuated markedly both above and below historical levels and that no long-term trend was easily discernible. While 1992 observations did not show the extreme depletions that had been predicted, a substantial decline appeared in early 1993.

The UV Index, also introduced by Environment Canada in 1992, provided the world's first daily UV forecast. It combined recently observed ozone levels with weather-forecasting models to predict daily UV intensity over major Canadian cities, rated on a scale from 0 to 10. An accompanying chart converted the UV rating into the average time a fair-skinned person will take to sunburn. The UV Watch is reported daily by news media with the weather forecasts, together with a set of simple warnings to protect against excessive exposure. WMO adopted the ten-point system as a standard for reporting UV exposure.

One objective of monitoring programs is to provide the data that decision makers need to take appropriate action. Since some increase in surface UV is bound to occur and people will face a change in risk that depends on their own protective actions, it is clear that for some decisions the people who need the monitoring information are members of the general public. These two programs were first attempts to provide this information, both to damp excessive concern about catastrophic risks and to provide people the information they need to manage their own risks. Polling results suggested that these programs were both increasing citizen's knowledge about UV risks and changing their behavior. Indeed, program organizers speculated that in Canada cancer reductions due to people reducing their exposure may exceed increases due to ozone depletion.

16.2.8 Monitoring of Acid Deposition in Europe

Wide networks of stations monitoring acid deposition have played an important role in the development of policies in response to acid rain in Europe. In particular the Cooperative European Monitoring and Evaluation Programme (EMEP) produced a long-term set of comparable data, which was used by policy makers in Europe in developing protocols to the Convention on Long-Range Transboundary Air Pollution (LRTAP).

In the mid-1940s a Swedish soil scientist, Hans Egnér, developed a systematic way to look at the fertilization of crops by nutrients from the atmosphere. This work led to the establishment of a network for the collection and chemical analysis of precipitation. Sampling buckets were set out at experimental farms all over Sweden, and the major chemical constituents deposited in these buckets (rain, snow, dust) were measured on a regular monthly basis. The acidity of precipitation was one of the chemical characteristics that was measured. Other agricultural scientists gradually expanded the network to Norway, Denmark, and Finland and later to most of Western and Central Europe. The expanded network was called the European Air Chemistry Network (EACN), and it provided the first large-scale and long-term data on the changing chemistry of precipitation and its importance for agriculture and forestry (Cowling 1982; Emanuelsson, Eriksson, and Egnér 1954; Egnér, Brodin, and Johansson 1955). In 1956 the International Meteorological Institute in Stockholm assumed responsibility for the coordination of the network. In 1957, as part of the International Geophysical Year, the network expanded to include Poland and the Soviet Union.

At the 1972 Stockholm Conference on the Environment Sweden presented a case study on *Air Pollution across National Boundaries: The Impact of Sulfur in Air and Precipitation* (Bolin et al. 1972). European nations agreed at the Conference that monitoring should be carried out to get more information on the issue of long-range transport of air pollutants. The OECD was selected as the forum for this monitoring program, and in 1972 the OECD launched a cooperative technical programme to measure the long-range transport of air pollutants. The objective was to determine the relative importance of local and distant sources of sulfur compounds in terms of their contribution to the air pollution over a region, with special attention being paid to the question of acidity in atmospheric precipitation (OECD 1979). The program was initiated after about two years of preparatory work supported by the Scandinavian Council for Applied Research. Eleven OECD members participated: Austria, Belgium, Denmark, the Federal Republic of Germany, Finland, France, the Netherlands, Norway, Sweden, Switzerland, and the United Kingdom.

The OECD program consisted of a set of monitoring stations collecting data on deposition. The Norwegian Institute for Air Research (NILU) acted as the Central Coordinating Unit and compiled the data.

In 1978 the OECD gave the program independent status as the Cooperative Programme for the Monitoring and Evaluation of the Long Range Transmission of Air Pollutants in Europe. The program became known by the acronym of its shortened name, European Monitoring and Evaluation Programme—EMEP. There were several reasons for making EMEP independent of the Organization for Economic Cooperation and Development (OECD), but most important was the need for participation of Eastern European countries (Dietrich 1992). The United Nations Economic Commission for Europe (UNECE), with the assistance of the World Meteorological Organization (WMO) and the United Nations Environment Programme (UNEP), assumed responsibility for EMEP.

EMEP was set up with three major tasks: pollutant measurement at monitoring stations, collection of emission estimates from countries, and modeling of air pollutant transport. The work was divided among four centers:

• The Chemical Coordinating Center (CCC) at the Norwegian Institute for Air Research (NILU),

• The Meteorological Synthesizing Center East (MSC-E) at the Institute of Applied Geophysics in Moscow,

• The Meteorological Synthesizing Center West (MSC-W) at the Norwegian Meteorological Institute in Oslo, and

• The Coordinating Center for Effects (CCE) at the National Institute for Environmental Health and Environmental Protection (RIVM) in the Netherlands.

EMEP expanded its data collection and analysis from one pollutant, sulfur, in the 1970s to a range of pollutants—including nitrogen dioxide, ammonia, volatile organic compounds, ozone, and heavy metals.

Within the framework of the UNECE the Convention on Long-Range Transboundary Air Pollution (LRTAP) was adopted in Geneva in 1979 and entered into force in 1983. In 1984 this Convention adopted a Protocol on Long-Term Financing of EMEP. This Protocol committed parties to mandatory annual contributions to the budget approved by the executive body. The Protocol entered into force in 1988 and has ensured the continuous operation of EMEP.

Interviews conducted in 1992 (Dietrich 1992) suggested that the most important contribution of EMEP was to develop information useful for policy makers, while constrained by limited resources and relatively simple methods and models. Scientific results were converted into practical, politically useful information, which was nearly universally accepted by European decision makers. A manual with recommended methods to sample air and precipitation and recommended analytical techniques was issued by EMEP in 1977 and regularly updated thereafter, so that by 1990 most European stations were using harmonized methods and techniques.

16.2.9   Forest Monitoring

By 1992 forest monitoring was being carried out in most countries. However, the monitoring was started at different times and for different reasons. The discovery of widespread forest dieback in Germany in the early 1980s led to increased monitoring of forest health in Germany and other European countries. This monitoring was a result of previous monitoring efforts and stimulated risk assessment, option assessment, and strategies to respond to the risk of acidification.

Forest monitoring started with episodical and localized surveys in Scandinavia (Odén 1967) and in North America (Likens and Borman 1974), where interest was in monitoring of the forest nutrient cycle. In Germany, monitoring of the nutrient cycle began in the late 1960s under the auspices of the International Biosphere Program, and this research played a major role in the forest-dieback and acid rain debate when it developed in the early 1980s.

The turning point in forest monitoring was in the years between 1983 and 1985, when systematic monitoring of forest health began in many countries. In Germany systematic forest monitoring began in 1983 through cooperation between the Federal Ministry of Nutrition, Agriculture, and Forests and the State Forest Administrations. Uniform sampling standards were introduced in 1984.

In the former Soviet Union forest monitoring with respect to the problem of acidification began in 1978 within a Soviet-American project on Forest Ecosystems and Pollutants Monitoring. The plan for a permanent system of forest monitoring was approved by the USSR government in 1984, with the objective of expanding the number of permanent stations to thirty-five to forty by 1990. The expanded system of joint monitoring of air pollution and forests began in Hungary in 1988. The U.S. forest monitoring system related to the acidification issue was established with the National Acid Precipitation Assessment Program in 1982.

In 1984 the Forestry Department of the Directorate General for Agriculture of the European Community started a feasibility study, and one of its major goals was the establishment of a monitoring network to look at forest health and acidification. In 1986 a Council Resolution on the protection of the community forest was adopted, and in the following year a report on forest monitoring was included in the fourth Environmental Action Plan.

In the United Kingdom the first annual survey of the impacts of acidification on forests was made by the Forestry Commission in 1984. The U.K. Tree Health

Group was set up in 1985 to make surveys of forests on a continuing basis.

Two major programs related to forest monitoring were adopted in the United States in the mid-1980s.[10] The Integrated Forest Study and, beginning in 1985, the Forest Response Program were established under the auspices of the National Acid Precipitation Assessment Program (NAPAP). Later, in 1990 the Environmental Monitoring and Assessment Program was initiated.

In summary, most systematic forest-monitoring systems were introduced in the 1980s after the discovery of the forest impacts of acid deposition. It is important to note, however, that the monitoring systems were changed over time, as knowledge of the problem and political interest in the issue increased. In the Federal Republic of Germany, for example, between 1982 and 1983 the reported areal extent of forest damage increased by a factor of 4, but the sampling method had also been improved during this interval. Subsequently, there were changes in the categorization of damage. For example, in 1983 the trees were categorized as weakly, moderately, or strongly damaged. For the strongly damaged trees there was no uniform estimate because often the dying trees had been removed from the forest. By 1984 the categories were changed to Step 0 (no sign of damage), step 1 (slight damage), and step 2 (medium), step 3 (strong damage), and step 4 (dead). In 1985 the official report (BELF 1985) also included information on how the extent of needle or leaf loss is used to classify the damage. In 1987 the Expert Council on Environmental Questions (SRU 1987) suggested that a smaller sample would suffice, since less damage was expected, so most German states (Länder) sampled a network of blocks that were 12 km by 8 km and not 4 km by 12 km or 4 km by 4 km as in a complete sample. In 1990 a complete sample was not possible because of the extensive forest damage caused by the storms during the year.

Thus, in the case of forest monitoring, systematic observations began in most countries after the risk of acid deposition had been discovered. After its introduction, however, the sampling and characterization of forest health were changed from year to year in response to various pressures.

16.2.10   Integrated Monitoring Systems

Despite the early recognition of the fact that monitoring of the environment requires more than isolated monitoring of individual components, the development of "integrated monitoring systems" was slow. In fact, it was only after the discovery of acid rain and stratospheric ozone depletion that systems were developed.

Integrated monitoring was designed to identify other problems that might require research efforts and to act as an early warning system of background environmental degradation. There was also the general objective of scientific advancement tied to the concept of integrated monitoring. This recognized the importance of earlier monitoring in providing the data for a general understanding of how pollutants were transported in air, which was to be extended to the terrestrial environment and water. Integrated monitoring also challenged the previous experience of single-discipline monitoring: climatologists monitored the climate, atmospheric chemists monitored precipitation chemistry, foresters monitored the forests, and aquatic chemists monitored aquatic systems. The design of an integrated monitoring network would be interdisciplinary, while the data would be collected by a single-discipline technician.

Already in the design of the Global Environmental Monitoring System (GEMS) (Munn 1973), it was recognized that a systems approach to monitoring and assessment would be necessary. It was realized that although subdividing the biosphere into components was often convenient, ultimately models of each medium must be coupled to improve understanding of the whole system. It was concluded that GEMS should be designed so that interactions between media (such as atmosphere and oceans) could be studied. At the same time it was noted that this would be difficult (Munn 1973) because physical understanding of global biosphere processes was limited by a lack of suitable data; and the feasibility of monitoring interface flux rates had not yet been demonstrated in many instances.

Clearly these difficulties slowed the development of truly integrated monitoring systems.[11]

In the early 1980s, monitoring was still being defined in terms of repetitive observing of one or more elements or indicators of the environment according to prearranged schedules in space and time, using comparable methodologies for environmental sensing and data collection as in the original definitions of the early 1970s. By the late 1980s, monitoring programs were expected to do much more: acid rain monitoring sites were being remanufactured to consider the issue of climate change. Monitoring sites were being designed to study at least all this as well as the evaluation and synthesis of complex data sets. Separate and uncoordinated observations of different elements were no longer considered sufficient for understanding environmental processes involved in the acid rain or any other atmospheric pollution issue.

This point was made by the International Joint Commission's (IJC) Air Quality Advisory Board's Expert Group on transboundary monitoring in 1987, which stressed that as the environmental response to pollutants becomes more complex, there needs to be a move toward integration of monitoring programs, so that data are collected simultaneously from many components of the environment.

Others quickly moved toward the concept of integrated monitoring networks. Sweden, among the world leaders in acid rain research, firmly believed that clarification of the consequences of different forms of environmental impact required multifaceted monitoring of the environment. Sweden operated an integrated monitoring network in the late 1980s of some twenty reference areas. Other nations followed, including the United States (LTER) and a pilot project through the United Nations Economic Commission for Europe (UNECE). Canada proposed an integrated monitoring network in its comprehensive environmental strategy, the Green Plan.

16.3 Patterns in and Motivations for Monitoring

16.3.1 Patterns over Time

There were various broad trends in monitoring of global environmental risks over the time period of this study. First, in general there was an increase in comprehensiveness in the temporal and spatial scales and in the parts of the causal chain that were monitored. The increasing comprehensiveness was, of course, partly a result of technological advances made over the past years for such techniques as remote sensing and data archival. However, some of the increase was motivated by concern about the risks. From a small monitoring network in Europe a network developed that was critical in validating models of atmospheric transportation and acidic deposition. After the discovery of the Antarctic ozone hole, European countries set up special expeditions to measure the stratospheric ozone concentration in the Arctic winter.

Another broad trend was in the increasing standardization of monitoring systems. This was a trend that had continued ever since systematic (as opposed to episodic) monitoring began and reflected the logical conclusion that standardization leads to comparability of data over space and time, which enhances the usefulness of the data collected. Even in the monitoring of forest health, standardized procedures were introduced within countries (such as Germany)[12] and within Europe (under the auspices of the U.N. Economic Commission for Europe).

The standardization of measurements across countries was balanced with the continuing guiding view that monitoring by other countries was inaccurate (the "not-invented-here" syndrome), where an indigenous system and an international-standard monitoring system might be operated side by side.

Looking at the actors involved in monitoring programs, government programs have played a predominant role in most countries. There are some examples of monitoring programs set up by industry. For example, the Central Electricity Generating Board set up a forest-health monitoring system in the United Kingdom. Likewise, industry programs recorded the statistics of CFC production and use (although the need to maintain the anonymity of individual companies for reasons of industrial competitiveness hampered the use of such data in risk and response assessment). Environmental NGOs did not generally set up monitoring programs, although they frequently interpreted and assessed the monitoring data of others. Examples of environmental NGOs setting up monitoring programs can be found, however: for example, in the United Kingdom Friends of the Earth and Greenpeace began monitoring tree damage in the 1980s.

There appears to be a trend toward "increasing politicization" of the monitoring function. That is, a shift was made from seeing monitoring primarily in the context of science assessments (with a tradition of and interest in open access) to seeing monitoring in the context of issues like blame allocation and compliance. In all three cases this had an influence on the development of the debate. In the climate case, the publication by the World Resources Institute (WRI) of a ranking of countries according to their greenhouse gas emissions was severely attacked by scientists from developing countries (Agarwal and Narain 1992), who felt that the ranking was biased and false. In the ozone case, tension grew between those who wanted to know the figures for production and use of CFCs and representatives of the industrial companies or in some cases individual countries who did not want to release the figures. In the case of industrial companies, the main argument for this reluctance was that of industrial competitiveness, and the solution was often found in the reporting of production figures to an independent office, which cumulated and published the figures anonymously.

16.3.2   Changes in Acid Rain Monitoring

As an example of the kinds of changes in monitoring of global environmental risks that have influenced the management of the risk, this section considers the monitoring of acid deposition in more detail. In response to a growing scientific and public concern about the long-range transport of acidity and its effects on segments of the environment, countries introduced more significant monitoring networks in the 1980s. These networks led to several changes in the monitoring capacity of countries.

Improved Qualitative, Temporal, and Spatial Coverage of Deposition Monitoring   While the spatial coverage of deposition monitoring was improved in all countries in response to the acid rain risk, coverage was still unequal and sometimes inadequate in the early 1990s. For example, Canada, Germany, the United Kingdom, and the United States had great variations in the number and density of their monitoring stations from region to region. In many cases, this was justified (that is, higher densities where risk is greatest), but sometimes whole regions of a country were inadequately monitored (such as in western Canada).

Some deficiencies were apparent in the comprehensiveness of the monitoring in North America, mainly due to the lack of interest in identifying and assessing other forms of acidic deposition. Both Canada and the United States continued to focus solely on the emissions and deposition of sulfur dioxide and nitrogen oxides but not on ammonia as most European countries (especially the Netherlands) had done.

North American countries also differed from those in Europe in their monitoring of wet sulfate deposition as a surrogate for total acidification. Canada and the United States used wet sulfate deposition because of the debates about the methods of measuring dry deposition.

Increased Efforts to Determine the Effects of Acid Deposition on Segments of the Ecosystem, Including Aquatics, Forests, and Agriculture   The focus on environmental effects monitoring depended primarily on the country of concern. Countries tended to focus their research and monitoring on a single receptor, one of primary concern to the general public. For example, Canada and the United States focused on aquatic effects, Germany highlighted forests, the Netherlands centered its attention on the natural terrestrial flora and fauna, Japan featured human health effects, and the United Kingdom concentrated on soil acidification. Counties established systems to monitor other ecological components, but their resources usually focused on a single receptor of concern.

Only one country committed the resources necessary to produce a comprehensive examination of the environmental effects due to acidifying substances—the United States. The United States examined aquatics, forests, agriculture, and human health in its National Acid Precipitation Assessment Program (NAPAP 1990). The acid rain issue and the so-called integrated-monitoring initiatives discussed earlier in this chapter prompted this comprehensive examination.

Better Attempts to Determine Emissions   The monitoring of acidifying emissions, while improving, still relied heavily by the mid-1990s on estimates through calculations based on energy, raw material, and production data—the so-called emission factors. There were very few direct and continuous monitoring programs for acidifying emissions; these were usually in place because of localized human-health effects.

A Sharing of Data through Internationalization
Monitoring of atmospheric emissions and deposition of acidifying substances was also effective in providing the necessary information on source-receptor relationships. This was especially true in the discussions that arose among northern European countries and between Canada and the United States.

An interesting public-information tool in Canada and Germany ensured that acid rain monitoring was both timely and relevant. In both countries, the print media were provided with daily and weekly summaries of pH concentrations of the precipitation falling in the relevant areas of concern. In Canada, back trajectories were used to indicate the source of the acid precipitation falling in that country, most of which originated in the United States. It had been argued that media reports communicated the government position that acid rain had to be addressed on both the national (Canada) and the international (United States) fronts.

In most countries, governments and industry were the sole participants of acid rain monitoring activities. This was not because other actors eschewed monitoring but because the high cost and required technical expertise limited the ability of most actors to participate. There were, of course, notable exceptions—for example, Japan and the United Kingdom. In Japan, a group of citizens and schools conducted a two-month survey of acid precipitation to challenge the more conservative Environmental Agency estimates. They found that 75 percent of the area surveyed had an acidity problem. In the United Kingdom, an NGO "imported" a German forest researcher to conduct a survey of damage to British forests, since the NGO did not trust the official status of risk assessment to forests as performed by the Forest Commission. Survey methods differed, but the German researcher's survey described signs of damage in thirtyone of forty-seven sites. These alternative monitoring activities have been scorned by scientists whose culture does not accept either the primitiveness of the surveys or the methodology of another country's scientists who do not adopt the same framing of the issue. (This is a good example of a barrier to cross-country social learning and of interest-based commitments and scientific culture.)

16.3.3 The Links between Monitoring and Other Risk Management Functions
In looking at monitoring in relation to global environmental risks, it is possible to identify a number of motivations for monitoring. In many cases these illustrate linkages between monitoring and other risk management functions.

Purely Scientific Some monitoring is carried out for purely scientific reasons to increase understanding about the world and with no concern about global environmental risks. There has been a long history (since the 1920s) of monitoring stratospheric ozone concentrations by atmospheric scientists. These data were reanalyzed in the 1980s to search for long-term global trends. Similarly, weather observations were recorded long before scientists began to look for global temperature trends. This kind of monitoring was weakly linked to risk assessment and other management functions inasmuch as the monitoring data become useful when a risk assessment suggests that a reanalysis of previously collected data would be useful.

There are some remarkable examples of monitoring for purely scientific reasons, which required reinterpretation after global environmental risks were discovered. For example, Munn (1973) discussed the monitoring of substances that he believed had no impact on human health and welfare but were useful indicators. An example cited by Munn was CFC-11, which he felt could be justified in terms of toxicity because the substance had no known effects at concentrations six orders of magnitude greater than the present level 10^{-10} by volume. For this reason it was suggested that routine monitoring would be invaluable, because most of the priority pollutants have very large naturally occurring global sources, making it difficult to isolate and follow the man-made components.[13]

Risk Management In a number of cases monitoring systems were motivated by the need to increase understanding of the risk—by looking for signals and trends, testing theories about the risk, validating, parameter estimating, and so on. This kind of monitoring can be clearly linked to the assessment functions, since it is used to explain and confirm assessments. There are good examples in the management of each of the global environmental risks studied here, in which monitoring efforts were established or intensified after a risk assessment had been made.

Verification Monitoring served an important role in providing verification that actors were doing what they were supposed to do or that environmental indicators were changing as predicted. This kind of monitoring was linked closely to the function of implementation, in particular with regard to monitoring of compliance. It can also be linked to goal and strategy formulation: in the climate change case a monitoring (reporting) system was established via the Climate Convention before a legally binding emission-reduction goal was agreed on. In the

European Union (EU) the Monitoring Directive was used to ascertain whether the stabilization goal could be achieved. The verification motivation for monitoring should also be closely related to the function of evaluation, since a self-conscious evaluation must consider whether the established goals were met.

Bureaucratic In some cases monitoring was performed only to fulfill existing regulatory or bureaucratic obligations—for example, the collection of trade, economic, and energy statistics. As shown earlier in this chapter, data on CFC production were collected for trade statistics for many years before the risk of stratospheric ozone depletion was discussed. Similarly, energy statistics were collected by the U.N. Statistical Office long before Keeling used them to show that the emissions of carbon dioxide had been increasing since the end of the nineteenth century. The link between monitoring of this kind and environmental risk management was thus in many respects fortuitous. When a risk assessment is made, data collected for other purposes may become useful.

Delaying Tactics In some cases, monitoring has been used as a substitute for taking other action in response to a global environmental risk. For instance, in the United States and United Kingdom monitoring of acid deposition was performed instead of goal formulation. In the climate case, a number of countries worked within the Intergovernmental Negotiating Committee for the Framework Convention on Climate Change to delay goal setting in favor of monitoring and research.

16.3.4 Is There Evidence That Monitoring of Global Environmental Risks "Improves" with Time?

Progress in monitoring global environmental risks is assessed here relative to two criteria: adequacy and effectiveness. As used in other chapters of this book, adequacy is a measure of the quality of inputs to the monitoring system, in particular the technical quality of the system. Effectiveness is an output measure of the impact of the monitoring system on the risk management. Did the monitoring lead to a change in the framing of the debate?

Ozone Monitoring Because the original purpose for monitoring ozone was meteorological interest, ozone monitoring has been in the hands of the international scientific community since the 1950s. Standards of technical adequacy have always been high. The increasing political salience of the ozone issue has in certain particulars brought both upward and downward pressure on adequacy of monitoring. International interest from UNEP in 1980 moved the U.K. Antarctic Survey team

not to ship its Dobson instrument back to England as a budget-cutting measure. But an increasing political profile has also been associated with measures in international circles to resist improving the adequacy of monitoring by promoting a more rapid shift from the present Dobson instrument to the technically superior Brewer and has in some instances led to political maneuvering to have monitoring stations in scientifically inappropriate places. The accessibility, comparability, and breadth of data have advanced continually through the increasing provision of services from the World Ozone Data Center in Toronto.

Effectiveness has been advanced by increasingly skillful and timely dissemination of new monitoring results to the news media. Monitoring of ozone concentrations over Antarctica was effective in reframing the debate on ozone depletion and in stimulating further steps in risk management (further risk assessments to confirm the origin of the depletion, goal and strategy formulation with regard to CFC emissions, and so on).

Acid Rain Monitoring The adequacy of acid rain monitoring of emissions, transport, deposition, and environmental effects has improved over time. When research efforts were beginning to assess the acid rain risk in the early 1970s, the paucity of data did not allow an adequate assessment of the potential risk. At the time, monitoring focused on deposition, and it was local or regional in scope, primarily at sites affected by nearby sources. The monitoring centered on sulfur dioxide and its potential effects on human health. Researchers held considerable reservations about the accuracy of the sulfur concentration data and the overall deposition rates.

The largest contributor to improving accuracy, however, was the internationalization of acid rain monitoring. First, monitoring required substantial resources—both capital and operating budgets—which was more easily acquired through shared funding arrangements. Adequate resources ensured an appropriate spatial and temporal complement. Second, cooperative monitoring also helped to extend the spatial coverage by allowing sites in other jurisdictions—that is, in several countries at a time. And third, shared responsibility for monitoring allowed cross-pollination of monitoring techniques, research initiatives, and monitoring results. Cooperative monitoring began in the mid-1950s with the EACN, in which a number of European countries participated. As described in section 16.2, this led to further development of EMEP, a European network for monitoring pollutants.

Monitoring has been very effective in framing and reframing the debate on acid rain. In fact, monitoring led to the original framing of the debate. For example, it was

Svante Odén of the Swedish University of Agricultural Sciences who in 1967, while analyzing the EACN data, reasoned that acidifying air pollutants could disperse over hundreds of kilometers—the first recognition of long-range transport of air pollutants in Europe. In Canada, Harold Harvey and Richard Beamish were monitoring fish populations and witnessed large mortalities whose only readily apparent cause seemed to be unusually acidic waters of many of the lakes they were monitoring. Harvey and Beamish were perplexed by the fact that they were observing acid deposition hundreds of kilometers away from any large polluting source.

The existence of the monitoring network, or monitoring activities for research, allowed these scientists to observe that something in the ecosystem was askew. Monitoring provided continuous long-term data to identify a problem and to suggest some diagnosis and understanding of this problem.

Climate Change Monitoring   The adequacy of monitoring of factors related to global climate change has to a large extent improved over time. The monitoring of concentrations of greenhouse gases in the atmosphere improved as the instrumentation was developed and as the number of measuring stations increased. On the other hand, however, the monitoring of greenhouse gas emissions did not improve dramatically over time. Especially in the case of widely dispersed sources, the emissions themselves (such as automobile exhaust) are not monitored: it is necessary to monitor related human activities and extrapolate. In particular, in the case of the emissions of carbon dioxide into the atmosphere as a result of tropical deforestation, the emissions have been very uncertain, and this uncertainty has played an important role over the past twenty years in the debate about the increasing greenhouse effect.

Improvements in the monitoring of climate change are closely related to the large advances made in meteorological observing systems. The spatial and temporal distribution of observations has improved, ocean observation systems have expanded, satellite systems with valuable new information have been introduced, and major international observation programs have been successfully carried out. However, while it is possible to produce accurate pictures of global and even regional changes of climate variables in the recent past, little has been done to monitor socioeconomic drivers and impacts of climatic change. Thus, the sensitivities of the socioeconomic variables are still often unknown.

There are several examples where monitoring has been effective in framing the debate. Keeling's first compilation of the concentration of carbon dioxide at Mauna Loa showed the rapidly increasing concentration and pointed to a need for action. In the case of the globally averaged earth surface temperature, there have been interesting developments. During the 1970s there were a number of scientists, not necessarily climatologists, who claimed that the globe (in particular the Northern Hemisphere) was cooling. The basic data used for making these assessments were the World Climate Data published once a decade. The data were mainly for stations on the continents of the Northern Hemisphere. Although data were available for some shipping routes and small islands, there were significant data problems and a generally poor distribution of observations. By the mid-1980s, however, scientists in the United Kingdom were able to clean up the data sufficiently to produce more truly global temperature data. These data showed that the cooling observed between 1940 and 1970 was restricted to the Northern Hemisphere land areas and was not a global phenomenon. The global temperature data showed that the globe as a whole had warmed by 0.3 to 0.7 degree Centigrade between 1880 and 1980. Several groups that were working on this basic data set but using different ways of interpolating and averaging confirmed the observation that the earth was warming, and this consensus was reflected in the Conference Statement of the 1985 Villach Conference, which led to an international response to the climate change issue.

One other good example of monitoring the data that were framing the debate is the use of ice-core data, especially that from the Vostok core. In the mid-1980s scientists were able to show on the basis of measurements made on air bubbles trapped in the deep ice of the Antarctic that over the past 150,000 years of earth's history there was a match between the changes in carbon dioxide and methane concentrations and the average temperature. This was another proof that there is a "greenhouse effect." The ice-core data also showed how human activities had dramatically changed the composition of the atmosphere in a very short part of the earth's history. This information was used in a number of countries to convince decision makers of the need to act.

16.4   Problems and Pitfalls in Monitoring

The stories told and patterns analyzed in this chapter and in chapters about individual arenas illustrate that over the period covered by this study a number of problems and pitfalls in the monitoring of environmental risks have materialized.

First, there appears to be a trend toward reducing the resources provided for monitoring. In the former Soviet Union, reductions in resources for scientific activities continued during the first half of the 1990s and threatened to lead to the shutdown of monitoring systems that had been in place for decades. Even in wealthier countries, such as Canada, the introduction of the concept of "integrated monitoring" can be viewed as a bureaucratic exercise of rationalization of monitoring sites across the country in the light of reduced resources.

The story of the discovery of the Antarctic ozone hole illustrates the dangers of either not trusting the data one has collected or of ignoring data because they fall outside the range that one would normally expect, leading to the delayed "discovery" of important environmental phenomena. In Germany a similar kind of story has been reconstructed with regard to the discovery of forest dieback (Ell and Luhmann 1995). Foresters in southern Germany had noticed that trees were dying in the 1970s and even earlier but felt that the phenomenon could be a result of their own poor management of individual forests, so there was no public acknowledgment of the observations. At the same time measurements were being made of the acidification of soils and its effects on heavy metal transportation in soils. After the results of this research were published, suggesting that one impact could be forest damage, the foresters realized that what they were observing was not necessarily "their fault" and that there was possibly a more general explanation. So the public discussion of the observations of forest damage began.

For each of the global environmental risks studied here, legitimacy and ownership problems have arisen with regard to monitoring data. In the case of CFCs, for example, to protect industrial competitiveness, data have been reported anonymously to independent offices, where the data have been cumulated, so that it was not possible to find out how much is produced by individual firms. The reporting of data has been unsatisfactory in the case of the Montreal Protocol. For example, at the Third Meeting of Parties to the Montreal Protocol in June 1991 the Implementation Committee noted that reporting was not satisfactory: of seventy-one parties (at that time), only thirty-one had reported complete data for 1986, which is the base year used to calculate the required phase-down of the original list of controlled substances. Of the remainder, nineteen parties had submitted incomplete data, six had reported no data available and/or requested assistance, two had reported that their data were included in those of another party, and thirteen parties, including four European Community members, had submitted no data at all (Bergeson and Parmann 1985). In the climate case, Agarwal and Narain (1992) claimed that a study published by the World Resources Institute in collaboration with the United Nations was an example of "environmental colonialism." In particular, Agarwal and Narain argued

that the figures used by WRI to calculate the amount of carbon dioxide and methane produced by each country are extremely questionable. For example, they found that the deforestation estimates made by WRI for Brazil and India were overestimated. Agarwal and Narain claimed that the data were used in the WRI report to blame developing countries for global warming and perpetuate the global inequality in the use of the earth's environment and resources. For each of the global environmental risks studied here, countries have displayed sensitivities about the publication and use of monitoring data by other countries or institutions in other countries. The U.N. Framework Convention on Climate Change deals with this problem by requiring the Conference of Parties to agree on methodologies for calculating emissions by sources and removals by sinks of greenhouse gases.

The importance of monitoring in the management of global environmental risks appears to be often underestimated. For each of the issues studied here, monitoring played a major role in reframing the debate and stimulating risk assessments, goal, and strategy formulation and implementation. The motivation for monitoring was often unrelated to the need to document environmental risks, which suggests that monitoring systems have not been designed to play a role in risk management, although in each case they have done so.

Appendix 16A. Acronyms

AFEAS	Alternative Fluorocarbon Environmental Acceptability Study
BAPMoN	Background Air Pollution Monitoring Network
BAS	British Antarctic Survey
CCC	Chemical Coordinating Centre at the NILU
CCE	Coordinating Centre for Effects at RIVM
CFC	chlorofluorocarbon
CH$_4$	methane
CMA	Chemical Manufacturers' Association
CMDL	Climate Monitoring and Diagnostics Laboratory
CO$_2$	carbon dioxide
EACN	European Air Chemistry Network
EC	European Community
EMEP	Cooperative European Monitoring and Evaluation Programme
EU	European Union

FAO	Food and Agriculture Organization	**UNEP**	United Nations Environment Programme
GAW	Global Atmosphere Watch	**UV**	ultraviolet
GEMS	Global Environmental Monitoring System	**VCI**	Verband der Chemischen Industrie (Trade Association of the Chemical Industry)(Germany)
GOS	Global Observing System		
IAEA	International Atomic Energy Agency	**WMO**	World Meteorological Organization
IAMAP	International Association of Meteorology and Atmospheric Physics	**WRI**	World Resources Institute

FAO   Food and Agriculture Organization

GAW   Global Atmosphere Watch

GEMS   Global Environmental Monitoring System

GOS   Global Observing System

IAEA   International Atomic Energy Agency

IAMAP   International Association of Meteorology and Atmospheric Physics

IGY   International Geophysical Year

IJC   International Joint Commission

ITC   International Tariff Commission

LRTAP   (Convention on) Long-Range Transboundary Air Pollution

LTER   Long-Term Ecological Research

MSC-E   Meteorological Synthesizing Centre East (at the Institute of Applied Geophysics in Moscow)

MSC-W   Meteorological Synthesizing Centre West (at the Norwegian Meteorological Institute in Oslo)

NAPAP   National Acid Precipitation Assessment Program

NASA   National Aeronautics and Space Administration (U.S.)

NGO   nongovernmental organization

NILU   Norwegian Institute for Air Research

NOAA   National Oceanic and Atmospheric Administration (U.S.)

ODG   ozone-depleting gas

OECD   Organization for Economic Cooperation and Development

OTP   Ozone Trends Panel

ppmv   parts per million volume

RIVM   National Institute for Environmental Health and Environmental Protection in the Netherlands

SBUV   solar backscatter ultraviolet instrument

SCOPE   Scientific Committee on Problems of the Environment

TOMS   Total Ozone Mapping Spectrometer

UBA   Umweltbundesamt (Federal Environment Agency)(Germany)

UNECE   United Nations Economic Commission for Europe

Notes

1. The authors are grateful for the contributions of all of the research teams that provided information on monitoring. For reviews of earlier drafts we also thank three anonymous reviewers.

2. National meteorological networks were in some cases established earlier than the international networks. For example, in the late eighteenth century the Earl of Palatinate (in Germany) set up the Societas Meteorologica Palatina in Mannheim, embracing stations like Hohenpeissenberg, Prague, Mannheim, and many others, some of which still exist today.

3. There are of course several other reasons for monitoring the environment, including general pollution surveys, regulatory pollution control, health effects assessments, vegetation damage assessments, modeling, trend analysis, and prediction.

4. The fact that the carbon dioxide concentration and temperature are highly correlated was important whether the CO_2 concentrations lagged the temperature trends or not, but these scientific discussions did not influence the impact of the correlations on the debate.

5. Brown was interested in the natural acidity of lakes and rivers, and his hypothesis was that carbon dioxide gas that is dissolved in the water is always in balance with the carbon dioxide in the air above the water (see Wiener 1990, 16).

6. BAPMoN started in the mid-1960s and was later incorporated into Global Atmosphere Watch.

7. Industry compliance with German regulations improved in the 1990s after the government passed the law banning CFCs and Halons (see chapter 3 on the German arena).

8. This discussion is drawn from Kawahire and Makino (1989) and Miura (1992).

9. During the organizing committee meeting of the 1984 Quadrennial Ozone Symposium, the committee originally intended to reject the paper by Chubachi, but one of the coeditors of the Proceedings of the Symposium insisted on including the paper as a poster paper, which was later published in the *Symposium Proceedings*.

10. Earlier than this, there was a Canadian program, and some effort was made to get the United States involved, according to sources in the Canadian Atmospheric Environment Services.

11. It should be noted, however, that the idea of integrated monitoring began to flourish in the late 1970s through the active support of the Soviet scientist Yuri Izrael, who organized several international workshops in the USSR.

12. German forest-damage reports (see chapter 3 on Germany).

13. At the same time as suggesting that CFC-11 would be a good indicator and has no impact on health or welfare, the SCOPE report

suggested that the gas is an intense infrared absorber in the 8-13 nanometer region and that a future rise in concentrations to above 10-9 by volume might be of concern in discussions of climate change. Thus, the climate change risk was recognized before the stratospheric ozone–depletion risk.

References

Agarwal, A., and S. Narain. 1992. *Global Warming in an Unequal World*. India: Centre for Science and Environment.

Bader, F. 1958. United States polar ice and snow studies in the International Geophysical Year. *Geophysical Monographs* 2: 177–181.

Barnola, J.M., D. Raynaud, Y.S. Korotkevich, and C. Lorius. 1987. Vostok ice core provides 160,000 year record of atmospheric CO_2. *Nature* 329: 408–414.

BELF. 1985. *Waldschadenserhebung 1985*. Bonn: Federal Ministry for Food, Agriculture, and Forests.

Bergeson, H.D., and G. Parmann, eds. 1995. *Green Globe Yearbook of International Cooperation on Environment and Development 1995*. Oxford: Oxford University Press.

Bolin, B., et al. 1972. *Swedish Case Study for the United Nations Conference on the Human Environment: Air Pollution across National Boundaries—The Impact of Sulfur in Air and Precipitation*. Stockholm: Norstadt.

Charlson, R., ed. 1973. *Stockholm Tropospheric Aerosol Seminar*. Report A.P. 14. Stockholm: Institute of Meteorology, University of Stockholm.

Chubachi, S. 1984. A special ozone observation at Syowa Station, Antarctica from February 1982 to January 1983. In C. Zeryos and A. Ghazi, eds., *Atmospheric Ozone*. Dordrecht: Reidel.

Cowling, E.B. 1982. Acid precipitation in historical perspective. *Environmental Science and Technology* 16: 110.

Delmas, R.J., J.M. Ascencio, and M. Legrand. 1980. Polar ice evidence that atmospheric CO_2 20,000 yr BP was 50 percent of present. *Nature* 284: 155–157.

Dietrich, W.F. 1992. Monitoring of long-range transport of air pollutants and the policy debate on acid deposition. Social Learning Project Archives, BCSIA, Kennedy School of Government, Harvard University.

Egnér, H., G. Brodin, and O. Johansson. 1955. Sampling technique and chemical examination of air and precipitation. *Kungl. Lantbrukshogskolans Annaler (Annals of the Royal Agricultural College of Sweden)* 22: 369–410.

Ell, R., and H.J. Luhmann. 1995. Von Scham, Schäden und Ursachen—Zur Entdeckung des Waldsterbens in Deutschland. In. G. Altner et al., eds., *Jahrbuch Ökologie 1996*. Munich: Beck Verlag.

Farman, J.C., B.G. Gardiner, and J.D. Shanklin. 1985. Larger losses of total ozone in Antarctica reveal seasonal ClO_x/NO_x interaction. *Nature* 315: 207–210.

Heath, D. E. 1986. Testimony before the Subcommittee on Health and the Environment of the House Committee on Energy and Commerce.

Hecht, A.D., W. Dansgaard, J.A. Eddy, S.J. Johnsen, M.A. Lange, C.C. Langway, C. Lorius, M.B. McElroy, H. Oeschger, G. Raisbeck, and P. Schlosser. 1989. Long-term ice records and global environmental changes. In H. Oeschger and C.C. Langway, eds., *The Environmental Record in Glaciers and Ice Sheets*. Chichester: Wiley.

International Atomic Energy Agency (IAEA). 1965. Methods of surveying and monitoring marine radioactivity. Safety Series No. 11. Vienna: IAEA.

International Tariff Commission (ITC). 1991. *Synthetic Organic Chemicals: U.S. Production and Sales, 1990*. Washington: International Tariff Commission.

Kawahire, Johi, and Makino Yukio. 1989. *Ozons. Gensho*. Tokyo: Yomiuri Shimbunsha.

Langway, C.C., and H. Oeschger. 1989. Introduction. In H. Oeschger and C.C. Langway, eds., *The Environmental Record in Glaciers and Ice Sheets*. Chichester: Wiley.

Likens, G.E., and F.H. Borman. 1974. Acid rain: A serious regional environmental problem. *Science* 184: 1176–1179.

Miura, Kaunori. 1992. Ozone observation pioneer: Chubachi Shigeru. In Ministry of Foreign Affairs, *The Forefront of the Environmental Movement in Japan*. Tokyo: Ministry of Foreign Affairs.

Munn, R.E. 1973. *Global Environmental Monitoring System (GEMS)*. SCOPE Report 3. Toronto: SCOPE.

National Acid Precipitation Assessment Program (NAPAP). 1990. *Acidic Deposition: State of Science and Technology*. Vol. 1, *Emissions, Atmospheric Processes, and Deposition;* Vol. 2, *Aquatic Processes and Effects;* Vol. 3, *Terrestrial Materials, Health, and Visibility Effects;* Vol. 4, *Control Technologies, Future Emissions, and Effects Valuation*. Washington: U.S. Government Printing Office.

National Aeronautics and Space Administration (NASA). 1988. Executive Summary of the Ozone Trends Panel, March 15. Washington: NASA.

Neftel, A., H. Oeschger, J. Schwander, B. Stauffer, and R. Zumbrunn. 1982. Ice core sample measurements give atmospheric CO_2 content during the past 40,000 yr. *Nature* 295: 220–223.

Odén, S. 1967. Nederbordens Forsurning *Dagens Nyheter*. October 24.

Organization for Economic Cooperation and Development (OECD). 1979. *The OECD Programme on Long-Range Transport of Air Pollutants: Measurements and Findings* (2nd ed.). Paris: OECD.

Pukelsheim, F. 1990. Robustness of statistical gossip and the Antarctic ozone hole. *The IMS Bulletin* 19: 540–545.

Shabecoff, Philip. 1988. DuPont halts chemicals that peril ozone. *New York Times,* March 25.

SRU. 1987. *Umweltgutachten 1987*. Stuttgart: Verlag W. Kohlhammer.

Tolba, M.K., O.A. El-Kholy, M. Holdgate, D. McMichael, and R.E. Munn, eds. 1992. *The World Environment 1972–1992*. London: Chapman and Hall.

Wiener, J. 1990. *The Next One Hundred Years*. New York: Bantam.

17
Option Assessment in the Management of Global Environmental Risks

William C. Clark, Josee van Eijndhoven, and Nancy M. Dickson with
Gerda Dinkelman, Peter M. Haas, Michael Huber, Angela Liberatore,
Diana Liverman, Edward A. Parson, Miranda A. Schreurs, Heather Smith,
Vassily Sokolov, Ferenc L. Tóth, and Brian Wynne[1]

17.1 Development and Analysis of Option Assessment

17.1.1 Outline of the Chapter

If the function of risk assessment is to ask, "What is the problem?" then the function of option assessment is to ask, "What are the possible solutions?" Option assessments may be formal, such as the Intergovernmental Panel on Climate Change's (IPCC) report on response strategies for climate change (IPCC 1991), or informal, such as the processes most people would go through if asked to advise a colleague on "What should we do about acid rain?" The goal of this chapter is to describe, understand, and evaluate changes in societies' assessments of their options for responding to global environmental risks. Section 17.1 defines option assessment as we use it here, provides an overview of its historical development in the cases reviewed for this study, and presents the normative and analytic frameworks we employed in our research. Sections 17.2 and 17.3 present our detailed findings on the products and processes of option assessment, respectively. Section 17.4 reviews possible explanations and interpretations of the findings and returns to the normative question of how well society is doing in the production of option assessments. Finally, Section 17.5 speculates on the policy and research needs highlighted by this study.

17.1.2 Option Assessment Defined

Options are particular measures that might be undertaken to help manage a problem or a risk.[2] Options can include the setting up of institutions (such as the Global Environment Facility—GEF), use of technologies (such as "clean-coal" processes), or the policy instruments employed in their implementation (such as carbon taxes). Other options include gathering information (through monitoring, research, or development) in preparation for action or doing nothing at all.

Options may be considered for adoption at many levels, including local choices over particular technologies, national strategies for emission reductions, and international conventions on risk management. We are concerned with all levels in this study. Indeed, one of our central tasks is to identify and explain any pattern that

may exist in the emergence of higher-level assessments through time or in particular arenas.

In general, an arena concerned with the management of an environmental risk will find itself actively debating several options at any given time. We call this set of actively debated measures the *option pool*. Some options enter the pool through formal option assessments. Others are injected by interested groups that wish to see particular measures adopted. In our studies, the pool of options often includes a relatively undifferentiated mix of simple measures and complex policies.

Option assessment is a process that involves systematic explication and analysis of measures that might be undertaken to help manage a risk.[3] Option assessments may be carried out by any actor group, including governments (e.g., German Bundestag 1989), industry (e.g., Confederation of British Industry 1989), nongovernmental organizations (e.g., Greenpeace 1992), and individuals. They can seek a balanced perspective or explicitly advocate specific measures.

Our primary approach to the study of option assessment has focused on written documents that report on the results of relatively formal assessment activities. We examined over 150 such documents, including most of the major publicly circulated option assessments produced in the arenas addressed by this study. These formal, public assessment reports clearly reflect major commitments of resources. Many of them occupied a prominent place in public and political debates over the management of global environmental risks. They also have the distinct advantage from an analytic perspective of constituting a reasonably well identified, bounded, and documented data set. The principal formal option assessments reviewed for this study are listed in table 17.1.

Formal assessments, however, capture only part of societies' historical efforts to determine what could be done about global environmental risks. Our preliminary investigations confirmed our expectations that some industries and governments were also producing private assessments. In many arenas informal option assessments communicated through internal memoranda or oral briefings seemed likely to play a more important role than formal ones. We therefore treated the formal

Table 17.1

Chronology of major option assessments (full citations in bibliography)

Date	Country	Assessment title	Actor group
A. Acid rain			
1967	Germany	*Stress on Forestry through Pollution Emissions and Their Technical, Silvicultural, Planning, and Legal Consequences*	Expert
1981	International Institutions (II)	*Proposals by the Working Party on Air Pollution Problems Regarding Control Techniques and Related Measures to Reduce Emissions of Sulfur Compounds in the Atmosphere*	Government
1982	United States and Canada	*Emissions, Costs, and Engineering: Final Report of the United States–Canada Work Group*	Government
1983	European Community	*Acid Rain: A Review of the Phenomenon in the European Economic Community and Europe*	Government
1983	Germany	*Last Chance for the Forest: The Avoidable Consequences of Acid Rain*	Expert
1984	United States	*Acid Rain and Transported Air Pollutants: Implications for Public Policy*	Government
1986	II	*Technologies for Controlling Nitrogen Oxide Emissions from Stationary Sources*	Government
1986	II	*Technologies for Controlling Nitrogen Oxide Emissions from Mobile Sources*	Government
1986	II	*Effective Enforcement Procedures*	Government
1987	United States	*NAPAP Interim Assessment: The Causes and Effects of Acidic Deposition*	Government
1988	European Community	*Acid Rain and Photochemical Oxidants Control Policies in the European Community*	Government
1988	Germany	*The Forest Dies from Stress*	Expert
1988	United States	*Polluted Coastal Waters: The Role of Acid Rain*	NGO
1989	Japan	*The First Results of a Survey of Acid Rain Policy*	Government
1990	II	*The RAINS Model of Acidification: Science and Strategies in Europe*	Expert
1990	II	*Emissions of Volatile Organic Compounds*	Government
1990	II	*Hydrocarbons and Ozone Formation: An Approach to Their Control Based on Likely Environmental Benefits*	Government
1990	United Kingdom	*Greenpeace Position on Acid Rain*	NGO
1991	II	*Exploration of Economic Instruments for Implementation of Cost-Effective Reductions of Sulfur Dioxide in Europe Using the Critical-Load Approach*	Government
1991	II	*The Critical-Load Concept and the Role of the Best-Available Technology and Other Approaches*	Government
1991	United States	*Acidic Deposition: State of Science and Technology*	Government
1991	United States	*The 1990 NAPAP Integrated Assessment Report*	Government
1991	United States	*Analysis of Alternative Sulfur Dioxide Reduction Strategies*	Industry
1992	Japan	*Regarding the Mid-Term Assessment of the Second Acid Rain Policy Survey*	Government
1992	Japan	*Green Aid Plan*	Government
B. Ozone depletion			
1975	United States	*Climatic Impact Assessment Report*	Government
1978	Germany	*Environment Opinion*	Government
1979	United States	*Protection against Depletion of Stratospheric Ozone by Chlorofluorocarbons*	Expert
1980	European Community	*Council Decision of 26 March 1980 Concerning CFCs in the Environment*	Government

(continued)

Table 17.1 (continued)

Chronology of major option assessments

Date	Country	Assessment title	Actor group
B. Ozone depletion (continued)			
1980	United States	*Economic Implications of Regulating CFC Emissions from Nonaerosol Applications*	Government
1980	Japan	*Regarding the Fluorocarbon Problem*	Government
1985	United States	*The Health Costs of Skin Cancer Caused by Ultraviolet Radiation*	Government
1986	II	*Report of the Second Part of the Workshop on the Control of CFCs*	Government
1988	II	*Report of the International Ozone Trends Panel*	Government
1988	United States	*Regulatory Impact Analysis*	Government
1988	Japan	*Regarding the Basics of a System for the Protection of the Ozone Layer*	Government
1988	Japan	*Fundamental Thoughts on the Regulation of Fluorocarbon Production for the Protection of the Ozone Layer*	Government
1989	II	*Technical Progress on Protecting the Ozone Layer*	Government
1989	II	*Economic Panel Report*	Government
1989	Germany	*Protecting the Earth's Atmosphere: An International Challenge*	Government
1991	II	*Scientific Assessment of Ozone Depletion*	Government
1991	Germany	*Protecting the Earth: A Status Report with Recommendations for a New Energy Policy*	Government
1991	United Kingdom	*Refrigeration and Air Conditioning CFC Phase Out: Advice on Alternatives and Guidelines for Users*	Government
1991	Japan	*Regarding Future Policies for the Protection of the Ozone Layer*	Government
1992	United Kingdom	*Making the Right Choices: Alternatives to CFCs and Other Ozone-Depleting Chemicals*	NGO
C. Climate change			
1970	II	*Man's Impact on Global Environment (Science)*	Expert
1971	II	*Inadvertent Climate Modification*	Expert
1971	II	*Global Environmental Monitoring*	Expert
1977	United States	*Energy and Climate*	Expert
1978	II	*Carbon Dioxide, Climate, and Society*	Expert
1979	Germany	*Our Threatened Climate: Ways of Averting the Carbon Dioxide Problem through Rational Energy Use*	Expert
1979	Germany	*Investigation of the Influence of Anthropogenic Factors on Climate*	Expert
1979	United States	*The Carbon Dioxide Problem: Implications for Policy in the Management of Energy and Other Problems*	Expert/government
1980	United Kingdom	*Climate Change: Its Potential Effects on the United Kingdom and the Implications for Research*	Government
1982	Japan	*International Responses to Global Environmental Problems*	Government
1983	United States	*Changing Climate*	Expert
1983	United States	*Can We Delay the Greenhouse Effect?*	Government
1986	II	*Report of the International Conference on the Assessment of the Role of Carbon Dioxide and of Other Greenhouse Gases in Climate Variations and Associated Impacts*	Expert
1986	II	*The Greenhouse Effect, Climatic Change, and Ecosystems*	Expert
1987	United States	*A Matter of Degrees: The Potential for Controlling the Greenhouse*	NGO
1988	II	*Developing Policies for Responding to Climate Change*	NGO/expert
1988	Japan	*Policy Recommendations Concerning Climate Change*	Government
1989	Germany	*Protecting the Earth's Atmosphere: An International Challenge*	Government

(continued)

52

William C. Clark, Josee van Eijndhoven, and Nancy M. Dickson et al.

Table 17.1 (continued)
Chronology of major option assessments

Date	Country	Assessment title	Actor group
C. Climate change (continued)			
1989	Netherlands	*To Choose or to Lose: National Environmental Policy Plan*	Government
1989	United Kingdom	*Getting out of the Greenhouse: An Agenda for United Kingdom Action on Energy Policy*	NGO
1989	United Kingdom	*The Greenhouse Effect and Energy Efficiency*	Industry
1989	Canada	*Study on the Reduction of Energy-Related Greenhouse Gas Emissions*	Government
1990	II	*Climate Change: The IPCC Response Strategies*	Government/expert
1990	European Community	*Energy in Europe: Energy for a New Century—The European Perspective*	Government
1990	United States	*Policy Options for Stabilizing Global Climate*	Government
1990	United Kingdom	*United Kingdom: A Case Study of the Potential for Reducing Carbon Dioxide Emissions*	Expert for U.S. government
1990	United Kingdom	*Transport and Global Warming: A Forward Look*	Industry
1990	Japan	*Action Program to Arrest Global Warming*	Government
1990	Canada	*National Action Strategy on Global Warming*	Government
1991	II	*Climate Change: Science, Impacts, and Policy*	NGO/expert
1991	European Community	*The Economics of Policies to Stabilize or Reduce Greenhouse Gas Emissions: The Case of Carbon Dioxide*	Government
1991	Germany	*Protecting the Earth: A Status Report with Recommendations for a New Energy Policy*	
1991	United States	*Changing by Degrees: Steps to Reduce Greenhouse Gases*	Government
1991	Netherlands	*Climate Change*	Government
1992	II	*Climate Change: The Supplementary Report to the IPCC Scientific Assessment*	NGO/expert
1992	European Community	*The Climate Challenge: Economic Aspects of the Community's Strategy for Limiting Carbon Dioxide Emissions*	Government
1992	United States	*Policy Implications of Greenhouse Warming*	Expert
1992	Japan	*Fourteen Proposals for a New Earth: Policy Triad for the Environment, Economy, and Energy*	Government and industry

assessments as only one possible source of the particular options in the active option pool at the time of their writing. We gathered additional evidence on actively discussed options from interviews with key assessment users, media reports, the academic literature, policy debates in parliaments and elsewhere, and lobbying efforts by interest groups. Our coverage of private or informal processes and products of option assessment is necessarily less systematic and spottier than our treatment of formal assessment reports. We nonetheless believe that the results summarized here—incorporating what we could discover about both formal and informal option assessments—provide a useful first reconnaissance of the role option assessment has actually played in the management of global environmental risks.

17.1.3 Option Assessment in Historical Perspective
Arena studies presented in part II of this book suggest that public debate over options for dealing with the risks

examined in this study was well developed before the risks themselves were defined as public issues. Existing policy debates on supersonic transport (SST), energy conservation, and local air pollution from coal burning seized on early scientific risk assessments of global environmental issues as additional reasons for advancing their respective arguments. In some cases, such as the SST opponents' use of ozone-depletion arguments, this initial framing of new problems by old solutions rapidly gave way to a debate centered on the risk itself. In others, such as the invocation of climate change risks by advocates of energy conservation, the initial "solution"-defined energy agenda continued to exert a significant influence on the evolution of the global environmental risk debate throughout its history.

The earliest option assessments explicitly focused on the global environmental risks studied here were highly informal affairs, often consisting of little more than a commonsense assertion by the scientists who first characterized the problem of the need for more research or emission reductions. In contrast to the development of risk assessment, few formal option assessments focusing on these problems were undertaken until after the problems were already well established on the policy agenda. As late as 1987, for example, the Netherlands government voiced a lament it shared with other nations that "no systematic evaluation of response options exists" for dealing with climate change. Formal option assessment of global environmental risks has therefore generally been a post-1980 phenomenon.

Even after global environmental issues emerged onto the policy agenda, however, option assessments were performed largely in a reactive rather than proactive mode. Most appeared in close association with particular national decisions or international negotiations. Indeed, as international schedules for protocol negotiation and amendment emerged in the mid-1980s, they became the pacesetters for option assessments in both domestic and international arenas. By the end of our study period in 1992, a surge of option assessments could be expected to appear in print shortly before each major international negotiating session on any of the global environmental risks. All the national arenas addressed in this study were increasingly incorporating formal international assessments into their policy deliberations and responding to international calls for formal national assessments of their own.

17.1.4 Questions about Option Assessment

This sketch of the development of options and their assessment in our three cases of global environmental risk, viewed against the larger issues raised in chapter 1, leads to two sets of questions that motivate our analysis in this chapter:

• *The development over time of the pool of options actively considered by society* What changes occurred in the set of options that constitute the contents of the option pool? Why were some types of options successfully incorporated into the social debate while others were not? What explains these patterns? To what extent, and in what ways, can the pool of actively considered options be said to have improved over the period addressed by this study? What, if any, was the role of formal option assessments in that improvement?

• *The development over time of the practice of option assessment* What changes occurred in the goals, scope, institutions, or use of option assessment? How did actual methods of assessment change? What changes occurred in who performed option assessments? To what extent, and in what ways, can the performance of option assessments be said to have improved over the period addressed by this study?

We describe the analytical approach used for investigating these two groups of questions in Section 17.2, before turning to a summary of our empirical findings in sections 17.3 and 17.4.

17.2 Analytic Approach

This section begins with a review of the conceptual issues involved in making sense of the role of option assessment in the management of global environmental risks. From this foundation, it then erects a descriptive classification of the options and assessments encountered in the cases we studied. It closes with the outline of a set of evaluative criteria that we then use for analyzing our data in sections 17.3 and 17.4.

17.2.1 Good Option Assessments?

We argue in this chapter that doing better option assessments—and even deciding what "better" might be—has proven to be substantially harder than doing better risk assessments. These difficulties reflect fundamental tensions latent in existing concepts of "usable knowledge"—that is, assessments to inform action or choice.

Issue Development: Stages versus Garbage Cans
The relationship between knowledge and action in the development of public-policy issues has been characterized in a variety of ways. At one extreme, issues are seen as developing through successive stages in which actors first identify problems and then seek solutions for them (e.g., deLeon 1999; Jones 1977). At another, it is solutions—in the guise of proposals to carry out particular actions—and their advocates that drive issue development and the policy process. In between, some scholars have argued that streams of problem definition and solution formulation develop independently, with only occasional joining in the "garbage can" of the public-policy arena (e.g., Cohen, March, and Olsen 1972; Kingdon 1995). Others have argued that coalitions of advocates play a conscious and influential role in joining particular problem and solution streams to advance their own agendas (e.g., Sabatier and Jenkins-Smith 1993, 1999).

From a stages perspective on issue development, the search for better option assessments might be expected to emphasize improved responsiveness to political goals, the promotion of "progressive problem shifts" in the sense of Lakatos (1970) and Majone (1980), or—most simply and ambitiously—the identification or design of more effective problem solutions. In contrast, from a garbage-can view of the policy process better option assessment is in large part simply the more effective partisan advocacy of particular solutions. An important metric of performance would be the extent to which assessments introduced

54

William C. Clark, Josee van Eijndhoven, and Nancy M. Dickson et al.

particular proposals into new policy communities and subsequently built familiarity with and acceptance of such proposals. Advocacy coalition views also focus on the persuasive uses of assessment but emphasize more the function of good assessments in connecting mutually supportive problem and solution streams.

Assessment: Comprehensive versus Incremental Rational actor models of choice hold that the best decisions will result from comprehensive or *synoptic* comparisons of all possible options. In practice, many of the assumptions of rational action are violated. Nonetheless, a common presumption is that good assessments should analyze as many options as possible, as thoroughly as possible, in ways that provide for rigorous comparisons. In a more practical vein, perceptive students of policy analysis have long observed that professional analysts may have contributed more to the social good by their occasional successes in increasing the number and quality of options debated in the political process than by their more numerous contributions to the theory of optimal choice among given options (Schelling 1983). By this reasoning, better option assessment should entail increasing the variety of options facing society and increasing the depth of society's understanding of those options. An equally powerful social science finding pushes option assessment in the opposite direction. Herbert Simon won a Nobel Prize for observing, among other things, that thinking about alternatives takes resources—time, effort, intelligence—that are always in short supply (Simon 1983). The more alternatives considered and the more deeply they are analyzed, the higher the cost in these crucial accounts. Real-world decision makers, in this view of the world, therefore "satisfice" rather than "optimize" in their treatment of options. Instead of trying to consider all possibilities, they rely on "successive limited comparisons" to find an option that seems "good enough" for the moment. They then stop and move on to other pressing tasks. Simon and his followers have argued that such "bounded rationality" is not only what real-world decision makers *actually* do but also what they *should* do for most effective performance. Numerous scholars have noted that public-policy making likewise does and indeed should reflect many of the same "satisficing" behaviors found in Simon's "boundedly rational" decision makers (Lindblom and Cohen 1979; Heclo 1977). By this reasoning, those performing better option assessment should consciously limit the variety of options given serious treatment and the depth of analysis to which those options are subject.

Science and Values: Shaping the Boundaries Questions of how science and values are and ought to be related in efforts to provide expert advice to the political

system have split the assessment community for decades. At one extreme, many argue that science can and should be separated from values and that science advice in policy contexts can and should seek to confine itself to "the facts" as strictly as humanly possible. In contrast, others have argued that science, values, and politics are inextricably intertwined—at least in cases such as global environmental risks—and should be frankly treated as such (e.g., Collingridge and Reeve 1986). Between these two extremes, there emerged over the period of our study a substantial body of institutional experimentation (such as government offices seeking to perform nonpartisan technology assessments) and academic scholarship (e.g., Jasanoff 1990) concerned with the negotiation and institutionalization of the *boundaries* between science and politics—boundaries that are selectively permeable in ways that permit the creation of what we have called here *usable knowledge.*

For those who view science as properly and pragmatically separable from politics, criteria for better option assessments tend to focus on the scholarly credentials of the assessors, the process of peer review, and the unbiased quality of results. Those impressed with the extent to which science and politics are intertwined have focused attention more on seeking improvements in the representation of various interests and stakeholders in assessment activities and on the explicit identification of whose interests any given assessment is promoting. The boundary-work school has struggled, not altogether successfully, with reforming assessment processes so that they replicate neither pure science nor pure politics but rather promote the construction of what Jasanoff (1990) has called "serviceable truth."

17.2.2 Descriptive Classification of Options and Assessments

The debates sketched above, even in their more subtle forms, yield no consensus on what might constitute good or even better option assessment in the management of global environmental risks. Our response to this dilemma in the present study has been to lean toward a positive rather than a normative analysis. In conducting our empirical work, we have attempted to take seriously the major concerns emerging from the usable-knowledge debate, not prejudging their respective merits. Fortunately, there turns out to be a good deal of overlap in those concerns. Virtually all parties to the usable-knowledge debate are interested in empirical data on such things as what options actually get assessed, how the assessments are carried out, and who the assessors turn out to be. The analytic approach we adopt in this chapter reflects our attempt to capture these common concerns in ways that will let us construct a solid empirical foundation on which a more satisfactory conceptual understanding can build concerning the role of option assessment in the management of global environmental change.

The first part of that approach is a descriptive framework for classifying the option assessments reflected in our empirical data. Many such classifications are possible. For the cases addressed in this study we found most useful the framework described in box 17.1, in which individual options and assessments are classified according to what we have called the *targets* and *means* they address. Table 17.2 uses this framework to present some of the options most commonly encountered in our study.

17.2.3   Analytic Criteria

From the general conceptual concerns reviewed earlier in this section, we derived a set of analytic criteria to use in bringing our specific data to bear on our general questions about the development of options and their assessments. These criteria are summarized in table 17.3 as relatively straightforward questions and described in more detail as we apply them in reaching the findings reported in sections 17.3 and 17.4. The criteria address both the assessment process and Table 17.2's pool of actively discussed options that emerges from assessments and other processes.

Perhaps optimistically, we believe that the evaluative criteria posed here should be relatively uncontroversial. That is, most scholars and practitioners of option assessment, other things being equal, seem likely to want to know about the properties addressed by those criteria. More problematic is the interpretations of the patterns we observe—whether "high" or "low" scores are more likely to be associated with "better" assessments or more "useful" option pools. We leave these interpretive questions to the final sections of this chapter.

17.3   Findings: The Option Pool

We begin here our analysis of how societies' understanding of their options for dealing with global environmental risks evolved over the period covered by our study. This section examines the pool of options actively debated by society. The next section turns to the development of the option-assessment process and its role in shaping the option pool.

Recall that we built the data set on the option pool from a variety of sources, including parliamentary proceedings, news accounts, and interviews as well as formal assessment documents. This means that options found in our sampling of the pool need not have originated in a formal assessment. One of our tasks is to discover the extent to which formal assessments in fact play an important role in shaping the option pool. Our analysis is structured according to the three relevant criteria introduced in table 17.3, examining in turn the dominance, richness, and scale of options actively debated in the public arena.

17.3.1   Dominance: Attention Devoted to Options

Dominance of options is the term we use to reflect the extent to which a few options or classes of options receive the bulk of attention in the risk management debate. Dominance is an important descriptor of the option pool because it allows us to see the impact of powerful interests, intellectual "frames," or other forces that push particular options forward and increase their visibility and potential for adoption. Analyzing the changing patterns of dominance in the option pool provides a good overview of how societies' response to the question "What can be done about global environmental risks?" has evolved.

Overall Patterns   Long before formal option assessments on global environmental risks were conducted, individual options related to these risks were advanced and discussed. Prior to the 1970s, the only sustained discussion of options for managing any of the global environmental risks addressed in this study focused on building the knowledge base through research programs. Other measures that can be seen in retrospect to have some bearing on management of these global risks were considered only in the context of debates about other problems including specific technologies such as the SST, local pollution, and regional climate variability. Only after the Stockholm Conference in 1972 did a widespread debate about possible options for dealing with risks of global environmental change begin to develop.

As a general rule, that debate rapidly became dominated by options that targeted emissions reduction through means of enhancing technical capacity or mandating performance standards. Options targeted on environmental remediation or impact reduction received much less attention—perhaps only a quarter—of that accorded emission reduction. As for *means* of risk management, societies exhibited a sustained recognition of the option of increasing knowledge through additional research and a growing interest in options that enhanced institutional capacity and the use of incentives. Nonetheless, even by the time of the Rio Conference in 1992 none of these alternative means had displaced those of technology development and command-based regulation as the dominant options under discussion for dealing with our global environmental risks.[4]

Table 17.2
Principal options addressed, 1957–1992 (see box 17.1 for definitions)

Means	Targets		
	Emissions	Environment	Impacts
A. Acid rain			
Capacity:			
Cognitive	—	—	—
Technical	Energy reduction Fuel mix for electricity (health) Fuel switching Scrub (sulfur dioxide) Energy efficiency Coal cleaning (desulfurization) Nuclear energy Fluidized bed combustion Catalytic converter Add-on technologies (nitrogen oxides, ammonia) Structural change in production and consumption	Liming Tall stacks Exhaust fans	Breed acid-resistant trees
Institutional	Differential reduction based on critical loads	—	Acidification fund
Incentive:			
Information	—	—	—
Market	Tax emissions Emission trading	—	Acidification fund
Command:			
Regulations	Product standards regarding sulfur dioxide content (gas, oil) Emission standards Mobil-source standards for nitrogen oxides Ambient standards based on health Speed limits Catalytic converter Standards for ammonia Standards based on "critical loads" Day without a car Emissions cap	—	—
B. Ozone depletion			
Capacity:			
Cognitive	Label CFCs	—	—
Technical	CFC substitutes Halon substitutes Non-CFC foams and packaging Recycling CFC conversion into safe and useful products	Increase ozone artificially	Hats, sunglasses, and sunscreen
Institutional	International agreements (Montreal) Targets for CFCs, halon, and bromine reduction Reliability of accounting Technology transfer Eliminate leakage Develop alternative application devices New refrigerants	—	—
Incentive:			
Information	—	—	—
Market	Tax CFCs Emissions trading	—	—

(*continued*)

Table 17.2 (continued)
Principal options addressed, 1957–1992

Means	Emissions	Environment	Impacts
		Targets	
B. Ozone depletion (continued)			
Command:			
Regulations	Ban or limit SST flights	—	—
	Ban or boycott aerosols		
	Freeze production or consumption of CFCs		
	Limit CFC production		
	Restrict CFC uses (foam, solvents, refrigerants)		
	Ban CFCs		
	Recycle CFCs		
	Regulation of non-CFC ODGs		
	Restrict trade in ODGs		
	Phaseout of HCFCs		
C. Climate change			
Capacity:			
Cognitive	Research on renewables	—	Warning systems to reduce storm damage
	Labeling appliances		Prepare for and anticipate climate change
Technical	Efficiency technologies (insulation)	Tree planting	Coastal defense
	Fuel switching	Carbon-fixation technologies	Plant breeding
	Renewable energy		Irrigation
	Nuclear power		
	Scrub carbon dioxide		
	Reduce non-carbon dioxide GHGs		
	CFC substitutes		
	Electric car		
	Advanced energy technology		
Institutional	Privatize electricity supply	—	Environmental aid funds
	International climate convention		
	Technology exchange		
Incentive:			
Information	—	—	—
Market	Tax carbon		
	Tax energy, fuels		
	Least-cost planning		
	Economic instruments		
	Low-interest loan and tax incentives for plant and equipment investment		
Command:			
Regulations	Ban synthetic fuels	—	Land-use control to reduce storm damage
	Auto efficiency standards		
	Transport-sector emission limits		
	Demographic policy		
	Reduce GHG emissions		
	Carbon dioxide emission quota		

Box 17.1
Options addressed in global environmental risk management: A framework for classification

To consider options and their assessments systematically we used a two-dimensional framework in which individual options and assessments are classified according to what we have called the *targets* and *means* they address. This enables us to delimit the full range of options that might be assessed or otherwise debated and to locate actual options and assessments within that range.

Targets

The *target* of an option or assessment can be defined as the portion of the causal chain of risk that it seeks to alter. We adopt the conventions for the causal chain model that is described in Chapter 1 of this volume, summarized in the introduction to Part III, and employed in the companion Chapter 15 on risk assessment. We overlay on that chain three corresponding targets for options and their assessments:

• *Emission targets,* for measures to reduce demand for the polluting activity (such as energy conservation) and measures to reduce the amount of emissions per unit demand (such as switching to cleaner fuels or installing technologies to capture or destroy pollutants before they are released to the environment);

• *Environment targets,* for measures to remove the pollutant from the environment (such as planting forests to absorb carbon dioxide) or to restore the environment to its prior condition (such as liming of acidified lakes); and

• *Impact targets,* for measures to protect from the changed environment people or things they value (such as employing sunscreens for protection from increased UV radiation), measures to introduce structures or species less vulnerable to a changed environment (such as drought-resistant crops), or measures to compensate people for impacts they experience (such as an acidification fund).

As noted in chapter 1 and the introduction to part III of this volume, these three targets have often been designated in the literature as *prevention, offset,* and *adaptation* options, respectively. We found in the course of our research, however, that strong connotations have developed around the use of these latter terms. Options characterized as targeted on *prevention* tended in many circles to be treated as though they were inherently good. In contrast, the labels *adaptation* and, especially, *offset* acquired for some people such negative connotations as to be virtually unmentionable. This situation developed to such an extreme that we found the label *prevention* being stretched by some to cover all options (for example, building sea walls to prevent impacts) merely to secure the perceived benefits of the classification. As an empirical finding, these naming battles are important and are discussed at the appropriate point later in this and other chapters. As an analytical matter, it seemed worthwhile to adopt a more neutral and descriptive terminology for our classification scheme; hence the three target labels given above.

We also found it helpful to add a fourth classification that we have called

• *Generic targets,* for measures that are intended to change the risk across the above categories (such as general research or institution building).

Means

The *means* dimension of our option classification is intended to address the different ways that things we do as a society can alter our relation to the risk at hand. In principle, most interventions or policies change what people *can* do (through changing capacity for research, monitoring, assessment, negotiation, and implementation), what they *want* to do (through changing concern, awareness, or incentives), or what they are *compelled* to do (through changing the social contract via commands embodied in prohibitions, requirements, or laws). Subdividing these logical categories to reflect certain distinctions that have proven important in the actual management of environmental risks, we developed six analytic categories of means:

• *Capacity: cognitive means,* for measures that address the knowledge base through commitment of resources to research, monitoring, or information dissemination;

• *Capacity: technical means,* for measures that address the physical capacity for responding to the risk (such as material substitutions, pollution control technologies, and tree planting);

• *Capacity: institutional means,* for measures that address the organizational capacity for responding to the risk (such as improving society's ability to perform assessments, forging international conventions, improving the contractual environment, and coordinating programs or activities);

• *Incentives: informational means,* for measures that spread awareness of the risk itself or the consequences of actions (such as research and monitoring programs, information campaigns, and labeling practices);

• *Incentives: market means,* for measures that change the perceived benefits and costs of alternative actions (such as taxes, subsidies, and various other market mechanisms); and

• *Command means,* for measures that commit, require, or prohibit changes in practices (such as bans, standards, limits, and voluntary agreements).

Table 17.3
Criteria for evaluating option assessments and their products

Criteria relevant to the option pool:

Dominance: Which options receive the most attention?

Richness: How great is the variety of options under discussion?

Scale: Which options are targeted at a local, national, or international level?

Criteria relevant to the assessment process:

Pluralism: Who performs the option assessments?

Breadth: What is the range of options assessed?

Locus: What kind of outcomes are assessed?

Mode: How are options scored in the assessment?

Issue-Specific Variation All three of the issues of environmental change covered in this study shared the basic feature of an option pool dominated by measures targeted on emission reduction. Instructive differences developed, however, in the relative emphasis accorded particular means of pursuing the dominant options in each issue. These differences in dominant means generally reflected differences in how the issues were framed. In particular, they corresponded to whether the offending emissions were seen as hazards to be eliminated, pollutants to be reduced to tolerable levels, or as inevitable by-products of an overconsumptive society.

The ozone issue was framed from the outset as a hazardous-technology problem (see table 17.2, panel B). As in the other issues, substantial discussion was carried out on the prospects for enhanced research capacity. But the option debate on ozone was distinctive in devoting most of its attention through the mid-1980s to means involving compulsory bans and phaseouts and only in the late-1980s turned to the development of technical alternatives that would make such draconian regulations palatable. For example, the SST debate was cast from its beginnings in the late 1960s in terms of whether the technology in question should be strictly constrained or even prohibited. Molina and Rowland's 1974 hypothesis of an ozone-depleting role of chlorofluorocarbons (CFCs) was followed quickly by moves to boycott or ban the use of CFCs in most aerosol uses. Research and development began quickly within the private sector on options that would substitute chemicals less dangerous to the ozone layer for CFCs in a variety of uses. This early attention to technical fixes waned in the early 1980s as the ozone issue fell from the public agenda. But when it returned in middecade, the option debate restarted where it had left off: with a focus on identifying and phasing out hazardous technologies, while promoting substitutes that were both feasible and safe.

The acid rain issue, in contrast, was born as a classic pollution problem—different from its predecessors primarily in that the pollutants in question had strayed beyond national borders. The most important options discussed can be found in table 17.2, panel A. For most of the time and arenas covered in this study, its option pool resembled that for other air pollutants, dominated by a succession of emission-reduction technologies. The transboundary character of acid rain led some "victim" nations—notably Canada and Sweden—to press for novel international regulatory frameworks. But concern remained focused on control of the acidifying pollutants. The overall option debate continued to be dominated by the same kinds of end-of-pipe technologies and mandatory emission standards familiar to the existing pollution-control community.

Particular technologies also shaped the early evolution of the option debate on climate change (see table 17.2, panel C), with nuclear-energy advocates in the United States and Germany playing important early roles. The dominant means for targeting emission reduction, however, were improvements in conservation and energy efficiency—emission-reduction measures that would reduce the demand for activities that could release harmful emissions. This entrained a much different coalition of actors and advocates than either the hazardous-technology or pollution-control orientation of the other issues we considered. In particular, it made climate change the global environmental problem of choice for those activists interested in promoting less consumptive or lower-energy lifestyles in the industrialized nations. The focus on demand also helped to promote the relatively early emergence of market incentives (such as carbon taxes) into the debate over management options for climate.

These issue-specific differences are, in retrospect, so obvious that their significance is easy to overlook. But the reduction of acid rain emissions *could* have been framed, as was climate, in terms of the need for reduction of demand. In fact, the policy debate remained dominated by existing supply-side options. The private-sector producers of CFCs *could* have joined the coal or oil interests in seeing themselves as purveyors of specific chemicals—CFCs—and in defending the unrestricted use of these chemicals to the bitter end. In fact, they came to see themselves as offering services rather than specific products and thus willing, under appropriate conditions, to search for alternative technologies that would provide the service at less risk (and, eventually, equal or greater profit). The climate debate *could* have taken seriously the early arguments that fossil fuel technologies were inherently hazardous to the global environment and aggressively promoted the development of alternative technologies, as was done in the ozone issue. In fact, debate on nonfossil technologies was discouraged by interests hostile to them—both pro-oil and antinuclear—and never reached a position of dominance in the option debate.

Arena-Specific Variation In contrast to the significant influence of issue-specific factors on the dominant means (though not targets) debated, few arena-specific differences in dominant options emerged. The major exceptions were those arenas preoccupied with generic capacity-building measures at the international level. Among these were the former Soviet Union, frequently Canada and the Netherlands, and not surprisingly the international institutions and European Community. As important as the ultimate contributions of these international capacity advocates may have been, however, the options they advanced generally remained in the shadow of debates on the dominant options for direct emissions reductions.

17.3.2 Richness: The Variety of Options Assessed

Richness of options is the term we use to reflect how many different kinds of options are available in the option pool. Richness can increase when discussion focuses on multiple targets, multiple means, or both. Richness is an important descriptor of the option pool because, as noted earlier, theory directs our attention to both potential social benefits and potential hazards of increasing the range of options actively considered in the risk management debate.

Overall Pattern The richness of the overall option pool increased throughout the study period. Most of this increase had occurred by the mid-1980s, however, with relatively little net enrichment in the late 1980s and early 1990s (for example, "bubbles" had entered the debate by 1986). In a somewhat broader perspective, we may conclude that most of the qualitative innovation in options occurred early in the issue cycle, while the subsequent debates over policy and implementation contributed rather little to the overall richness of the option pool.

The dominant option categories—technical and command means of emission reduction—accounted for the largest increase in richness over the study period. Nonetheless, these dominant options represented only about half of the total richness of the option pool. At least three other groups of options also made substantial contributions to the pool and, ultimately, contributed some of the most important innovations in policy discussions that had emerged by the time of the Rio Conference: research capacity, institutional capacity, and economic incentives.

Research Capacity Among the earliest options considered and pursued in all issues and most arenas was the development of a knowledge base to better delineate the risks and the prospects for dealing with them. Discussion about the need to enhance research capacity continued throughout the study period, though, as we argued above, even by the mid-1970s it had become a secondary rather than dominant option in most arenas.[5] Throughout our study period there occurred sporadic exhortations that the research should be made "relevant" or "responsive" to policy. In practice, however, these exhortations were usually transformed into a basic research agenda consistent with a science-driven rather than a policy-driven view of knowledge seeking. Such policy-driven funding tended to be restricted to the development or evaluation of basic technological capacity to reduce emissions.

Institutional Capacity The need to enhance institutional capacity for international negotiation of responses to global environmental risks was stressed in Sweden's submission on acid rain to the Stockholm Conference. The United Nation's Environment Programme's (UNEP) founding and development of an "outer-limits" program in the wake of the Conference provided an early international venue for largely scientific discussions on global issues. But despite support from a few national arenas, international options for responding to our global environmental risks had barely begun to receive sustained and widespread attention by the early 1980s (e.g., Kay and Jacobson 1983; Caldwell 1991). Over the next decade, however, the range of such options under active discussion expanded rapidly. By the time of the Rio Conference, a rich array of options to enhance institutional capacity had been proposed, were being actively debated in policy as well as academic circles, or had moved beyond the talking stage to implementation. These included formal international protocols and conventions with associated compliance-enhancement mechanisms, alternatives for organizing their associated secretariats, mechanisms for financial and technology transfers, proposals for international efforts to build assessment capacity in the developing countries, and the Rio process itself.

Economic Incentives A further increase in the richness of the pool of options was due to the introduction of economic-incentive mechanisms into the debate. Although relatively high fuel taxes had long exerted a strong influence on energy conservation and efficiency measures in Western Europe and Japan, by the 1970s only the United States had formally explored (without adopting) taxation as an option for managing global environmental risks. Wider discussion of taxes as a means of emissions reduction surfaced only in the late 1980s, primarily in the context of the climate change issue but with a presence in the ozone and acid rain debates as well. By the time of Rio and its aftermath, a wide range of economic incentives had become a standard part of the option menu discussed by national and international policy makers concerned with the management of global environmental risks.

Finally, it is worth noting the relative absence from the overall option pool of measures targeting the environment or impacts on society. To be sure, measures such as lake liming, sunscreen lotions, and sea walls were discussed and even featured in individual option assessments. But with some exceptions in the climate case noted below, such alternatives seldom became incorporated in the mainstream policy debates over the prospects for coping with global environmental risks. In some cases, such as the Netherlands' attention to dikes, this was because the option in question was treated as part of another issue area that was only marginally affected by global environmental concerns. In others, such as Canada's skepticism over liming of acidified lakes, the neglect reflected an objective evaluation of the option's practical limitations. And in still other instances, such as the ridicule heaped on the U.S. government's "hats and sunglasses" response to the ozone risk, alternatives were actively repressed as efforts to deflect attention from what many viewed as the "real" problem of emission reduction. Regardless of cause, the question remains of whether the relative neglect of options targeting the environment and impacts has enhanced or undermined society's efforts to cope with global environmental risks. We defer its consideration to the conclusion of this chapter.

Issue-Specific Variation In general, differences among issues in the richness of options actively debated are small relative to the similarities noted above.

Acid rain, alone of our issues, shows a clear *decline* in the richness of options being actively considered at the time of the Rio Conference relative to the situation a decade earlier. This eventual decline in richness occurs in virtually every option category with the exception of economic incentives for emission reduction, where the range of actively debated alternatives was increasing in anticipation of negotiations on the second Sulfur Protocol. Recall that acid rain was the first of our issues to emerge onto the policy agenda and the first to pass on to a phase of well-institutionalized implementation and review. The decline in richness associated with this latter institutionalized implementation phase of acid rain management raises the question of whether we might expect declines in the richness of options actively debated in the ozone

and climate issues as they eventually mature. At the time of the Rio Conference, however, neither of these declines was yet in evidence.

Another issue-specific difference in the richness of options considered involves the attention paid to economic incentives. Such options ultimately played a major role in the climate debate but a relatively minor one elsewhere. As noted above, the possibility of taxing carbon emissions to reduce demand for carbon dioxide–releasing fuels had already been raised in the 1970s. By the mid-1980s, more than half of the arenas in our study were seriously discussing the use of carbon and energy taxes or least-cost energy planning as plausible means for addressing the risk of climate change. In contrast, options involving economic incentives were virtually absent from the debate on stratospheric ozone depletion and acid rain in the 1970s. They subsequently made modest inroads into acid rain policy in the 1980s as a means for promoting broad-based improvements in energy efficiency and for improving the efficiency of how emission-reduction efforts were allocated among regions. By the late 1980s, ozone-policy discussions were increasingly invoking taxes as a possible means for smoothing mandated phaseouts and for recovering associated windfall profits. At the time of the Rio Conference, however, the climate debate continued to encompass a far greater variety of incentive measures than was evident in either acid rain or ozone.[6]

Finally, it is worth pointing out that while options targeted at changing the environment (rather than emissions or social impacts) have never played a major role in any policy debate, they have been most widely discussed for the climate issue. The most popular such option in the climate case concerned planting trees to remove carbon dioxide (CO_2) from the atmosphere. This option received positive and sustained attention from the 1970s onward in a number of arenas, was given substantial play in the IPCC's 1990 report on response strategies, and became a central element of several arenas' subsequent action plans for climate change.[7] No other environment-targeting option in any of the issues has achieved such prominence in the policy debate. We suspect, however, that the popularity of the tree-planting option in the climate debate has more to do with trees than with climate. Tree planting emerged as the ultimate no-regrets option: attractive on its own merits, with great symbolic value and little cost, and in most debates closely linked with the equally attractive emission-reduction measure of "saving the (rain) forest." The widespread reaction against all other *geoengineering* (or, in our terms, *environment-modifying*) options raised in the climate debate supports the interpretation that pro-forest rather than pro-CO_2 removal sentiments created

62

William C. Clark, Josee van Eijndhoven, and Nancy M. Dickson et al.

this exception to the normal focus on emission-reduction options during the period of our study.

Arena-Specific Variation Most options discussed anywhere were eventually discussed everywhere. In this sense, there is little variation in the option pool among arenas. When we turn, however, to an analysis of which arenas have been most responsible for enriching the pool—either by introducing options to the pool for the first time or by keeping them there when others' attention lagged—some interesting patterns do emerge in particular option categories:

• *Technical capacity for changing emissions* Perhaps not surprisingly, Japan, Germany, the United Kingdom, the United States, and the Netherlands have been the principal sources of innovations in these options, while international institutions[8] have functioned as disseminators.

• *Institutional capacity for international cooperation* The former Soviet Union, Canada, the United States, and, predictably, the family of international organizations have repeatedly taken the lead in pushing for the development of international institutional capacity. The Netherlands has also had an aggressive record except for its relative absence from the ozone debate.

• *Research and monitoring capacity for better understanding* The push for more research to inform the policy debate has been almost universal, but leadership and sustenance have often been provided by the United Kingdom, the United States, the former Soviet Union, and the international institutions.[9]

• *Economic incentives* for emission reduction have been a specialty of the United States and, to a lesser extent, Germany and Japan. Debate in the United States first raised the options of carbon taxes and emission trading in the 1970s. These were picked up and further developed as policy instruments in the late 1980s and 1990s in Europe and elsewhere, but analysts in the United States have led the way in showing how economic incentives could be applied to each of the issues addressed in this study.

17.3.3 Scale: The Jurisdictional Scope of the Option Pool

Scale of options is the term we use to distinguish the jurisdictional scale, from local to global, at which a particular measure might be undertaken and in terms of which its desirability is evaluated. As a practical matter, we attempted to distinguish the extent to which the option pool was filled with local alternatives, such as liming proposals for particular lakes; national alternatives, such as strategies for meeting a country's CFC reduction quota; or international measures, such as provisions for joint

implementation of greenhouse gas reduction targets. We are concerned with all levels in this study. Indeed, one of our central tasks is to identify and explain any pattern that may exist in the emergence of higher-level, internationally conceived options through time or in particular arenas. Again, we do not presume that higher-level assessments are better ones: a simple comparison of two pollutant-scrubbing technologies may be exactly the assessment that some local decision makers most need.

We must emphasize at the outset that our research methodology introduced a substantial bias into our efforts to address questions of level. As noted earlier, our ability to sample the option pool considered by the private sector and subnational governmental decision makers turned out to be much weaker and less systematic than our ability to sample at the national and international levels. When lower-level decision makers appeared at the national or international levels—as in Germany's Enquete Commission or the Technology Assessment Panels set up under the Montreal Protocol—we captured their views (German Bundestag 1989, 1991; UNEP 1987, 1989a). When they did not, neither did we. This means that a substantial but unknown fraction of local-level options have remained invisible to us, almost certainly distorting our analysis. Until further research can evaluate and correct this distortion, we must therefore present our observations about the level of options considered in the debate over global environmental risks even more tentatively than we do for other dimensions of this initial reconnaissance. With these reservations in mind, however, our preliminary findings are relatively straightforward.

Overall Patterns The level of options actively debated in society generally increased from local to national or global in the course of the study period. At least in the public arena, this seems to have included something of a displacement phenomenon in which more debate on national options meant less debate on local ones. For example, national strategies for the reduction of acidifying emissions were not publicly debated in Germany until 1982, by which time the relative merits of particular emission-reducing technologies had been discussed for at least a decade. Alternatives for multinational emission reductions entered the public discourse shortly thereafter, while discussion of local options (such as a comparison of particular technologies) fell from public view.

In most arenas, the elevation of the option debate from the local to the national or international level commonly occurred only *after* an active debate on relevant national or international goals or obligations had begun.[10] For example, as described in chapter 18 (on goal and strategy formation), the so-called Toronto goal of a 20 percent reduction in greenhouse gas emissions precipitated a widespread debate on possible means of meeting such an objective. Politically derived goal statements set the scale of the option debate. The converse possibility—that prospective exploration of national policy options would drive the emergence of political goals—seems not to have occurred with any frequency in the histories reviewed in this study.

Arena-Specific Variation   Two major exceptions to the general patterns described above stand out. First, as a complement to the overall trend toward discussion of options at the national and higher levels, international institutions began in the 1990s to provide a forum for the discussion of *local* options within a broader context. Perhaps the clearest examples of this function are the various Technology Panels growing out of the Montreal Protocol.[11] As described in more detail in chapter 13 (on international institutions), the Technology Panels analyzed costs, benefits, and constraints of options for reducing emissions of various ozone-depleting gases. On the basis of these local assessments, the Panels evaluated the feasibility of alternative global phaseout schedules. Private-sector users and producers were intimately involved in the work of the Panels, thus helping to ensure a strong linkage between global and local perspectives. Discussions conducted under the auspices of the Response Strategies Working Group of the IPCC and various Working Groups of the Executive Body of the Long-Range Transboundary Air Pollution (LRTAP) Convention play a similar if somewhat less well developed role.

A second important exception to the general pattern noted above is that in the United States national- and international-level options were under active discussion well before comparable goals had been articulated in the public debate. Local-level comparisons of individual technologies were certainly carried out and reported in the United States, often to excess. But from the first serious option assessments published from the late 1970s onward, the emphasis was on national comparisons of alternative action programs. The options evaluated were often very few or narrow—for example, fuel switching versus sulfur scrubbing as a means of meeting national sulfur emission-reduction goals. However narrow, they were uniformly evaluated at the level of national policy and not of individual plant or process. The generalization advanced above about the role of political goals in elevating the level of option discussion thus seems not to hold for the United States. Subject to further investigation, we suspect that this is a reflection of the relatively large, independent, and proactive policy-analysis community in the United States and its exceptional inclination to push analysis in the absence of any perceived political demand.

An Open Question   This review of the changing scale of the option debate leaves us with an important question that our data do not allow us to resolve: How well is the flourishing debate on national and international options linked back into an increased appreciation of the options available to governmental and private-sector decision makers operating at the subnational or local level? Little evidence of a critical debate on local options shows up in our data, though, as previously noted, this might be due to bias in our collection procedures. Clearly, however, there was less illumination of local options in the public debate on our global environmental risks in the early 1990s than there had been a decade before. We know that various efforts conducted through international institutions had begun to attempt a recoupling of global- and local-option assessments in the early 1990s. We lack, however, any evidence to tell us whether these activities have affected the perceptions of available options for a significant fraction of relevant local decision makers.

17.4   Findings: The Assessment Process

In this section we analyze the option assessments conducted during our study period according to the four process criteria introduced in table 17.3: pluralism, breadth, locus, and mode. The first of these criteria lets us ask how open the assessment process is to multiple interests in society. The next two characterize the degree of "integration" in the assessment process. The last explores different approaches to scoring or comparing options within an assessment. In each case, we begin with an effort to identify the broadest possible generalizations before turning to variation associated with particular cases and arenas.

17.4.1   Pluralism: Who Performs Option Assessments?

Pluralism is a term we use to reflect the difference between situations in which the process of option assessment is dominated by single political actors or sectors (such as one government ministry) and ones in which access is open to a wider range of participants. Participation could, in principle, entail roles in either the performance or the review of assessments. High participation could come about either through participation of multiple actors in each assessment or through a process in which multiple actors were able to conduct and disseminate assessments of their own. We are interested in this criterion because of the inherently political (that is, interest-laden) character of option assessments and the possibility of using option assessment to influence the option pool.

As discussed in section 17.2, the first option assessments to be focused on global environmental risks were generally performed on an ad hoc basis by individual scientists in the context of their personal appraisals of the problems at hand. Even into the mid-1980s, some of the most influential links between knowledge and action occurred through scientists' assessments that raised concern over global environmental risks without directly addressing policy options. The World Meteorological Organization Blue Books on atmospheric ozone, published in 1986, exemplify such influence (WMO et al. 1986). The role in option assessment of engineers, economists, and professional policy analysts grew throughout the period, however. This trend developed earliest and most energetically in the United States, which by the late 1970s had already fielded a significant number of option assessments based on expertise outside of the natural sciences. More inclusive approaches became pervasive among our study arenas throughout the 1980s, as exemplified by the representation of political stakeholders in the 1987 international Bellagio assessment of policies for responding to climate change (Jäger 1988) and Germany's Enquete Commission on Protecting the Earth's Atmosphere (German Bundestag 1991). By the time of the Rio Conference, experienced policy analysts within the context of formally institutionalized international assessments were performing many of the most visible and influential option studies. Representative of these latter approaches were the International Institute for Applied Systems Analysis's (IIASA) Regional Acidification Information and Simulation (RAINS) models of acidification, the Technology Assessment Panels set up under the Montreal Protocol, and the IPCC Response Strategies Working Group.

For our study period as a whole, governmental and intergovernmental institutions were the principal sources of option assessments. In some arenas, such as the former Soviet Union and Japan, their dominance was overt: governmental bodies provided the only formal venue for assessment of alternative options and responses. In other arenas where significant nongovernmental assessment activity took place (for instance the United States and European Community), governmental bodies nonetheless stood out in terms of the sheer number of assessments they conducted or commissioned. In still other arenas—notably Canada and Germany—governmental institutions played more of a synthesis role, gathering assessments performed by other actors and cataloging or comparing them. Finally, in the acid rain case the Dutch government actively promoted the development of assessment methods for use at the national and international levels.

The international arena was also populated predominantly by intergovernmental rather than nongovernmental assessments. These often functioned in the synthetic style noted above.[12] The history and form of this intergovernmental ascendancy varied significantly by issue. For ozone and European acid rain, a formal intergovernmental framework for policy negotiations and supporting assessment existed prior to the emergence of these issues into the political limelight. As a result, intergovernmental assessments rapidly came to dominate the international political discussion even as they made substantial use and synthesis of nongovernmental analyses.[13] In contrast, for the case of climate change, the effectively nongovernmental assessments emerging from the Villach and Bellagio processes in the mid-1980s played a significant role in bringing the issue to the attention of political actors.[14] Once alerted, however, these governments promptly intervened to replace this nongovernmental assessment process with the intergovernmental IPCC.

As suggested by these international examples, while governments may have generally dominated the option-assessment process, they had no monopoly on it. Many, if not most, of the governmental and intergovernmental reports we reviewed engaged representatives of the industry, expert, or nongovernmental organization (NGO) communities in their design, analysis, or review efforts. In some cases, this inclusiveness was formal and overt, as in the corporatist character of Germany's Enquete Commissions or the Montreal Protocol's Technology Assessment Panels. Even in the government-dominated cases of Japan and the former Soviet Union, however, outside advisory bodies including representatives of academia and industry were substantially involved in governmental option assessments. Independent production of option assessments by nongovernmental groups—experts, industry, and "green" organizations—was common in the United States, the United Kingdom, and Germany but less so elsewhere. Finally, it is worth noting that we found an intriguing number of instances in which consultants—professional policy analysts—simultaneously performed similar option assessments for multiple actors. (The consulting firm ICF, for example, performed separate acid rain option assessments for governmental, industry, and NGO groups in the United States in the mid-1980s.) The potential implications that such professionalization of assessment holds for improving practices or homogenizing outcomes would seem to be substantial but require further investigation.

Government assessments seem to have had a preferred position in setting the agenda of options to be debated. In several of our arenas where a substantial number of assessments were performed by groups other than government, those groups were in fact responding to and critiquing government assessments. This is most evident in the material submitted to parliamentary (or, in the United States, congressional) committees but can be seen in a close reading of some other assessments. It should be emphasized that some important options were nonetheless brought to the broader debate through the activities and assessments of actors outside of the government. And if government reports and hearings provided the focus for much of the assessment debate, they also provided a visible forum that other actors used to advance their views. Some of the most influential assessment processes (such as the IIASA RAINS effort, performed in support of the LRTAP negotiations) were effective in part because they were nongovernmental and international and thus provided a relatively neutral forum for analysis and debate.[15]

We are left with the impression that, in general, the process of option assessment for global environmental risks was a relatively pluralistic one, though with substantial variation among arenas. Assessments from government were most plentiful. But the voices of other social actors became increasingly omnipresent, persistent, articulate, and influential.

17.4.2 Breadth: What Is the Range of Options Considered?

Our analysis seeks to capture the degree of *integration* in option assessments and the processes that produce them through two terms: *breadth* and *locus*. *Breadth* is the term we use to reflect the range of options systematically considered in particular assessments. We want to distinguish relatively straightforward exploration of (or advocacy for) a single option, from efforts to discuss in some formal way, within some formal framework, a larger number of alternatives. This is important because of the tension between the views of assessment that emphasize the theoretical value of *comprehensive* analysis of options on the one hand and views that emphasize the practical utility of *successive limited comparisons* on the other.[16]

We classified an assessment as *narrow* in breadth when it considered only a single option. We also applied this classification if all the options considered by an assessment were within a single cell of the target-by-means classification defined in table 17.2—for example, a discussion of alternative technologies for providing aerosol propulsion without the use of CFCs. We assigned a *medium* breadth designation to assessments that explored multiple means of addressing a single target—for example, a comparison of mandated technologies versus voluntary compacts in reducing sulfur emissions. We also designated as medium breadth those assessments that

explored multiple targets addressed through a single means—for example, an evaluation of the role of market mechanisms for limiting and adapting to climate change. Finally, we described as *broad* those (few) assessments that assessed a range of options involving multiple means and targets.

Our central, if unsurprising, finding is that broad comparisons in a single assessment are rare. More than half of the assessments we reviewed were of narrow breadth. Many of these considered only a single option, scenario, or policy, implicitly contrasted with a base case or no-action option. The remainder addressed a very limited range of similar fixes for the problem at hand (that is, options within a single cell of table 17.2). Typical examples, spanning our history, include

• One of the earliest assessments in our collection (the German government's 1975 *Systems Analysis of Desulphurization Processes*);

• Some of the most model-intensive (such as a large number of U.S. assessments based on models of how energy price changes, including taxes, would affect carbon dioxide emissions) (including Edmonds and Reilly 1983); and

• One of the latest assessments in our study period (the Montreal Protocol's Technology Review Panel assessment of 1989) (UNEP 1989a).

Nearly all the other assessments we studied were of medium breadth, usually comparing alternative means (such as standards versus incentives) for changing a single target: emissions. (We documented earlier in section 17.3, Findings: The Option Pool, our finding that the dominant focus of the option debate by far was on the target of emission reductions, via means of either technological substitution or command-and-control regulation.) Typical of these assessments were

• The comparisons of emission-reduction strategies for sulfur dioxide performed with IIASA's RAINS model in the late 1980s (Alcamo, Shaw, and Hordijk 1990; Amann 1992; Hordijk 1991);

• The European Community's comparison of economic incentives versus command-and-control policies in its 1990 "Community Action Program to Limit EC Carbon Dioxide Emissions and to Improve the Security of Energy Supply" (CEC 1990); and

• The report of UNEP's 1986 Leesburg workshop on the control of chlorofluorocarbons (UNEP 1986).

Broad assessments ranging over multiple means and targets were rare. Those that did exist were as likely to be performed by individuals as by large international

66

William C. Clark, Josee van Eijndhoven, and Nancy M. Dickson et al.

committees. Typical of the individual efforts was Wentzel's 1967 prescient, if inevitably flawed, assessment of the ways in which German forest damage due to air pollution could be combatted by an integrated strategy. This involved emission-reduction technologies, tall smokestacks to disperse pollutants, land-use planning, and better silviculture practices, including the breeding of less vulnerable trees. In a similar vein, comprehensive and structured overviews of options for addressing the climate change problem were put forward by Thomas C. Schelling in his contribution to the U.S. National Research Council's 1983 report *Changing Climate* and by Swart, de Boois, and Vellinga (1989) in the Netherlands. Examples of high-breadth, committee-based assessments include the early Climatic Impact Assessment Program (CIAP) study of possible SST impacts on the stratosphere, the Dutch National Environmental Plan of 1988, and the initial work of the IPCC on global climate change (U.S. DOT 1975; Netherlands Ministries of Housing, Physical Planning, and the Environment 1989; IPCC 1991).

Over time, the proportion of narrow assessments generally declined as the issues of acid rain, ozone depletion, and climate change matured and entered the policy arena. By the time of the Rio Conference there was simply less of an audience for simple advocacy of particular options or detailed cost comparisons of closely related technologies than there had been a decade or more earlier. This trend was not universal, however. At least for the ozone issue, what one observer has described as "the happiest group of regulated industries in the world" had by the early 1990s embraced the extremely narrow Montreal Protocol Technology Assessment Panels as an unusually useful option-assessment exercise. Comparably narrow, decision-focused option assessments were just beginning to take shape in support of the Global Environment Facility (GEF) decision making and Joint Implementation negotiations as our study period drew to a close, but their ultimate fate cannot be evaluated with the data we have reviewed here.

The overall decline in narrow option assessments was matched by an increase in medium-breadth comparisons. From the late 1980s to the end of our study period these most often compared market incentives with command-and-control approaches to emission reduction. This trend toward assessing a range of means did not, however, entrain a comparable increase in the range of targets assessed. Comparisons of what could be achieved with comparable investments in emission reduction, environmental restoration, and social vulnerability reduction remained as rare—and apparently as unwanted—at the time of Rio as they had been at the time of Stockholm 20 years earlier.

The time trend in assessment breadth is most confused at the broad end of the spectrum. On the one hand, there has been a clear move away from would-be comprehensive option comparisons toward more limited assessments on a number of fronts. The first big ozone option assessment—the U.S. CIAP study of 1975—is still among the broadest ever attempted. But its very breadth seems to have entrained such controversy that the model was shied away from in later studies. A decade later, UNEP's Rome and Leesburg meetings were intended to provide a broad assessment of the same issue but again failed to produce either consensus or influence. By 1992, as noted above, the preferred form of option assessment on the ozone issue had evolved to the extremely narrow form adopted by the highly successful Technology Assessment Panels. For the climate issue, in contrast, Rio took place during a period of rising demand for and production of broad option assessments. On the international front, Sweden's Beijer Institute had organized a study of *The Full Range of Responses to Anticipated Climate Change* (UNEP and Beijer Institute 1989), followed by the broad survey of the first IPCC response strategies report (1991) and even more ambitious plans for a second round of option assessments published after our study period drew to a close (IPCC 1996a, 1996b).[17] Nationally, despite the difficulties encountered by Germany in adding policy detail to its highly acclaimed Enquete Commission Report, a number of countries (including Japan, the Netherlands, and the United States) had launched substantial integrated modeling efforts better to assess the options for dealing with the risk of climate change.[18] The contrast with the ozone case remained as striking as it was unexamined. The acid rain issue presents a somewhat intermediate situation. IIASA's RAINS assessments, for example, did eventually compare command and incentive means of achieving emissions reductions but explicitly rejected any effort to assess interventions targeted directly on mitigation of environmental or social impacts.

17.4.3 Locus: What Kind of Outcomes Are Assessed?

Locus is our second criterion for characterizing the degree of integration in option-assessment processes. We use *locus* to indicate where on the *causal chain* of risk management the consequences of an option or policy intervention are assessed. In both principle and practice, an option assessment can adopt one or more loci for its characterization of option consequences. Locus can thus distinguish end-to-end assessments that go all the way down the chain to characterize options in terms of their consequences for social systems from narrower assessments that characterize options only in terms of their consequences for emissions or the atmospheric environment. Locus should be important because the logical rationale for assessing options in terms of their consequences for society rather than for emissions collides with the relatively greater demands on uncertain data and analytic capability made by the assessment of environmental or social impact consequences.

All possible loci were present in our sample of assessments. For example:

• *Emissions locus* assessments compared possible actions in terms of their consequences for properties related to emissions such as carbon dioxide releases or CFC production rates;

• *Environment locus* assessments compared possible actions in terms of their consequences for properties of the environment such as acid-deposition loadings, rates of temperature increase, or the amount of ozone depletion;

• *Impact or society locus* assessments compared possible actions in terms of their consequences for and impacts on society through changing the amount of harvestable forest, skin cancer, or deaths from heat stroke;

• *Multiple locus* assessments combined two or three of the individual loci in the same assessment.

Our principal empirical finding is that for the vast majority of assessments conducted over our study period, the locus for characterizing the consequences of options was emissions. Much more rarely did assessments include characterizations of options in terms of their consequences for mitigating damage to the environment or impacts on society. This pattern in the locus of option assessments has not been uniform through time, however. Greater emphasis on consequences for the environment and society occurred in assessments conducted at both the beginning and the end of our study period.

A significant fraction of the earliest assessments on our global environmental risks *did* adopt comprehensive approaches, attempting end-to-end treatments of the science, policy, and its environmental or social impact consequences. For example, the 1967 German study mentioned above examined the relative capacity of several options to reduce impacts on forests from air pollution (Wentzel 1967). The U.S. government's CIAP report on SSTs and stratospheric ozone included an entire volume of results on economic impacts of ozone depletion and its prevention (U.S. DOT 1975). And the first international assessment of *Carbon Dioxide, Climate, and Society,* conducted under the auspices of the World Meteorological Organization (WMO), UNEP, and the Scientific Committee on Problems of the Environment (SCOPE) in 1978,

presented a quantitative evaluation of the impact of alternative energy options on the rate of global temperature increase and commented on likely social consequences as well (Williams 1978).

This early enthusiasm did not last, however. As formal assessments of global environmental risks became more plentiful in the late 1970s and early 1980s, the fraction addressing consequences of policy options at anything other than an emission locus declined substantially. This was especially true in the United States, where the greatest number of formal assessments was carried out. There, criticism of option assessments was disproportionately directed at their characterizations of management options in terms of consequences for environmental and, especially, social impacts.[19] Several factors doubtlessly contributed to this criticism, including both the greater technical uncertainty and the greater political saliency entailed in option assessments with a locus on environment or social impacts. Whatever the cause, however, the ensuing controversy not infrequently spilled over from social impact or policy questions to the basic science in ways that threatened to undermine its credibility as well. This problem was already evident in responses to the United States' 1975 CIAP report and became more pressing as interested parties exploited the inevitable uncertainties for partisan advantage.[20] By 1985, end-to-end assessment efforts had virtually disappeared from the scene. The most important studies published at that time either dropped the assessment of options entirely from their scope of work or confined option assessment to a limited emissions locus. This was true for each issue we studied, as suggested by the dominant thrust of SCOPE's climate assessment reviewed at Villach, the WMO/NASA Blue Books on ozone, and the United Nations Economic Commission for Europe (UNECE) assessments supporting the Thirty Percent (emissions reduction) Club signatories of the Sulfur Protocol to LRTAP.[21]

Many in the core scientific community were reluctant to move assessment beyond a science-only approach or, at most, an emission-locus characterization of alternative policy scenarios. But as the issue dynamics of our global environmental issues moved increasingly toward serious policy development, an increasing number of analysts and policy advisors argued that option assessments restricted to an emissions locus provided less than adequate foundations for informed decision.

Three potentially complementary responses to the dilemma of producing credible but useful assessments evolved in the late 1980s. The first agreed to conduct wide-ranging assessments that included attention to the environmental- or even social-impact consequences of possible options. It also attempted to construct a clear

68

William C. Clark, Josee van Eijndhoven, and Nancy M. Dickson et al.

organizational separation between what were thought to be these inevitably controversial option assessments and the underlying consensual scientific assessments of basic cause and effect. By 1988, variants of this strategy had been adopted by the international assessment process supporting both the Montreal Protocol (with its separate panels and reports for scientific, environmental, economic, and technology assessment) and the IPCC (with its separate working groups and reports for science, impacts, and response strategies).[22]

A second response to the assessment dilemma of the mid-1980s was to extend science-based option assessments beyond their existing emissions locus to incorporate environmental consequences—but to define *environmental consequences* in a sufficiently restricted way that a scientific consensus on the results could be attained and defended. By 1988, this approach had been implemented through the adoption of a *critical (acidification) loads* approach to option assessments supporting the ongoing LRTAP negotiations.[23] The idea of *chlorine loading* had also been introduced as a common environment-locus metric in ozone assessments (WMO et al. 1986). Attempts to define a useful and defensible critical load for carbon dioxide and its radiative equivalents were continuing in the context of negotiations on a climate convention at the close of our study period.[24] Despite its apparent attractiveness and initial acceptance in acid rain applications, however, the critical-load approach to extending the locus of assessment met serious criticism on both scientific and policy grounds.[25]

Finally, a third response sought to tackle the dilemma of scientific credibility and practical utility directly through the careful construction of truly integrated end-to-end assessments within which options would be evaluated in terms of their consequences for emissions, environment, and social impact loci. While NGOs in North America and Europe had carried on in the old tradition of comprehensive assessments through the 1980s, formal efforts to create rigorous integrated assessments gathered momentum only in the early 1990s, primarily through the efforts of modeling groups in the Netherlands and the United States. In essence, these groups proposed to do better what they believed had been done badly in the CIAP and other end-to-end assessments of twenty years before. The much less ambitious but apparently successful model-based integrated assessments carried out under IIASA's RAINS project had taken almost a decade to move from concept to application. At the close of our study period, it was therefore far too early to judge whether the newest integrated assessments of global environmental risks were likely to prove effective.

17.4.4 Mode: How Are Options Scored in the Assessment?

Mode is a term we apply to distinguish assessments that go little beyond naming or describing an option from assessments that go into substantially more detail. We find that assessments of global environmental risks have at times scored or evaluated the options they consider in terms of feasibility (technological or administrative), efficiency (option cost, cost effectiveness, or cost-benefit measures), equity, and combinations of the above. Mode is important because it addresses what many see to be the utility—even necessity—of formal criteria for comparing or even ranking options. As with our other criteria, however, we explore here the modes in which our sample of global environmental assessments have been conducted without presuming that one mode is better than another.

The majority of the studies reviewed for this analysis use option-cost calculations to characterize possible interventions. Many others merely describe options or focus on the technical performance characteristics of the option under consideration. Relatively few characterize options in terms of the damages that society would avoid (benefits it would gain) by implementing them. Formal evaluations of the political or administrative feasibility of alternative options are even rarer.

There are few significant time trends in the mode of characterization employed in the option assessments we reviewed. Some of both the earliest assessments we have reviewed (such as Wentzel's 1967 study on forest damage and air pollution in Germany) and the latest (such as the Montreal Protocol Technology Assessment Panel Reports) characterize their options in detailed technical performance terms with minimal cost information. Of the few benefit evaluations that have ever been carried out, as many were published before 1985 as after. Nor have they become obviously less controversial with time. CIAP's 1975 assessment of the benefits of avoiding ozone depletion attracted much the same kind and amount of criticism as the European Community's 1987 decision analysis of the benefits of acid rain reduction and Nordhaus and Yohe's 1983 attempt to estimate the global costs of allowing climate change to proceed (U.S. DOT 1975; CDA 1987; Nordhaus and Yohe 1983). The cost or cost-effectiveness mode of assessment was dominant throughout the study period.

Arena differences are also not obviously significant. The United States may stand out somewhat for its exceptional reliance on economic modes of assessment. Almost no major American option assessment of the period failed to calculate costs for the options it considered. Perhaps its most ambitious option study of the latter half of our study period—the U.S. National Research

Council's (NRC) 1992 *Policy Implications of Green-house Warming*—was a tour de force in the sustained application of a consistent cost-effectiveness mode of assessment. Benefits assessments, in contrast, became less popular with time in the United States and, with the exception of late work on the greenhouse issue, had virtually vanished by the mid-1980s.[26] The mode of assessment elsewhere was somewhat more varied, with many major studies not moving beyond the naming of possibilities and the characterization of technical feasibility. The most catholic of all arenas in their mode of assessment, perhaps not surprisingly, were the international institutions. The IPCC response strategies report of 1991 was in that sense typical, displaying a bit of all known assessment modes, none of which were pursued consistently enough to provide grounds for useful comparison.

One additional observation regarding assessment mode in our sample of global environmental assessments deserves mention. Though we did not systematically code for scenarios as a mode of option assessment, our evidence suggests that we should have. For a scenario mode of analysis and comparison is clearly—in retrospect—common to a number of the option-assessment approaches that seem to have been most used and useful during our study period. This includes Edmonds and Reilly's (1983, 1985) energy-modeling assessments of U.S. greenhouse gas emissions, IIASA's RAINS model of European acidification (Alcamo, Shaw, and Hordijk 1990), the Dutch IMAGE model (Rotmans 1989), and (perhaps) the family of chlorine-loading models used for evaluations of progress in implementing the Montreal Protocol. Some of these approaches employ a cost-effectiveness or even cost-benefit mode of analysis; others essentially estimate what is technically feasible. What they have in common is that they can be used reasonably transparently and flexibly to provide somehow comparable scorings of multiple options of interest. Their apparent popularity suggests that it may be the ability to respond consistently to "What if" questions about possible options, rather than any particular mode of comparison, that makes for usefulness in option assessments (Blinder 1988).[27]

17.5 Emergent Patterns and Possible Explanations

In this section we reach beyond the detailed empirical findings outlined previously to suggest several emergent patterns concerning the relation of options and their assessment to the management of global environmental risks. We also advance a number of possible explanations for those patterns that could profitably be subject to critical testing in future work.

We believe that the general patterns and explanations advanced here encompass most instances of option-assessment, in most arenas, for most of our issues, during most of our study period. If asked to speculate on the most likely role of future option assessments in the management of global environmental risks, we would expect a continuation of these patterns and the processes that shaped them. Nonetheless, our data show important exceptions to the generalizations advanced here. In this section, we seek to acknowledge such outliers but not to pursue their interpretation or implications. The latter task is reserved for the final section of this chapter, where we explore the prospects for improving option assessment's contribution to the management of global environmental risks.

17.5.1 Emergent Patterns

The option debate has remained focused on a small subset of the potential alternatives for managing global environmental risks. In particular, it has been dominated by measures to reduce, by means of technology mandates or emission standards, releases of a few selected chemicals. Capacity-building and incentive-creating measures for emission reduction received far less attention. Measures targeted on environmental damages or social impacts were rarely treated as serious options for responding to global environmental risks. Indeed, their consideration was not infrequently actively repressed. There was no significant decrease with time or issue maturity in the proportion of assessments with a narrow focus. The few assessments that did attempt to expand significantly the range of options considered had not obviously altered the subsequent course of issue evolution by the time of the Rio Conference in late 1992. For the host of new experiments in increasing the breadth of option assessments that had been launched in the context of the Climate Convention, the jury was still out at the close of our study period.

Assessments of options for managing global environmental risks were largely restricted to limited comparisons of feasibility or cost among similar alternatives. Benefits were seldom appraised. More comprehensive assessments attempting to integrate feasibility, cost, and benefit considerations were attempted, found wanting, and succeeded by more limited efforts throughout the study period. Even the most significant apparent exceptions to the general pattern described here went little further than to provide cost-effectiveness comparisons for alternative means of achieving specified goals of emission or reduction. The surge of interest in integrated climate assessments that swept many of our

70

William C. Clark, Josee van Eijndhoven, and Nancy M. Dickson et al.

study arenas in the early 1990s had no contemporary parallel in the communities of ozone or acid rain assessors. By the time of the Rio Conference in 1992 the effectiveness of the newest round of integrated climate assessments was yet to be demonstrated.

Option assessments were driven more by developments in the world of action than in the world of knowledge. The emergence of new knowledge about global environmental problems did not directly precipitate new assessments of the options for dealing with them. Indeed, (risk) assessments of problems and (option) assessments of solutions evolved as quite separate functions for most of the issues and arenas we studied. The emergence of new management goals rather than new scientific findings precipitated most new option assessments. Once the risks were firmly established on the policy agenda, the schedule of formal policy decisions or negotiations set the pace for the conduct of subsequent option assessments. Likewise, the needs of such negotiations set the agenda of assessments. This was particularly true of international negotiations that, once established, tended to homogenize the content and synchronize the conduct of option assessments across participating arenas for all of our issues.

Option assessments served to deepen and spread awareness of existing options, rather than to bring new options into the policy debate or remove old ones from it. Instead, most options ever considered for an issue were introduced through advocates and entrepreneurs within a couple of years of the issue emerging onto the policy agenda. Subsequent assessment activity mostly "stirred the soup," bringing particular items to the surface and consequent attention but introducing little in the way of new ingredients. Perhaps more surprisingly, despite their emphasis on feasibility analysis, formal option assessments rarely managed to remove existing alternatives from active consideration except through the cumulative effects of neglect. Indeed, crucial assessments that effectively differentiate alternative options for policy seem at least as rare as crucial experiments that effectively differentiate competing theories for science.

Option assessments generally failed to generate a cumulative body of reliable knowledge concerning alternative responses to global environmental risks. Successive option assessments effectively started from scratch, except in the narrowest areas of specialized concern noted above. They seldom built on or even acknowledged the methodological innovations or empirical findings of their predecessors except on the occasion of attacking them. This was true even within the body of assessments applied to single issues within particular arenas. Even rarer were assessments built on findings and experience in other issues or arenas. As a result, our history of option assessments was shaped more by habit, methodological fads, and solution advocacy than by the growth of critically validated knowledge in anything that might be called a maturing field of option assessment. Some encouraging exceptions to this finding exist, however—for example, RAINS assessments for acid rain and the Technology Assessment Panels for ozone. As noted above, we speculate on their significance in the final, forward-looking section of this chapter.

None of these conclusions regarding the role of option assessment in the management of global environmental risks is without important exceptions. But neither is any particularly unlikely or unexpected from the perspective of the broader theoretical debates introduced at the beginning of this chapter. Returning to those perspectives and the tensions they encompass helps to suggest some possible explanations for the patterns and exceptions we have observed.

17.5.2 Issue Development

Our findings in this study seem best illuminated by a conceptual view of issue development that includes elements of Kingdon's (1995) independent problem and policy streams; that delineates developmental stages in terms of the agendas on which the issue resides, rather than a functional sequence running from problem identification through implementation; and that stresses the significance of periods of instability when issues are moving up from their early place on the agendas of a few specialists onto more broadly visible public agendas.[28] In particular, our work suggests that the roles played by options, their advocates, and assessments in issue development differ systematically according to an issue's current place on the social agenda.

The most unambiguous and distinctive criterion for demarcating stages of development for our global environmental issues is the period of rapid increase in public and political attention that all the issues experienced in the mid- to late 1980s (see chapter 14 on issue attention). During the extended period preceding this emergence of our issues onto public agendas, the options and advocates that most influenced the framing of the debate were imported from other, more mature issue areas: the SST controversy, the energy crisis, the regulation of local air pollution. Consistent with garbage-can views of issue development, well-organized advocates from these well-developed issue areas seized on the emerging scientific concerns for global environmental risks as additional justifications for advancing their own agendas. Formal option assessment played nearly no part at this stage, partly because no one really wanted options assessed and partly because no one had an independent mandate to assess them.

As described elsewhere in this volume, the rapid emergence of our issues onto the policy agendas in the mid- to late 1980s was precipitated by a number of factors. But neither the introduction of new management options into the policy stream nor the promotion of old ones through option assessments seems to have played much role in destabilizing the old order or drawing public attention to the issue.[29] Once the global environmental problems were established on the public agenda, however, and goals for managing their risks began to be debated publicly, the variety of options under active consideration rose rapidly.[30] These reflected not only established interests that saw in the increased attention a new opportunity for advancing their causes, but also a new array of potential solutions pushed by new advocates, many of whom were turning their attention to the now highly visible global environmental issues for the first time. The continued dominance of both the initial and enhanced option pools by emission-limiting technologies and standards is consistent with expectations that alternatives bestowing concentrated benefits will receive more attention from advocates than those that do not. An increase in formal option assessments followed rather than led these additions to the option pool. These assessments largely reflected the pools they encountered and did not seek to change the distribution of options considered.

The period for innovation in and enrichment of the option pool was of limited duration. Few additional new ideas were introduced to the option pool during the succeeding years when domestic policies and international agreements were being negotiated and implemented. The demand for assessment remained relatively high, however, and was met by a wide array of government, NGO, industry, and academic studies. By and large, the effect of these studies was consistent with expectations from the broader literature portraying issue development in terms of competing policy streams. Assessments promoted wider diffusion and deeper understanding of existing proposals for the management of global environmental risks but rarely had decisive influence in either elevating particular options to a position of dominance or removing particular options from the debate.[31] Most exceptions to this general pattern were narrow feasibility studies conducted in situations where an agreement on policy goals was already well established (such as the findings of the Technology Assessment Panels established under the Montreal Protocol). Only in the rarest of cases—most notably the 1989 German Enquete Commission Report on *Protecting the Earth's Atmosphere*—did these new option

assessments directly feed back into a reassessment of the politically enunciated goals from which they originated.

17.5.3 The Assessment Process

Our finding that most option assessments consisted of narrow comparisons among a few similar alternatives would appear to be wholly in line with the expectations of the bounded-rationality models of analysis in the policy process. And we found little evidence in the historical record that the theoretical virtues of comprehensive comparisons stressed in some rational-actor literatures had been realized in practice.

The explanation for our findings cannot, however, be quite as simple as this broad-brush overview suggests. For example, bounded-rationality perspectives are as consistent with narrow comparisons among impact-reduction options as with narrow comparisons among emission-reduction options. In fact, the assessments we reviewed were heavily biased toward the latter from the beginning of our study period and remained that way throughout. This initial dominance of the option pool by emission-reduction measures is consistent with both the general value preference of many of our actor groups for prevention over remediation and the likelihood (noted earlier) that the options bestowing concentrated benefits (such as on the manufacturers of scrubber technologies or on the regulators of emissions) will attract relatively many and forceful advocates. Once the initial dominance of emission-reduction measures was established, several factors may have reinforced it.

One possible reinforcement is similar to the charmed-circle problem in natural science: the tendency for researchers to identify uncritically a small subset of the possible problems as worthy of further attention and to disregard and indeed denigrate work on alternatives. (An example from our history is the continued focus of acid rain assessments—even those performed by disinterested individuals and institutions—on scrub versus fuel-switch alternatives for reducing sulfur dioxide emissions, long after other options had emerged.)

A second is what Harvey Brooks has called the pitfall of "technological monocultures" (Brooks 1973). This is the tendency of one or a few initially popular options to crowd out others in early stages of the policy debate through founder effects or the benefits of scale, leaving society with an unnecessarily narrow selection of alternatives when new scientific findings or goal and strategy changes alter the terms of the debate later on. We emphasize, however, that these and related anchoring hypotheses would need a more critical analysis than we have been able to give them before they should be entertained as more than probabilities.

72

William C. Clark, Josee van Eijndhoven, and Nancy M. Dickson et al.

Returning to our competing views of assessment, one common set of explanations for the "satisficing" behaviors associated with bounded-rationality perspectives is that the users of assessments do not want to consider a wide range of alternatives. In fact, however, the policy makers and advisors in our study called frequently for comprehensive assessments, most of which were then not performed. Likewise, the "limited successive comparisons" described by Lindblom and others are often said to be the result of the limited resources available to analysts. In fact, of the few broad comparisons that do appear in our histories, most were not particularly resource intensive. They were just ignored. This situation is reminiscent of the dilemma with which Lindblom and Cohen open their 1979 monograph on the problems of *Usable Knowledge: Social Problem Solving*. Observing the strained relationships in U.S. social-policy circles between professional analysts and policy makers, they observed that analysts were dissatisfied "because they are not listened to," while policy makers were unhappy "because they do not hear much they want to listen to." The authors go on to attribute this situation to fundamental misconceptions on the part of both performers and users of assessment as to what analysis can and should be expected to accomplish.

Similar misunderstandings almost certainly pervade the communities involved with option assessments for our global environmental issues. This is not surprising when we recall that until very late in our study period there were few occasions and fewer institutions in which users and performers of assessments could come together regularly and informally enough to begin learning about and adjusting to one another's perspectives. Some such mutual accommodation doubtless occurred in those of our arenas where most assessment was informal, personal, and high level (such as in the former Soviet Union). But the greatest change came with the schedule of continuous policy review and revision that emerged under the convention, protocol, amendment system pioneered by the LRTAP Convention and developed more fully in the Montreal Protocol for the ozone case. This linkage created a regular, sustained demand for assessments that could discover and serve specific limited policy needs. It also helped to create for the first time identifiable and stable communities of option assessors functioning at both the international and national levels. For reasons that we turn to in the next section, the emergence of these assessment communities has almost certainly been central to some of the most valuable contributions that option assessment has made to the management of global environmental risks.

17.5.4 Boundary Work

The literature on boundary work that is reviewed at the beginning of this chapter highlights the difficulties of achieving credibility and legitimacy for scholarly analysis deployed in policy contexts. In particular, the literature emphasizes that partisan interest in the conclusions of such analysis creates pressures to deconstruct scientific consensus and that the boundary between what is treated as *disinterested science* and what as *value-laden policy* in any particular case needs to be negotiated. These perspectives help to make sense of a striking feature of our issue histories: the evolution of assessment practices in directions that increased the degree of scholarly consensus over their results but at the apparent cost of decreasing the policy relevance of those results.

This tendency toward *design for consensus* shows up in our data in three ways. First, as already noted, risk assessment and option assessment were systematically separated in the vast majority of the cases and arenas we studied. In many cases, the risk assessments simply became stand-alone documents, with no institutional connection to any options studies whatsoever. Even ostensibly end-to-end studies such as the IPCC's 1991 report placed its science and response assessments in separate volumes and did everything possible to ensure that criticism of the latter did not undermine the consensus on the former. Some of this separation doubtless reflected decisions by political leaders not to give a public platform to any assessment that would address policy issues. But our interviews suggest that the scientists responsible for the risk assessments were often active proponents of separation as well. They simply saw option assessment as a relatively unscientific and readily politicized activity that posed an unwelcome threat to their search for consensual findings (Tolba 1994).

Also consistent with the notion that assessment practices are shaped by a search for consensus are our findings on assessment locus. Recall that despite the often expressed desire of assessment users for appraisals of policy options in terms of changed impacts on society, the proportion of assessments performing such appraisals declined and the proportion restricting their attention to changes in emissions increased throughout most of our study period. Again, other factors may have contributed to this trend. But there is little question that appraisal of options in terms of ultimate impacts on society is likely to be both more politically sensitive and more scientifically uncertain than appraisal in terms of changes in emissions. The shifting of assessment locus leftward on the causal chain illustrated in the introduction to this section thus simultaneously reduces both the motivation and the potential for political deconstruction of assessment results.

An analogous explanation is consistent with our admittedly less stark findings on the preferred mode of scoring option assessments. Given the choice of characterizing options in terms of their feasibility, cost-effectiveness, or benefits, practicing assessors almost never opted for the latter. Instead, they preferred cost-effectiveness or feasibility assessments despite the persistent demands by both potential political users and potential scholarly contributors for more benefit studies (e.g., U.S. NAPAP 1991). In so doing, they rejected the more controversial mode of analysis in favor of approaches less vulnerable to either technical or political attack.

Pending further research, we remain agnostic on whether most option assessors had explicit objectives of conflict reduction or consensus seeking in mind when they designed their studies. But whatever their conscious motivations, the premium placed on consensus by assessors and policy makers alike, together with the dissent and criticism associated with many of the assessment designs that society increasingly came to avoid, make the evolutionary account of assessment practice we have sketched above worth considering.

The problem, of course, is that many of the features that made assessments more likely to secure a scholarly consensus also tended to make them less relevant to the needs of policy makers and other assessment users. This may be part of the reason that we see so little satisfaction with option assessment on the part of the policy-making community and so little impact of option assessment on the policy debate through much of the history we have studied. We return to this tradeoff and its implications for efforts to improve option assessments in the concluding section of this chapter.

Exceptions to this pattern exist, however, and are potentially instructive. The literature on the boundary work that is needed to produce usable knowledge emphasizes the importance of institutional settings and practices that allow assessment users and producers to feel each other out—to negotiate and adjust over time a balance between the scientifically defensible and the policy relevant. For most of our arenas and most of our history, institutions of this sort were nonexistent or of extraordinarily limited scope or duration. Such boundary negotiations as did occur took place at a distance, with producers and users sporadically reacting to each other. Alternatively, as in the case of the German Enquete Commission assessment, intensive boundary negotiations were carried out on an ad hoc basis over a limited period but then ceased. Producers and users of assessments were left again at a distance from one another, interacting principally through sporadic pronouncements of mutual dissatisfaction or miscomprehension.

In the later years of our study period settings for more direct, sustained, and productive negotiations of the boundaries between science and policy began to emerge. This occurred largely through institutionalization of the continuing (re)assessment processes introduced to support international deliberations on our global environmental risks. At least for the more mature issues we studied—acid rain and ozone—the producers and consumers of option assessments who were brought together in these settings overtly engaged in a continuous process of negotiating pragmatic boundaries between the technical and political dimensions of their concerns. The resulting agreements on "serviceable truth" had by the early 1990s helped to promote some of what were arguably the most effective option assessments encountered in this study: IIASA's RAINS support for development of the LRTAP Sulfur Protocols and the UNEP Technology Assessment Panels' inputs to successive amendments of the Montreal Protocol. Whether the technical complexities and political stakes of the climate case would leave room for comparably effective boundary work was still open to debate in the wake of the Rio Conference.

17.6 Toward Better Option Assessment

To what extent, and in what ways, can the practice of option assessment in the management of global environmental risks be said to have improved over the period addressed in this study? What are its prospects for getting better in the future?

As noted in many other chapters of this book, any effort to address such normative issues requires an explicit set or system of critical standards. Ideally, we would ground our evaluation in standards widely accepted by users and producers of option assessments. We argued in section 17.1 of this chapter that no widely shared standards existed for the conduct of option assessments in general. We left open the possibility that there would have emerged in the field of global environmental management at least an ad hoc consensus on what constituted good or at least better practice. Our empirical finding, not altogether unexpected, is that such a consensus does not in fact exist.

We therefore adopt here (as elsewhere in this study) the broad criteria for the appraisal of "scientific inquiries with policy implications" developed by Clark and Majone (1985) from their review of other assessment studies. These criteria are discussed in the introduction to part III.[32] The inevitable limitations of any such evaluative criteria notwithstanding, we suspect that much could be gained were the community dealing with option assessment in the management of global environmental

74

William C. Clark, Josee van Eijndhoven, and Nancy M. Dickson et al.

risks to spend somewhat more time reflecting on the themes they address. Our remarks here are intended to promote the process by suggesting some of the evaluative considerations arising from our study that might be addressed in an extended, communitywide discussion on improving option assessment.

17.6.1 Effectiveness

Our evidence clearly shows that if *effectiveness* were to be defined in terms of the tangible impact of individual option assessments on particular decisions, then most option assessments of global environmental risks undertaken during the period studied here would have to be judged as ineffective in the extreme. (So would virtually all other intellectual activity undertaken to shape social action.) In keeping with the arguments regarding effectiveness advanced earlier, we believe it is more helpful to examine the impact of assessments on the development of the policy or option *agenda* rather than on individual decisions themselves. In particular, we borrow from Lakatos (1970), Boothroyd (1978), and Majone (1980) to propose that option assessment be judged progressive or effective to the extent that it "succeeds in disposing of issues, i.e., in moving them from the stage of contention to a class of issues which the actors in the policy process judge to be in a state of satisfactory, if temporary, resolution" (Majone 1980, 15). In practice, this suggests that effective assessment practices would be able to accommodate any new option proposal and to provide authoritative illumination of its merits relative to existing options. It would mean that the option pool would be systematically drained of particular options via assessments that showed them to be inferior to their competitors. Finally, effective assessments would offer other options at least some systematic, merit-based protection from random or ideological elimination from the pool. While an extended defense of this view is beyond the scope of this chapter, we note that it is consistent with much of modern thinking on the nature of both effective science and effective politics.[33] We argue here only that it provides perspectives that we have found helpful in critically appraising our history of option assessment for the management of global environmental risks.

Even by this more forgiving criterion, however, option assessment for the management of global environmental risks cannot be said to have been a particularly effective enterprise over the period of this study. As we have shown, option assessment almost never managed to give "daylight and fair play" to certain kinds of options—most notably those targeted on environmental restoration. It was therefore unable to contribute much to an informed debate on where and under what conditions such options

might be attractive. Likewise, option assessment generally failed to stand in the way of the ideology-based removal of such options from the pool of actively debated alternatives. Exceptions to this pattern—for example, the 1992 study by the U.S. National Research Council on *Policy Implications of Greenhouse Warming*—showed that improving the "effectiveness" of assessments at providing "daylight and fair play" to such options was not impossible, just difficult (U.S. NRC 1992).

It would also seem that neither single assessments nor the assessment process as a whole were particularly successful in removing options from the debate. Instead, options generally accumulated. Many assessments offered strong arguments against particular options—for example, economic analyses showing the inferiority of mandated technology options for acid rain reduction. Few were effective in harnessing rational argument to remove those options from active consideration by the political process. One potentially significant exception to this pattern comes in the latter period of the acid rain case where, as we have shown, the richness of the option pool systematically decreased. If this decrease were due in substantial part to authoritative findings of option assessments that certain options were inferior to others, it would represent a plausible case for increased effectiveness. This may have occurred. But our data do not allow us at this stage to reject the alternative hypotheses advanced earlier that political and bureaucratic factors irrelevant to the technical merits of the case pushed in the same direction.

As we understand the history reviewed here, it does not allow us to muster more than one tentative cheer for the effectiveness of option assessment in the management of global environmental risks. We turn next to consideration of other criteria of appraisal that might help us to understand why this is so.

17.6.2 Adequacy

Widespread agreement on the adequacy or appropriateness of particular methodologies and data sets was achieved only over highly restricted domains of option assessment. Typical examples include assessments of the technical feasibility of alternatives to particular ozone-depleting chemicals, the cost-effectiveness of similar sulfur scrubbers, or the implications of fuel switching for carbon emissions. In such cases, analysts came to share basic assumptions about what constituted quality work. Reliable knowledge accumulated. In contrast, for the integrated assessments, multitarget or multimeans comparisons, and benefit analyses that were in demand from critics of the narrow and limited assessment approaches, there was little agreement on appropriate methods, processes, or data. Predictably, little resembling the accumulation of reliable facts had emerged from such ambitious assessments by the end of our study period. Rather, each study tended to start from scratch with attention to its predecessors consisting, at most, of passing swipes at their presumed methodological failings.

On the positive side, the option-assessment community has not been unaware of the enormous challenges presented by the relative immaturity of its methodological development. As we have shown, the vast majority of actual assessment activity conducted on global environmental risks consisted precisely of the sorts of focused studies that—whatever their limitations—*could* create reliable knowledge. They did this by relying heavily on established disciplinary approaches (such as cost-effectiveness analysis) for which criteria of methodological adequacy were already well developed. Moreover, we have evidence from a few assessments (such as IIASA's RAINS effort and the work of the Energy Modeling Forum) that a serious long-term commitment to the incremental development *and structured criticism* of more ambitious and interdisciplinary approaches can indeed contribute to cumulative and reliable knowledge about options across broader and broader domains. Finally, the increased attention to systematic development and comparison of integrated assessment methods evident toward the end of our study period can be viewed as a step in a propitious—if historically perilous—direction.

17.6.3 Value

The role of criteria of value in option assessment is to help ensure some correspondence between the things that assessors can productively study and the things that users (really) want to know from them. This is a daunting task, as evidenced by the frequent complaint in fields ranging from economic to energy policy that most analysis is never used by anyone (Blinder 1988). Valuable option assessments for the management of global environmental risks would appear to have been as scarce as those anywhere else. As we have shown, policy and decision makers in the global environmental realm have been underwhelmed by the value of most assessments they have received. And most assessments have been undertaken with no clear understanding of whose choices or decisions they were seeking to illuminate. With no generally accepted criteria of value to draw on, it is hardly surprising that assessments have focused on a charmed circle of emission reduction options defined more by their analytical tractability and advocates' energy than by any articulate notion of social need. Reciprocally, with no general appreciation among potential users of the "art of the soluble" in option assessment, it is hardly surprising that

their sporadic demands for more useful or policy-relevant analysis have often pushed the assessment community beyond its depth. Overall, there has developed little in the way of a critical tradition reflecting on the choice of which problems experienced by which policy makers assessment ought to address. And there has been little progress in the assessment community—despite a decade of exhortations—in making more explicit and open to debate the value assumptions latent in the methods they adopt.

There are some bright lights in this generally bleak picture. Many of these are relatively pedestrian, such as the growing recognition—shared with other substantive areas of policy analysis—that assessment models encouraging "What if?" questions from potential users are a particularly valuable tool for linking science and policy concerns.[34] More profound understanding is emerging as well, especially in the growing recognition that sustained "boundary negotiations" between the producers and users of assessment can make an enormous difference in generating value in the assessment process. We have evidence that such value-adding boundary work can be relatively ad hoc, as in the German Enquete Commission, or permanently institutionalized, as in the support provided by IIASA's RAINS assessment for the LRTAP Convention negotiators. The crucial factor seems to be that option assessments produced through an intimate collaboration of analysts and specific users are more likely to prove worthwhile than those conducted for generic support of policy development.

17.6.4 Legitimacy

As noted in the introduction to this chapter, option assessment resides in a precarious position on the cusp of knowledge and action. The perceived legitimacy of its products could plausibly depend on how it handles both the scientific challenge of providing "fair play to new ideas" and the political challenge of protecting minority views from the potential tyranny of the majority. The political dimension of option-assessment legitimacy has occasionally been thrust to the fore, as in the Indian criticism that the IPCC had a rich-world bias or the damage so frequently wrought to the credibility of U.S. assessment efforts by editorially minded political appointees.[35] In general, however, most formal option assessments of global environmental risks have been treated as primarily scientific or technical exercises rather than as mixed scientific and political ones.[36] Within this prevailing narrow scientific framing of option assessment, legitimacy or credibility became almost entirely equated with the ability to articulate a consensus finding of the assessment community.

76

William C. Clark, Josee van Eijndhoven, and Nancy M. Dickson et al.

This has had two consequences, neither obviously salutary. First, it has made the necessarily sticky problem of including the perspectives of (political) users in the design of option assessments even more difficult. This, in turn, has exacerbated the problem, noted above, of ensuring that the option assessments are valuable to someone in particular. Second, the emphasis on consensus may well have pushed assessments in the same direction as the considerations of methodological adequacy noted above. Limited studies of similar alternatives, compared in conventional disciplinary terms, are those most likely to survive peer review and support a consensus opinion. Strong pressures developed to frame option assessments in ways that would let them pass such tests. "Monster barring"—a term applied in the history of science to repression of assaults on the conventional wisdom—was commonplace, and technically difficult or ideologically troublesome considerations such as adaptation or geo-engineering or equity calculations rarely made it into the mainline option debate (Lakatos 1970). When for various reasons they did, the assessments that carried them frequently paid a steep price in terms of perceived legitimacy.

The more encouraging moments in this story are again concentrated where Jasanoff has suggested we would find them: in studies with intense boundary-work negotiations on what to include in assessments and what would constitute "serviceable truth" in the conclusions emanating from them (Jasanoff 1990). The starkest example is perhaps the Montreal Protocol's Technology Assessment Panel, which managed to include stakeholders in the assessment studies without falling victim to accusations of interest-group "capture" or political censorship.

17.6.5 Prospects for Improvement

Option assessment for the management of global environmental risks emerges from the story we have told here as what Ravetz has called an "immature" field of inquiry. *Immaturity* is a common phenomenon encountered in all manner of knowledge-based endeavors. Its use as a descriptive classification is no more pejorative than would be its use in other contexts to characterize a young, juvenile, or immature organism. Well before any of the major option assessments studied here were conducted, Ravetz (1971, 368) observed:

Watching the activity of [an immature] field over a period of years, one does not witness the steady cumulation of new facts, perhaps superseding but never completely destroying the old. Instead, there is a succession of leading schools, each with a manifesto that is more impressive than its achievements, and each passing into obscurity when its turn on the stage is over.

What are the prospects for improvement of an immature field such as option assessment? Drawing on experience from throughout the history of basic and applied science, Ravetz (1971, 368) cautioned against the

hope that the condition [of immaturity] is easily remedied. Most such claims, and the research strategies organized around them, involve a concentration on one aspect of the work which is seen to be deficient, and whose improvement should produce full maturity. Heroic attempts have been made to amass empirical data; to apply mathematical and computational tools for the production of information . . . [and] to develop methods and a methodology appropriate for the discipline. In each case, the attempt is to reproduce what is believed to be the crucial feature of an established science . . . . But almost all these one-sided efforts fail utterly . . . . For the condition of [immaturity] is not an accidental deficiency in some component of the materials of a field, but is a systematic weakness in those materials and in the social activity whereby they are produced.

Option assessment for the management of global environmental risks, in the first quarter century of its existence that we have studied here, fits the older and broader pattern of disciplinary immaturity to an unnerving degree. And Ravetz's warning about the futility of "heroic" fixes casts the IPCC experience that closes our study period in a not wholly encouraging light. Nonetheless, as Ravetz's own phrasing suggests, immature fields can grow up. And there is some evidence that in the later years of our study period option assessment for global environmental risks was beginning to come of age. We close this chapter by highlighting four signs of adolescence, if not yet maturity, in option assessment that would almost certainly reward special encouragement in coming years.

International Institutionalization Most of the recent progress in growth toward maturity for option assessment of global environmental risks has been due to the advent of regularized international assessments performed in support of international policy deliberations. This international institutionalization is arguably the most significant change to have occurred in the option-assessment landscape over the period of our study. As we have seen, it has taken a variety of forms ranging from the conspicuous involvement of nongovernmental groups providing ongoing professional assessment support to the LRTAP negotiations, through the overtly intergovernmental structure of the IPCC. More significant than formal structure has been the unprecedented opportunities that international institutionalization has established for sustained interaction among the users and producers of assessments. These interactions are in no sense sufficient to produce better option assessments. But they have contributed significantly to the emergence of self-aware and reflective communities of individuals around the world that are committed, inter alia, to making better option assessments for the management of global environmental risks. Such community building in and of itself is surely a positive step along the long road to maturity described by Ravetz and therefore deserves encouragement. But it has also contributed to more tangible and immediate benefits in several areas to which we now turn.

Critical Standards We noted earlier that no widely shared set of critical standards for "good" assessments of options to manage global environmental risks developed during our study period. Except in the narrowest areas of specialized concern, successive option assessments effectively started from scratch. They seldom built on, or even acknowledged, the methodological innovations or empirical findings of their predecessors except on the occasion of attacking them. The lack of common critical standards contributed significantly to the manifest lack of cumulative progress in, and "immaturity" of, the option debate. The exceptions to this general pattern have occurred primarily within the recently internationalized assessment communities noted above. Though still somewhat inarticulate about the standards they employ, the professional assessment communities associated with later revisions of protocols for stratospheric ozone depletion and (European) acid rain have been able to build cumulative, reliable knowledge and to mount effective quality-control procedures. The community of climate assessors has further to go, but even there the intense and sustained methodological discussions that have grown up around the challenges of "integrated assessment" offer some hope of increasing maturity and effectiveness. The tasks ahead are many but surely include efforts to make more articulate and formal the discussions of critical standards that have begun to emerge in these international communities of assessment professionals. More daunting, but surely worthwhile, will be organized efforts to share experiences, insights, and lessons across the existing issue-oriented communities, thereby opening the door to the development of useful critical standards for global environmental assessment.

User Partnerships Our study suggests that option assessments can benefit substantially from close and continuing negotiations between assessment producers and users over appropriate objectives and methods. Such negotiations have been rare or restricted to the most stylized and formal exchanges throughout most of our study period. In their absence, we have shown that the uncritical assumption that assessments should seek "consensus" may result in purchasing legitimacy for an option assessment at the cost of reducing its value to decision making. But the historical experience also suggests that this undesirable tradeoff can be mitigated through institutional arrangements that bring together users and producers in a joint negotiation of assessment objectives and methods. As suggested by the examples of the German Enquete Commission and the UNECE's relations with IIASA in support of the LRTAP process, such negotiation focuses on defining the boundary between the scientific and political dimensions of assessment. The outcome, when negotiations are successful, is an agreement on what kinds of findings, produced through what analytical and quality assurance procedures, will produce "serviceable truth" for the case at hand. Such boundary negotiations, and the partnerships between assessment producers and users that they create, should almost certainly be encouraged in a wider range of assessment circumstances.

Historical Perspective Finally, our study emphasizes the need for option assessment to see itself in historical perspective. Option assessment plays very different social roles at different stages in the development of global environmental issues. For assessments to be helpful or effective—indeed for them to avoid stoking unproductive controversy—they must be undertaken with more regard to what they can reasonably contribute *given the current state of issue development* than has generally been the case. Early on, before the issues have attracted significant public or political attention, formal assessment seems to serve primarily an advocacy function for options pushed by existing interests. Even the most professionally disinterested efforts to conduct detailed comparisons among such options are almost guaranteed to be rendered ineffective by unavoidable weaknesses in the needed data. Moreover, efforts to conduct fair comparisons of alternatives at this stage find few enthusiastic users. The best direct contribution of option assessment to the management of global environmental risks at this stage of the debate may well be to help illuminate broad, qualitative questions of whether the risk in question is likely to be relatively easy, or relatively hard, to control.

Option assessments seem to have little to do with opening the "windows of opportunity" through which global environmental risks finally climb onto the policy agenda. The demand for their production nonetheless rises rapidly with issue emergence. Formal option assessments created in response to this demand will generally be completed too late to contribute much to the intense political debate associated with initial rise in interest; informal, rapid assessments are more likely to be more important and influential at this point.

78

William C. Clark, Josee van Eijndhoven, and Nancy M. Dickson et al.

If a consensus on goals for the management of a global environmental risk eventually develops in the political arena, the formal assessments initiated during and subsequent to the initial rise in issue attention may play a substantial role in subsequent management of the issue. *Until* goal consensus is reached, however, option assessment is likely to remain peripheral to the policy debate or to be used only in narrow advocacy contexts. In such circumstances, option assessments tend to be at best irrelevant to the management process. More often, they become surrogate battlegrounds for unresolved value conflicts, undermining the credibility of the option-assessment process and any assessment of underlying risk to which the option study is closely linked. We conclude that extensive use of formal option assessments is likely to be most useful and effective when focused on issues (or parts of issues) where political goals are already agreed on.

Transcending the importance of attuning the ambitions of specific option assessments to their historical circumstances, however, is the need for viewing the activity of option assessment on a historical time scale. Building the institutions, professional communities, critical standards, and political trust that we have emphasized here as central to the improvement of option assessment all have taken time—time on the order of at least a decade for the cases we have studied. More useful option assessment in the future will require professionals who are willing to commit to it as a career and societies willing to support it on comparable time scales. Better option assessment for global environmental risks will require, above all, patience.

Appendix 17A.   Acronyms

CFC	chlorofluorocarbon
CIAP	Climatic Impact Assessment Program
CO₂	carbon dioxide
DoT	Department of Transportation (U.S.)
EC	European Community
EPA	Environmental Protection Agency (U.S.)
GEF	Global Environment Facility
IIASA	International Institute for Applied Systems Analysis
IMAGE	Integrated Model for Assessment of the Greenhouse Effect
ICSU	International Council for Scientific Unions
IPCC	Intergovernmental Panel on Climate Change
LRTAP	(Convention on) Long-Range Transboundary Air Pollution
NAPAP	National Acid Precipitation Assessment Program (U.S.)
NASA	National Aeronautic and Space Administration (U.S.)
NGO	nongovernmental organization
NRC	National Research Council (U.S.)
RAINS	Regional Acidification Information and Simulation (model)
SCOPE	Scientific Committee on Problems of the Environment
SST	supersonic transport
UNECE	United Nations Economic Commission for Europe
UNEP	United Nations Environment Programme
WMO	World Meteorological Organization

Notes

1. Many people contributed to the research on which this chapter is based. We wish to acknowledge the contributions of the following participants in the Summer Workshops of the project: Jeroen van der Sluis, Rodney Dobell, Renate Ell, Elena Nikitina, Claudia Blumhuber, Vladimir Pisarev, Karen Svensson, and Jill Jäger. For reviews of working papers on which this chapter is based, we thank David Fairman, James Hammitt, Clark Miller, Edward Parson, and David Victor.

2. The following terms are sometimes used by others as synonyms for what we term *options: actions, interventions, measures, practices, policies, responses.* We have chosen *option* in preference to other possible terms because it seems the most general and least judgmental of those available. In particular, we preferred it over *response* because of the latter's implication that the debate on possible measures or actions necessarily follows *as a response* to the identification of a risk. As we show later in this chapter, there are both theoretical and empirical reasons to reject this presumption. We preferred *option* over *policy* because of the latter term's restriction in many circles to measures considered by governments. Our treatment, in contrast, is interested in options considered by other actors as well.

3. For the purposes of this study, we treat at least some of the activities carried out under the following labels as encompassing the process of option assessment: *environmental impact assessment, policy analysis, response assessment,* and *technology assessment.* As noted above, the label *risk assessment* is often used to include the two processes that we have in this project distinguished as *risk assessment* and *option assessment.*

4. Technology and command means between them represented perhaps 60 percent of the options debated through the period.

5. The former Soviet Union is the one arena in which the research option arguably remained dominant across all cases in the 1980s.

6. The most notable exception is the United States, where taxes on CFCs were introduced early and, subsequent to 1988, the use of market incentives took on the characteristics of a crusade in government and NGO circles.

7. In 1989 environmental ministers from sixty-eight countries meeting in the Netherlands formulated a plan of action to mitigate global warming. It included a goal of a net increase of world forest cover of 30 million acres per year by 2000 (Climate Conference Secretariat 1989). In 1990 the Bush administration took a supportive position on tree planting through the America the Beautiful program with a goal to plant 1 billion trees per year during the 1990s using state-sponsored and private-sector involvement at the local level.

8. We use the term here to include both the European Union and the organizations discussed in Chapter 13 on international institutions.

9. Of course, the three countries named here were also often branded as among the worst laggards in implementing policy measures. This correlation lends some credence to the notion that demands for more knowledge can constitute a strategy for delaying action. But the arena studies in this book also present compelling evidence for these and other arenas that knowledge was frequently pursued in good faith and was intended as a complement to, rather than substitute for, action. A more sophisticated explanation for the exercise of scientific leadership by these countries is needed.

10. In some cases, such as the Canadian discussion of options for dealing with the ozone problem, foreign (U.S.) goals provided the motivation for a domestic debate on national policy alternatives.

11. The original Technology Review Panel was set up in 1989 (UNEP/OzL.Pro.WGII(1)/4*, December 6, 1989). By 1991, an expanded agenda was developed by the Technology and Economic Assessment Panel.

12. North American acid rain assessments and climate assessments prior to the creation of the IPCC are important exceptions.

13. The ozone problems' technology-assessment panels enlisted substantial industrial input, while the LRTAP process made unique use of IIASA's nongovernmental RAINs model as a decision-support tool.

14. For the United States' engagement with the climate issue, what mattered most was the development under the auspices of the World Climate Program of a series of international scientific assessments of the climate issue held in Villach, Austria, during the period from 1980 to 1987. The joint sponsorship of these meetings by WMO, UNEP, and the International Council of Scientific Unions (ICSU) gave them a peculiar quasi-governmental, quasi-nongovernmental character. The 1980, 1983, and 1985 meetings were sponsored by UNEP, WMO, and ICSU (WMO et al. 1981; WMO 1984; UNEP, WMO, and ICSU 1985). The project to organize the 1987 workshops (Jäger 1988) was initiated by the Beijer Institute (Stockholm), the Environmental Defense Fund (New York), and the Woods Hole Research Center (Massachusetts).

15. Many arenas eventually became active promoters of international assessments—for example, the Netherlands in the case of the RAINS model.

16. Note that the *kind* of analysis or comparison undertaken in the assessment is not at issue here—only the variety of options somehow analyzed or compared. The metric of comparison is dealt with later in this section, under the *mode* criterion.

17. This second round, when published as the *Second IPCC Assessment* in 1996, did little to resolve the question of where option assessment was, or ought to be, headed. In particular, the reports of Working Groups II and III that addressed options for dealing with climate change, while reporting much of substance, were reduced in many cases to addressing methodological debates and philosophical controversies. The similari-

ties to the dilemmas surrounding the CIAP integrated assessment of more than twenty years earlier were striking, leading many to question whether the field of option assessment was, even in the mid-1990s, up to the task of undertaking broad, comprehensive option assessments of global environmental risks.

18. The German efforts were published as German Bundestag (1989, 1991). The Japanese studies were being carried out under the umbrella of the government's ambitious New Earth 21 Program (Japanese Energy and Industrial Technology Council et al. 1992). The Dutch work included use of its Integrated Model for Assessment of the Greenhouse Effect (IMAGE) to support the international climate negotiations (Rotmans and den Elzen 1993). The United States reentered the field of broad option assessments with the National Research Council's 1992 assessment of *Policy Implications of Greenhouse Warming* (U.S. NRC 1992). It also launched a major program of "integrated assessment" under its Global Change Research Plan, with efforts underway at MIT and Carnegie Mellon University.

19. See, for example, Glantz, Robinson, and Krenz (1982) for a summary of the critiques leveled at the CIAP report and its contemporaries.

20. This is a more general phenomenon, as indicated by the account of assessment experience in the United Kingdom given by Collingridge and Reeve (1986).

21. In 1985, SCOPE was finalizing the draft climate assessment that was to be the focus of discussion at the Villach International Assessment Conference in September (Bolin et al. 1986). The best-known recent national assessments on which SCOPE and Villach drew had also truncated their options assessments to an emissions locus (e.g., U.S. NRC 1983). Under WMO auspices and NASA leadership, 1985 saw the preparation of what would be published as WMO et al. (1986).

22. The first approach, characterized by UNEP's Executive Director M.K. Tolba as the most important lesson to be learned from the history of global environmental assessment, was to separate organizationally the ostensibly objective, consensual assessments of the basic science from the presumptively manipulable, contentious assessments of impacts and, especially, policy (Tolba 1994). A corresponding retreat by the scientific community from the early experiments in end-to-end, integrated assessments can be clearly seen in the evolution of the United States' and then the international community's work on the ozone question. These developed a successively *less* extensive locus—and also became less controversial—through the late 1970s and early 1980s. The portion of the 1975 CIAP study that elicited the greatest criticism was its economic-benefit assessments—in our terminology, an option assessment with a "social impact" locus for its appraisal of policy consequences. When the National Research Council (NRC) conducted its first assessment of "The Environmental Impact of Stratospheric Flight" in 1975, it simply dropped the social-impact component (U.S. NRC, 1975). As the NRC returned to the ozone-depletion issue in subsequent years, it further disintegrated its assessment approach by placing its relatively uncontroversial discussion of causes in separate volumes from its more controversial analysis of impacts, options, and their downstream effects (e.g., U.S. NRC 1976, 1979, 1984). By the time assessments were being prepared to support the emerging Montreal Protocol process in 1986, the separation was complete: NASA led an international effort involving the elite scientific community to produce a consensual and influential account of the basic causation of depletion. The U.S. Environmental Protection Agency (EPA) ran a separate effort, drawing heavily on private-sector consultants, to assess downstream effects. The assessment of policy options was handled in yet another set of reports that focused almost exclusively on an emissions locus. The

ozone debate during the preparations for Montreal was backed by at least three separate international assessments. One was a review of the basic science—WMO's *Atmospheric Ozone: 1985* Blue Books—that included no analysis of impacts or control options at all (WMO et al. 1986). A second, emerging from a UNEP and U.S. EPA initiative, addressed health and environmental impacts while ignoring the science of depletion or the impacts of policy (Titus 1986). A third, emerging from a quite different UNEP meeting at Würzburg in 1987, took the science and impacts as given and assessed the consequences for future ozone depletion of various options for the control of ozone depleting gases (UNEP 1987).

When the post-Montreal assessment process was formalized in 1988 and 1989, the disintegration was complete: separate panels were established for scientific assessment, environmental assessment, economics (impact) assessment, and technology (option) assessment.

The last high-profile international ozone assessment of our study period—*Scientific Assessment of Ozone Depletion: 1991* (WMO et al. 1992)—included an evaluation of the impact of alternative phaseout scenarios on chlorine loading of the stratosphere (that is, it adopted an environment locus to characterize alternative scenarios) but did not report on downstream impacts, policies that might bring about the scenarios, or policies to restore the environment or reduce the impacts of ozone depletion on society (WMO et al. 1992). The generally accepted success of this strategy led to the adoption of a similar approach for the IPCC in 1988, which established essentially independent working groups and assessment reports for science, impacts, and policy dimensions of the climate problem.

23. The acid rain case, which moved earliest into the international policy realm, was also the first to see a concerted effort to reintroduce an environment locus into its option assessments. This effort focused on the concept of critical loads. The idea of a scientific threshold concentration of pollutants, above which adverse environmental effects would be manifest, had been established in principle at the time of the Stockholm Conference. By 1980 Canada and the United States had cast their Memorandum of Intent on acid rain in terms that put "appropriate deposition rates"—an environment locus—at the center of the policy discussion. The same idea, now cast explicitly in terms of critical loads, surfaced in Europe at the 1985 Helsinki Conference, was pushed thereafter by the Nordic Council of Ministers, came under consideration of the Executive Board of LRTAP by 1987, and had been formally proposed as the basis for the second Sulfur Protocol of LRTAP in 1992. In tandem with this growing demand, a number of domestic assessment groups and the international RAINS assessment at IIASA incorporated the critical-loads concept as an environment locus for their policy and option studies (Amann 1992).

24. An example is the effort in the Netherlands to define ceilings for environmental loadings compatible with "sustainable development" (Weterings and Opschoor 1992).

25. Policy objections in the acid rain case were posed by at least Germany, Austria, and Switzerland, which argued that a critical-load approach, which in a sense authorizes some degree of pollution, clashes with the preventive principle. In the United States, objections have at least in part reflected historical problems of implementation with the somewhat analogous air-quality ambient standards. Scientific objections highlight the simplistic character of the concept, citing the heterogeneity, nonlinearity, and complexity of ecosystem response. These latter objections are particularly serious, given the hope that critical loads would provide a scientifically sound foundation for extending option assessments to an environment locus. See Levy (1993), Amann (1992), Dietrich (1995), Staff (1992), and Rotmans (1989).

26. In the United States there was by the early 1980s a growing conviction among many associated with the environmental assessment enterprise that it was not helped much by efforts to estimate the benefits of policy options. For example, in the late 1970s and early 1980s the U.S. government had sponsored several unproductive efforts to estimate benefits of air-pollution reduction. The last of these concluded that "rather than the prolific breeding of benefits numbers," assessments should evaluate options in terms of the contribution they make to the variety of choices available (Crocker 1983).

27. Blinder (1988) has made a similar observation about option assessments for the design of macroeconomic policy.

28. Kingdon (1995) refers to the transition from a *governmental* agenda that includes those items under discussion by people in and around government to a *decision* agenda of hot issues ripe for policy development. Since our scope of actors was somewhat broader than his, and his research methods for ascertaining what was on the agendas at what time could not be directly applied over the long periods and multiple arenas we studied, this study does not provide a rigorous test or evaluation of Kingdon's concepts. We have used a similar but somewhat different terminology to avoid implying more direct comparisons than we are prepared to make in this chapter (see chapter 22 on conclusions and implications for the future).

29. The "discovery" by CFC manufacturers in 1986 that they might be able to produce safe substitutes for a price has been interpreted by some as a crucial factor in moving the ozone issue onto the policy agenda. Evidence developed in the U.S. and international institutions chapters of this book (chapters 11 and 13) suggests that this view is an oversimplification and does not constitute a significant exception to the generalization we advance here.

30. In our case, a central feature accompanying the opening of the policy window was the emergence of politically authoritative goals for management. These political goals for managing the risks became the demand pull that had previously left option assessments without an audience. Once a goal of, for example, "reducing emissions by 20 percent" had focused the debate, there was a great incentive for politicians to ask, "How?" and for advocates to answer the question. Option assessments—some exploratory, some persuasive—suddenly had a raison d'être. In addition, the emergence of authoritative goals for managing our global environmental risks created a supply push for new options and assessments. Higher issue visibility and a heightened prospect for political action brought more and more groups previously indifferent to the issue to see the choice among management alternatives as affecting their interests. Preferred options from these newly awakened interests were pushed into the policy arena, often backed with advocacy assessments of their own and further increasing the demand for additional, comparative assessments.

31. Kingdon (1995), for example, notes a similar role for technical debate on policy options in a number of U.S. domestic issue areas and links this function back to garbage-can views of the policy process.

32. These criteria in turn are based on Jerome Ravetz's (1971) seminal work on *Scientific Knowledge and Its Social Problems* plus Clark and Majone's (1985) study "The Critical Appraisal of Scientific Inquiries with Policy Implications."

33. For the scientific case, see Lakatos (1970). For the politics case, note Seymore Martin Lipset's claim that "moderation is facilitated by the system's capacity to resolve key dividing issues before new ones arise. If the issues . . . are allowed to accumulate, they reinforce each other," leading to unproductive conflict (Lipset 1963, 79).

34. Examples include the Edmonds-Reilly family of models for assessing the relationship between energy growth and carbon dioxide emissions, the RAINS modeling framework for supporting LRTAP, and the Würzburg modeling workshop on the chlorine-loading implications of alternative CFC regulations (UNEP 1987). See also Ravetz (1997).

35. The U.S. case study reported elsewhere in this book (in chapter 11) records examples from each of the issue areas covered in this story.

36. The German Enquete Commission study of 1991 is an important exception. Moreover, as noted in earlier sections of the text, a significant but poorly documented number of informal assessments seem to have operated under a more balanced view of the scientific and political dimensions of legitimacy. The dominance of the scientific perspective on option assessment reflects the relatively large volume of assessments generated or shaped by the United States and thus influenced by that country's exceptional appetite for science-based policy development.

References

Agren, Christer. 1992. Interview with William Dietrich. Gothenburg, Sweden, June 3.

Alcamo, Joseph, Roderick Shaw, and Leen Hordijk, eds. 1990. *The RAINS Model of Acidification: Science and Strategies in Europe.* Dordrecht: Kluwer.

Amann, Markus. 1992. Trading of emission reduction commitments for sulfur dioxide in Europe. Status Report 92-03 Laxenburg, Austria: IIASA.

Blinder, Alan. 1988. Economic policy and economic science: The case of macroeconomics. In K.T. Newton, T. Schweitzer, and J.-P. Voyer, eds., *Perspective 2000: Proceedings of a Conference Sponsored by the Economic Council of Canada, December 1988.* Ottawa: Canadian Government Publishing Centre.

Bolin, Bert, et al., eds. 1986. *The Greenhouse Effect, Climatic Change, and Ecosystems.* SCOPE 29. Chichester: Wiley.

Boothroyd, Hilton. 1978. *Articulate Intervention: The Interface of Science, Mathematics, and Administration.* London: Taylor and Francis.

Brooks, Harvey. 1973. The state of the art: Technology assessment as a process. *International Social Science Journal* 25(3): 247–256.

Bull, K.R. 1991. The critical loads/levels approach to gaseous pollutant emission control. *Environmental Pollution* 69: 105–123.

Caldwell, Lynton Keith. 1991. Law and environment in an era of transition: Reconciling domestic and international law. *Colorado Journal of International Law and Policy* 2: 1–24.

Cambridge Decision Analysts (CDA) and Environmental Resources Limited. 1987. Acid rain and photochemical oxidants control policies in the European Community: A decision analysis framework. Prepared for DG XI of the European Community Commission and the Netherlands Ministry of Public Housing.

Clark, William C., and G. Majone. 1985. The critical appraisal of scientific inquiries with policy implications. *Science, Technology, and Human Values* 10(3): 6–19.

Climate Conference Secretariat. 1989. *The Noordwijk Declaration on Climatic Change.* Ministerial Conference on Atmospheric Pollution and Climatic Change. Organized by the Minister of Housing, Physical Planning, and Environment of the Netherlands in cooperation with UNEP and WMO. Held in Noordwijk, The Netherlands, November 6–7, 1989. Leidschendam: Climate Conference Secretariat.

Cohen, Michael, James March, and Johan Olsen. 1972. A garbage-can model of organizational choice. *Administrative Science Quarterly* 17: 1–25.

Collingridge, D., and C. Reeve. 1986. *Science Speaks to Power.* New York: St. Martin's Press.

Commission of the European Communities (CEC). 1990. Community action program to limit EC carbon dioxide emissions and to improve the security of the energy supply. Communication of November 28.

Confederation of British Industry. 1989. *The Greenhouse Effect and Energy Efficiency.* London: Association for the Conservation of Energy.

Crocker, T.D. 1983. What economics can currently say about the benefits of acid deposition control. In H.M. Trebing, eds., *Adjusting to Regulatory Pricing and Marketing Realities.* East Lansing: Michigan State University.

deLeon, Peter. 1999. The stages approach to the policy process: What has it done? Where is it going? In Paul A. Sabatier, ed., *Theories of the Policy Process* (pp. 19–32). Boulder: Westview Press.

Dietrich, William. 1995. The challenge of selecting goals: Case studies regarding the use of critical levels. CSIA Discussion Paper 95-05, Cambridge: Kennedy School of Government, Harvard University.

Edmonds, J., and J.M. Reilly. 1983. A long-term global energy-economic model of carbon dioxide release from fossil fuel use. *Energy Economics* 5(2): 74–87.

————. 1985. *Global Energy: Assessing the Future.* Oxford: Oxford University Press.

German Bundestag, ed. 1989. *Protecting the Earth's Atmosphere: An International Challenge.* Interim report of the Study Commission of the Eleventh German Bundestag. Bonn: German Bundestag.

————. 1991. *Protecting the Earth: A Status Report with Recommendations for a New Energy Policy.* Bonn: German Bundestag.

Glantz, Michael H., J. Robinson, and M.E. Krenz. 1982. Climate-related impact studies: A review of past experience. In W.C. Clark, ed., *Carbon Dioxide Review, 1982.* New York: Oxford University Press.

Greenpeace. 1992. *Making the Right Choices: Alternatives to CFCs and Other Ozone-Depleting Chemicals.* London: U.K. Greenpeace.

Griesshammer, Rainer. 1983. *Letzte Chance für den Wald? Die Abwendbaren Folgen des Sauren Regens (Last Chance for the Forest? The Avoidable Consequences of Acid Rain).* Freiburg: Dreisam-Verlag.

Heclo, Hugh. 1977. *A Government of Strangers.* Washington: Brookings Institution.

Hordijk, Leen. 1991. Use of the RAINS model in acid rain negotiations in Europe. *Environmental Science and Technology* 25(4): 596–602.

International Council of Scientific Unions (ICSU), UNEP, and WMO. 1986. *Report of the International Conference on the Assessment of the Role of Carbon Dioxide and of Other Greenhouse Gases in Climate Variations and Associated Impacts.* October 9–15, 1985. Villach, Austria. WMO Report No. 661. Geneva: World Meteorological Organization.

Intergovernmental Panel on Climate Change (IPCC). 1991. *Climate Change: The IPCC Response Strategies.* Report from Working Group III. Washington: Island Press.

Intergovernmental Panel on Climate Change (IPCC), J. Bruce, Hoesung Lee, and E. Haites, eds. 1996a. *Climate Change 1995: Economic and Social Dimensions of Climate Change.* Working Group III Contribution to the Second Assessment Report of the Intergovernmental Panel on Climate Change. New York: Cambridge University Press.

———. R.T. Watson, M.C. Zinyowera, and R.H. Moss, eds. 1996b. *Climate Change 1995: Impacts, Adaptations, and Mitigation of Climate Change—Scientific-Technical Analyses.* Working Group II Contribution to the Second Assessment Report of the Intergovernmental Panel on Climate Change. New York: Cambridge University Press.

IPCC (*See* Intergovernmental Panel on Climate Change).

Jäger, Jill. 1988. *Developing Policies for Responding to Climatic Change.* A report to the UNEP and WMO, WMO/TD-No. 225. Geneva: WMO.

Japanese Energy and Industrial Technology Council. 1992. *Fourteen Proposals for a New Earth: Policy Triad for the Envionment, Economy, and Energy.* Tokyo: Energy and Industrial Policy Council.

Jasanoff, Sheila. 1990. *The Fifth Branch: Science Advisors as Policy-makers.* Cambridge: Harvard University Press.

Jones, Charles O. 1977. *An Introduction to the Study of Public Policy.* North Scituate, Mass.: Duxbury Press.

Kay, David A., and Harold K. Jacobson, eds. 1983. *Environmental Protection: The International Dimension.* Totowa, N.J.: Allanheld, Osmun.

Kingdon, John W. 1995. *Agendas, Alternatives, and Public Policies.* New York: Harper Collins.

Lakatos, Imre. 1970. Falsification and the methodology of scientific research programs. In I. Lakatos and A. Musgrave, eds., *Criticism and the Growth of Knowledge.* Cambridge: Cambridge University Press.

Levy, Marc. 1993. European acid rain: The power of tote-board diplomacy. In Peter M. Haas, Robert O. Keohane and Marc A. Levy. *Institutions for the Earth.* Cambridge: MIT Press.

Lindblom, Charles E. 1990. *Policy-Making Process.* Englewood Cliffs, N.J.: Prentice Hall.

Lindblom, Charles E., and D.K. Cohen. 1979. *Usable Knowledge: Social Problem Solving.* New Haven: Yale University Press.

Lipset, S.M. 1963. *Political Man: The Social Basis of Politics.* Garden City, N.Y.: Anchor Books.

Majone, G. 1980. Policies as theories. *Omega* 8: 151–162.

Molina, Mario, and F.S. Rowland. 1974. Stratospheric sink of chloro-fluoromethanes: Chlorine atom catalysed destruction of ozone. *Nature* 249: 810–812.

Netherlands Ministries of Housing, Physical Planning, and the Environment, Agriculture and Fisheries, Transportation and Public Works, and Economic Affairs. 1989. *To Choose or to Lose: National Environmental Policy Plan.* The Hague: Ministry of Housing, Physical Planning, and Environment.

Nordhaus, William D., and Gary W. Yohe. 1983. Future paths of energy and carbon dioxide. In National Research Council, *Changing Climate.* Washington: National Academy Press.

RAND. 1980. Economic implications of regulating chlorofluorocarbon emissions from nonaerosol applications. R-2524-EPA. Santa Monica: RAND.

———. 1981. Economic impact assessment of a chlorofluorocarbon production cap. N-1656-EPA. Santa Monica: RAND.

———. 1981. Allocating chlorofluorocarbon permits: Who gains, who loses, and what is the cost? R-2806-EPA. Santa Monica: RAND.

Ravetz, Jerome R. 1971. *Scientific Knowledge and Its Social Problems.* Oxford: Clarendon Press.

———. 1997. The science of "what if?" *Futures* 29(6): 533–539.

Rotmans, Jan. 1989. *IMAGE: An Integrated Model to Assess the Green-house Effect.* Dordecht: Kluwer.

Rotmans, J., and M.G.J den Elzen. 1993. De Rol van Het IMAGE-model in de klimaatproblematiek. *Milieu* 8(2): 49–56.

Sabatier, Paul A., ed. 1999. *Theories of the Policy Process.* Boulder, CO: Westview Press.

Sabatier, Paul, and Hank C. Jenkins-Smith, eds. 1993. *Policy Change and Learning: An Advocacy Coalition Approach.* Boulder: Westview Press.

Sabatier, Paul A., and Hank C. Jenkins-Smith. 1999. The advocacy coalition framework: An assessment. In Paul A. Sabatier, ed., *Theories of the Policy Process* (pp. 117–166). Boulder, CO: Westview Press.

Schelling, Thomas C. 1983. Climatic change: Implications for welfare and policy. In National Research Council, *Changing Climate.* Washington: National Academy Press.

Simon, Herbert A. 1983. *Reason in Human Affairs.* Stanford: Stanford University Press.

Staaf, Hakan. 1992. Interview with William Dietrich, Solna, Sweden, June 2.

Swart, R., H. de Boois, and P. Vellinga. 1989. Targets for climatic change. In *The Full Range of Responses to Anticipated Climatic Change.* Stockholm: United Nations Environment Programme and the Beijer Insitute.

Titus, James G., ed. 1986. *Effects of the Changes in Stratospheric Ozone and Global Climate* (4 vols.). Washington: UNEP and U.S. EPA.

Tolba, Mostafa K. 1994. Interview with William C. Clark and Nancy M. Dickson at Harvard University, Cambridge, Mass., October 24.

United Nations Environment Programme (UNEP). 1986. Ad Hoc Working Group of Legal and Technical Experts. 1986. Report of the second part of the Workshop on the Control of Chlorfluorocarbons. Leesburg, VA., September 8–12. Nairobi: UNEP.

———. 1987. Ad hoc scientific meeting to compare model-generated assessments of ozone layer change for various strategies for CFC control. UNEP/WG.167/INF.1. Würzburg, April 9–10. Nairobi: UNEP.

———. 1989a. *Economic Panel Report.* Nairobi: UNEP.

———. 1989b. *Technical Progress on Protecting the Ozone Layer.* Nairobi: UNEP.

———. 1991. *Technology and Economic Assessment Panels. Reports.* Nairobi: UNEP.

———. 1994. *Technology and Economic Assessment Panels Reports.* Nairobi: UNEP.

UNEP and the Beijer Institute. 1989. *The Full Range of Responses to Anticipated Climatic Change*. Stockholm: Beijer Institute.

U.S. Department of Energy (DoE). 1985. *Projecting the Climatic Effects of Increasing Carbon Dioxide*. DOE/ER-0237. Washington: DoE.

U.S. Department of Transportation (DoT). 1975. *Climatic Impact Assessment Program* (6 vols.). Washington: DoT.

U.S. National Acid Precipitation Assessment Program (NAPAP). 1991. *The Experience and Legacy of NAPAP*. Report of the NAPAP Oversight Review Board to the Joint Chairs Council of the Interagency Task Force on Acidic Deposition. Washington: NAPAP.

U.S. National Research Council (NRC). 1975. Climatic Impact Committee. *Environmental Impact of Stratospheric Flight*. Washington: National Academy Press.

U.S. National Research Council (NRC). 1976. *Halocarbons: Environmental Effects of Chlorofluoromethane Release*. Committee on Impacts of Stratospheric Change. Assembly of Mathematical and Physical Sciences. Washington: National Academy Press.

————. 1979. *Protection against Depletion of Stratospheric Ozone by Chlorofluorocarbons*. Committee on Impacts of Stratospheric Change, Committee on Alternatives for the Reduction of Chlorofluorocarbon Emissions, Commission on Sociotechnical Systems, Assembly of Mathematical and Physical Sciences. Washington: National Academy Press.

————. 1983. *Changing Climate*. Washington: National Academy Press.

————. 1984. *Causes and Effects of Changes in Stratospheric Ozone: Update 1983*. Washington: National Academy Press.

————. 1992. *Policy Implications of Greenhouse Warming*. Washington: National Academies Press.

Wentzel, D.F. 1967. Die Belastung der Forstwirtschaft durch Immissionen und ihre technischen, waldbaulichen, raumplanerischen und rechtlichen Folgerungen (Stress on forestry through pollution emissions and their technical, silvicultural, planning and legal consequences). In *Generalversammlung der CEA, Veroffentlichungen der CEA*. Heft 35 Schweiz: Brugg.

Weterings, R.A.P.M., and J.B. Opschoor. 1992. *De milieugebruiksruimte als uitdaging voor Technologieontwikkeling*. Raad voor het Milieu- et Natuuronderzoek, Rijswijk.

Williams, Jill, ed. 1978. *Carbon Dioxide, Climate, and Society*. Oxford: Pergamon Press.

World Meteorological Organization (WMO). 1984. *Report of the Study Conference on Sensitivity of Ecosystems and Society to Climate Change*. Villach, Austria, September 19–23. WCP-83. Geneva: WMO.

WMO, UNEP, and ICSU. 1981. *On the Assessment of the Role of CO_2 on Climate Variations and Their Impact*. Proceedings of a Meeting of Experts held in Villach, Austria, November 1980. Geneva: WMO.

WMO, U.S. NASA, FAA, NOAA, UNEP, CEC, and BMFT. 1986. *Atmospheric Ozone 1985: Assessment of Our Understanding of the Processes Controlling Its Present Distribution and Change* (3 vols.). WMO Report No. 16. Geneva: WMO.

WMO, U.S. NASA, U.S. NOAA, U.K. DOE, and UNEP, 1992. *Scientific Assessment of Ozone Depletion: 1991*. Washington: U.S. NASA and U.S. NOAA; London: U.K. DOE Nairobi: WMO.

Bibliography of Major Option Assessments

Alcamo, Joseph, Roderick Shaw, and Leen Hordijk, eds. 1990. *The RAINS Model of Acidification: Science and Strategies in Europe*. Dordrecht: Kluwer.

Bach, Wilfrid. 1979. Untersuchung der Beeinflussung des Klimas durch Anthropogene Factoren. (Investigation of the influence of anthropogenic factors on climate). Forschungsstelle für Angewandte Klimtologie und Umweltstudien, Institute fur Geographie, Universitat Munster.

————1984. *Our Threatened Climate: Ways of Averting the CO_2 Problem through Rational Energy Use*. Dordrecht: Reidel.

Bolin, Bert, et al., eds. 1986. *The Greenhouse Effect, Climatic Change, and Ecosystems*. SCOPE 29. Chichester: Wiley.

Canadian Council of Ministers of the Environment. 1993. In F.K. Hare, ed., *The Canadian Climate Program*. Toronto: Atmospheric Environment Service.

Commission of the European Communities (CEC). 1983. *Acid Rain: A Review of the Phenomenon in the EEC and Europe*. Prepared by Environmental Resources Limited for the CEC, Directorate-General for Environment, Consumer Protection and Nuclear Safety. London: Graham and Trotman.

————.1988. Acid rain and photochemical oxidants control policies in the European Community: A decision analysis framework. Prepared by Cambridge Decision Analysts (CDI) and ERL for the CEC, Directorate General for Environment, Consumer Protection and Nuclear Safety (DGXI), and the Netherlands Ministry for Public Housing, Physical Planning, and the Environment. Wiltshire, U.K.: Leopard Press.

————.1990. Directorate-General for Energy. *Energy in Europe: Energy for a New Century—The European Perspective*. Special issue. Brussels: EU DG XVII (Energy).

————.1991. The economics of policies to stablize or reduce greenhouse gas emissions: The case of CO_2. Prepared by Matthias Mors. No. 87.————Brussels: Economic Papers of EU DG II (Economic Affairs).

————.1992. Directorate-General for Economic and Financial Affairs. The climate challenge: Economic aspects of the Community's strategy for limiting CO_2 emissions. No. 51. Brussels: ECSC-EEC-EAEC.

Commission on Monitoring of the Scientific Committee on Problems of the Environment (SCOPE) of the International Council of Scientific Unions (ICSU). 1971. *Global Environmental Monitoring*. A report submitted to the U.N. Conference on the Human Environment, Stockholm 1972. SCOPE Report 1. Stockholm: ICSU/SCOPE.

Confederation of British Industry. 1989. *The Greenhouse Effect and Energy Efficiency*. London: Association for the Conservation of Energy.

DPA Group. 1989. *Study on the Reduction of Energy-Related Greenhouse Gas Emissions*. Toronto: DPA.

Economic Commission of Europe (ECE). 1986. Working Group on Nitrogen Oxides. Technologies for controlling NO_x emissions from stationary sources. Geneva: United Nations.

————.1986a. Working Group on Nitrogen Oxides. Effective enforcement procedures. Geneva: United Nations.

————.1986b. Working Group on Nitrogen Oxides. Technologies for controlling NO_x emissions from mobile sources: Light- and heavy-duty

84

William C. Clark, Josee van Eijndhoven, and Nancy M. Dickson et al.

diesel-fueled vehicles and heavy-duty gasoline-fueled vehicles. Geneva: United Nations.

————.1990a. Working Group on Volatile Organic Compounds. Emissions of volatile organic compounds (VOCs) from stationary sources and possibilities of their control. Geneva: United Nations.

————.1990b. Working Group on Volatile Organic Compounds. Hydrocarbons and ozone formation: An approach to their control based on likely environmental benefits. Submitted by the delegation of the United Kingdom. Geneva: UNEP.

————.1991a.Working Group on Abatement Strategies. The critical load concept and the role of the best available technology and other approaches. Geneva: United Nations.

————.1991b.Working Group on Abatement Strategies. Exploration of economic instruments for implementation of cost-effective reductions of SO_2 in Europe using the critical loads approach: Draft report prepared by governmental designated experts. Geneva: UNEP.

Electric Power Research Institute (EPRI). 1991. Analysis of alternative SO_2 reduction strategies. EN/GS-7132. Palo Alto: EPRI.

Environmental Defense Fund (EDF). 1988. *Polluted Coastal Waters: The Role of Acid Rain.* NewYork: EDF.

European Community. 1980. Council decision of 26 March 1980 concerning CFCs in the environment. OJL 90, 3.4.1980, 45.

German Bundestag, ed. 1989. *Protecting the Earth's Atmosphere: An International Challenge.* Interim Report of the Study Commission of the Eleventh German Bundestag Preventive Measures to Protect the Earth's Atmosphere. Bonn: German Bundestag.

————, ed. 1991. *Protecting the Earth: A Status Report with Recommendations for a New Energy Policy.* Bonn: Dt. Bundestag, Referat Offentlichkeitsarbet.

German Rat von Sachverstaendigen fuer Umweltfragen. 1978. Environment Opinion. Umweltgutachten 1978. BT-DRS 8/1938. Bonn: Bundestagdrucksachen.

Greenpeace. 1990. Memorandum submitted by Greenpeace. April 18, 1990. Evidence before the U.K. Energy Committee.

————. 1992. *Making the Right Choices: Alternatives to CFCs and Other Ozone-Depleting Chemicals.* London: U.K. Greenpeace.

Griesshammer, Rainer. 1983. *Letzte Chance fur den Wald? Die abwendbaren Folgen des Sauren Regens (Last Chance for the Forest. The Avoidable Consequences of Acid Rain).* Freiburg: Dreisam-Verlag.

ICSU, UNEP, and WMO. 1986. Report of the international conference on the assessment of the role of carbon dioxide and of other greenhouse gases in climate variations and associated impacts. WMO Report No. 661. Villach, Austria, October 9–15, 1985.

IEB for LRTAP. 1981. Proposals by the working party on air pollution problems regarding control techniques and related measures to reduce emissions of sulphur compounds in the atmosphere. Geneva: United Nations.

Intergovernmental Panel on Climate Change (IPCC). 1990. *Climate Change: The IPCC Response Strategies.* Report from Working Group III. Washington: Island Press.

————.1992. *Climate Change 1992: The Supplementary Report to the IPCC Scientific Assessment.* Cambridge: Cambridge University Press.

Jackson, Tim, and Simon Roberts. 1989. *Getting out of the Greenhouse: An Agenda for U.K. Action on Energy Policy.* London: Friends of the Earth.

Jäger, Jill. 1988. Developing policies for responding to climatic change. A report to the UNEP and WMO. WMO/TD-No. 225. A summary of the discussion and recommendations of the workshops held in Villach, Austria, September 28–October 2, 1987, and Bellagio, November 9–13, 1987, under the auspices of the Beijer Institute, Stockholm.

Jäger, Jill, and H.L. Ferguson, eds. 1991. *Climate Change: Science, Impacts, and Policy.* Proceedings of the Second World Climate Conference. Cambridge: Cambridge University Press.

Japanese Ad Hoc Group on Global Environmental Problems. 1982. International responses to global environmental problems. Tokyo: Ad Hoc Group on Global Environmental Problems.

Japanese Central Council on Environmental Pollution Control. 1988. Regarding the basics of a system for the protection of the ozone layer. Submitted to the Environment Agency. Tokyo: Central Council on Environmental Pollution Control.

Japanese Council of Ministers for Global Environmental Conservation. 1990. Action program to arrest global warming. Tokyo: Council of Ministers for Global Environmental Conservation.

Japanese Environment Agency. 1980. Regarding the fluorocarbon problem. Environmental Health Division, Health Survey Office. Tokyo: Environment Agency.

————. 1988. Policy recommendations concerning climate change. Interim report of the Advisory Committee on Climate Change. Tokyo: Environment Agency.

————. 1989. The first results of a survey of acid rain policy. Tokyo: Environment Agency.

————. 1992. Regarding the midterm assessment of the second acid rain policy survey. Tokyo: Environment Agency.

Japanese MITI. 1988. Fundamental thoughts on the regulation of fluorocarbon production for the protection of the ozone layer. Chemical Substances Committee. Tokyo: MITI.

————. 1991. Chemical Products Advisory Committee. Regarding future policies for the protection of the ozone layer. Midterm report. Tokyo: MITI.

————. 1992. Green Aid Plan. Tokyo: MITI.

Japanese Special Committees on Energy and Environment: Industrial Structure Council, Advisory Committee for Energy, and Industrial Technology Council. 1992. *Fourteen Proposals for a New Earth: Policy Triad for the Environment, Economy, and Energy.* Tokyo: Energy and Industrial Technology Council.

Mintzer, Irving. 1987. *A Matter of Degrees: The Potential for Controlling the Greenhouse Effect.* Washington: World Resources Institute.

Netherlands Ministry of Housing, Physical Planning, and the Environment. 1989. *To Choose or to Lose: The National Environmental Policy Plan.* The Hague: Ministry of Housing, Physical Planning, and Environment.

————. 1991. Briefing to the Second Chamber of the State General. No. 22232, September 4.

Schutt, Peter. 1988. *Der Wald Stirbt an Stress (The Forest Dies from Stress).* Frankfurt/M-Berlin: Ullstein Sachbuch.

Skea, J.F. 1990. *United Kingdom: A Case Study of the Potential for Reducing Carbon Dioxide Emissions.* Prepared for U.S. EPA with U.S. DOE. Richland, Wash.: Pacific Northwest Laboratory.

Study of Critical Environmental Problems (SCEP). 1970. In C.L. Wilson and W.H. Matthews, eds., *Man's Impact on the Global Environment.* Cambridge: MIT Press.

Study of Man's Impact on Climate (SMIC). 1971. In W.H. Matthews, W.W. Kellogg, and G.D. Robinson, eds., *Inadvertent Climate Modification.* Cambridge: MIT Press.

U.K. Department of Trade and Industry. 1991. Refrigeration and air conditioning CFC phase out: Advice on alternatives and guidelines for users. London: Her Majesty's Stationery Office.

U.K. Interdepartmental Group on Climatology. 1980. Climate change: Its potential effects on the United Kingdom and the implications for research. London: Her Majesty's Stationery Office.

United Nations Environment Programme (UNEP). Ad Hoc Working Group of Legal and Technical Experts. 1986. Report of the second part of the workshop on the control of chlorfluorocarbons. Leesburg, USA, September 8–12.

———. Economic Panel. 1989. *Economic Panel Report.* Nairobi: UNEP.

———. Technology Review Panel. 1989. *Technical Progress on Protecting the Ozone Layer.* Nairobi: UNEP.

U.S.-Canada Work Group 3B. 1982. *Emissions, Costs, and Engineering: Final Report.* Washington: EPA.

U.S. Department of Transportation (DoT). 1975. *Climate Impact Assessment Program* (6 vols.). Washington: DoT.

U.S. Environmental Protection Agency (EPA). 1980. *Economic Implications of Regulating Chlorofluorocarbon Emissions from Nonaerosol Applications.* Rand Report prepared for EPA, R-2524-EPA. Santa Monica: RAND.

———. 1983. *Can We Delay the Greenhouse Effect?* Edited by S. Seidel and K. Deyes. Washington: EPA.

———. 1985. *The Health Costs of Skin Cancer Caused by Ultra-Violet Radiation.* Rand Report prepared for EPA, N-2538-EPA. Santa Monica: RAND.

———. 1988. Office of Air and Radiation, Office of Program Development, and Stratospheric Protection Program. *Regulatory Impact Analysis.* Washington: EPA.

———. 1990. *Policy Options for Stabilizing Global Climate.* Edited by Daniel Lashof and Dennis Tirpak. EPA 21P-20003.1.Washington: EPA.

U.S. National Acid Precipitation Assessment Program (NAPAP). 1987. *Interim Assessment: The Causes and Effects of Acidic Deposition.* Washington: U.S. Government Printing Office.

———. 1991. *Acidic Deposition: State of Science and Technology.* Washington: U.S. Government Printing Office.

———. 1991. 1990 *Integrated Assessment Report.* Washington: U.S. Government Printing Office.

U.S. National Aeronautics and Space Administration (NASA), U.S. National Oceanic and Atmospheric Administration (NOAA), U.S. Federal Aviation Administration (FAA), World Meteorological Organization (WMO), and U.N. Environmental Program (UNEP). 1988. *Report of the International Ozone Trends Panel: 1988.* Washington: NASA, NOAA, FAA; Geneva: WMO; and Nairobi: UNEP.

U.S. NASA, U.S. NOAA, U.K. DOE, UNEP, and WMO. 1992. *Scientific Assessment of Ozone Depletion: 1991.* Washington: NASA and NOAA; London: DOE; Nairobi: UNEP; and Geneva: WMO.

U.S. National Research Council (NRC). 1977. *Energy and Climate.* Geophysics Study Committee, Geophysics Research Board, and Assembly of Mathematical and Physical Sciences. Washington: National Academy Press.

———. 1979. *Protection against Depletion of Stratospheric Ozone by Chlorofluorocarbons. Committee on Impacts of Stratospheric Change.* Committee on Alternatives for the Reduction of Chlorofluorocarbon Emissions. Commission on Sociotechnical Systems. Assembly of Mathematical and Physical Sciences. Washington: National Academy Press.

———. 1983. *Changing Climate.* Washington: National Academy Press.

———. 1992. *Policy Implications of Greenhouse Warming.* Washington: National Academy Press.

U.S. Office of Technology Assessment (OTA). 1984. *Acid Rain and Transported Air Pollutants: Implications for Public Policy.* Washington: U.S. Government Printing Office.

———. 1991. *Changing by Degrees: Steps to Reduce Greenhouse Gases.* OTA-0-482. Washington: U.S. Government Printing Office.

Wentzel, D.F. 1967. Die Belastung der Forstwirtschaft durch Immissionen und ihre technischen, waldbaulichen, raumplanerischen und rechtlichen Folgerungen in Generalversammlung der CEA, Veroffentlichungen der CEA (Stress on forestry through pollution emissions and their technical, silvicultural, planning and legal consequences). Heft 35 Schweiz: Brugg.

Williams, Jill, ed. 1978. *Carbon Dioxide, Climate, and Society.* Oxford, N.Y.: Pergamon Press.

Woodwell, George, Gordon MacDonald, Roger Revelle, and C. David Keeling. 1979. *The Carbon Dioxide Problem: Implications for Policy in the Management of Energy and Other Resources: A Report to the Council on Environmental Quality.* Manuscipt, Harvard University Library, Cambridge, Mass.

18
Goal and Strategy Formulation in the
Management of Global Environmental Risks

Marc A. Levy, Jeannine Cavender-Bares, and William C. Clark
with Gerda Dinkelman, Elena Nikitina, Ruud Pleune, and Heather Smith[1]

18.1 Introduction

With this chapter we begin to shift our attention away from the *knowledge* that societies assembled regarding global environmental risks and toward the *action* they took regarding those risks. In particular, we turn from the "What *could* be done?" of option assessment to the "What *should* be done?" of goal and strategy formulation.

The chapter is organized as follows. Section 18.1 outlines our concept of "goals and strategies" and criteria for their evaluation. It also maps our analytic approach, describing the empirical materials from which we extracted goals and strategies of particular actors and the way in which we classified the results. Sections 18.2, 18.3, and 18.4 illustrate the richness of cause-and-effect linkages involving goal setting in the management of global environmental risks for the cases of ozone depletion, acid rain, and climate change, respectively. More general findings regarding patterns in goals and strategies are presented in section 18.5. Finally, section 18.6 highlights the principal conclusions and questions suggested by our study.

18.1.1 Definitions

Goals are statements of objectives or of conditions that an actor professes a wish to bring about. Often multiple goals will be related hierarchically, with lower-level goals facilitating the achievement of higher-level ones. Representative goals expressed in the cases covered by this study include

- Protecting the environment,

- Reducing total sulfur dioxide (SO_2) emissions by 30 percent,

- Stabilizing greenhouse gas concentrations in the atmosphere at a level that would prevent dangerous anthropogenic interference with the climate system, and

- Marketing an electric car by 2000.

Strategies are plans for how goals will be achieved; they specify combinations of policies or response options to be implemented to obtain the desired results. A (hypothetical) example of a strategy relevant to the climate change case is to tax gasoline at $50 per gallon, provide $500 million per year in funding for renewable-energy research and development (R&D), revise construction codes to require greater insulation and more efficient lighting, and spend 0.7 percent of gross national product on foreign development assistance.

The analytical distinction between goals and strategies is not hard and fast. What is a goal to one actor may be a strategy to another. "Market an electric car by 2000" could be a goal for General Motors and a strategy for the U.S. government. A statement that appears to one observer as a strategy in service of some overarching goal may appear to another as a goal in itself. Some strategy statements are made with the overarching goal never stated explicitly. For these reasons, this chapter does not attempt to distinguish finely between goals and strategies.

Espousing a certain goal may at times appear to be an act of implementation. If the act of formulating a goal weren't so often perceived to be as important as action, there wouldn't be such vigorous fights over what goals to espouse. Likewise, there is also apparent overlap with the function of assessing policy or response options. An option assessment, for its part, may attempt to identify the feasibility of achieving a specific goal. A distinction remains, however. Option assessment is primarily a technical exercise concerned with comparing alternative policies or probing the consequences and feasibility of a single intervention. But the specification of which options *ought* to be chosen is the task of goal and strategy formulation. Option assessment may thus be thought of as appraising the "building material" of management (its physical, aesthetic, and economic characteristics), goals as dealing with architectural design, and strategy formulation as choosing particular materials and drawing up a schedule by which they will be assembled.[2]

In the research reported here, we have assumed that goal and strategy statements could be generated by the entire range of actor groups under consideration. Some goal statements, of course, carry more political authority than others: when a head of state espouses a goal for society, it means something different than when an environmental NGO does so. But whether this difference in

88

Marc Levy, Jeannine Cavender-Bares, and William C. Clark et al.

political authority makes a difference in how society responds is an empirical question.

18.1.2 Expectations

The academic literature on the role of goals and goal formulation in the evolution of social issues is varied and contentious. At one extreme, rational actor, decision-analysis, and planning perspectives often accord goals (or objectives) a central but unproblematic position, treating them as immutable and exogenous to the processes involved in seeking solutions for the problem at hand. At the opposite extreme, some scholars seem to treat goals and options for achieving them as equally malleable and problematic terms in the policy debate. In this view, ends and means are always proffered tentatively and influence one another in a dynamic process of mutual adjustment (e.g., Hirschman and Lindblom 1962, 218) or amplification (e.g., Kingdon 1995). Between these extremes, a number of analysts who have focused on questions of learning allow that goals can be reconsidered in light of experience or other forms of new information. But nearly all who have empirically studied such possibilities emphasize the difficulty and rarity of the reflective adjustment of ends relative to the more common response of adapting new means (Argyris and Schon 1978; Wildavsky 1979, 39; Hall 1993; Lee 1993).

Despite their disagreements, these various perspectives generally share a recognition that many sorts of goals appear in public-policy debates, variously specifying direction of change, desired end-states, and norms to be attained or maintained. Moreover, most writers accept that the use of goals in public discourse on issues can be instrumental or symbolic—or both at the same time, even for the same actor. Finally, virtually all who have reflected on the status of goals in the evolution of public issues agree that, in one way or another, goals matter (Michaels 1973, 147–148; Vickers 1965, 31–35). The goal of this chapter is to investigate how they have mattered, and to what effect, in the evolution of social responses to the challenges posed by our cases of global environmental change. In particular, given our focus on the prospects for learning in the response to such risks, we are interested in how to evaluate specific goals in terms of their contribution to long-term issue development.

18.1.3 What Makes a Good Goal or Strategy?

The most intuitive response to the question of how to evaluate goals and strategies is that good ones help to solve the problems they are aimed at and bad ones don't. Unfortunately, such a perspective is not adequate for assessments of goals and strategies aimed at managing complex social problems that have many dimensions and that are not likely to be "solved" in the sense that they disappear from social consideration. Most global environmental risks fall into this category: they can be managed well or badly, but they seldom can be solved once and for all.

The merits of goals and strategies can therefore be judged fairly only according to how well they promote effective management—the "getting on" with the problem at hand. This distinction between judging goals according to whether they solve the problem and whether they promote more effective management is roughly analogous to the distinction Imre Lakatos made for the evaluation of scientific theories. Lakatos (1970, 91–138) argued that scientific theories could be evaluated narrowly according to their ability to survive falsification tests and more broadly according to their contribution to a progressive research program. Lakatos's ideas were subsequently applied to good effect in the critical evaluation of long-term evolution in the realms of public ideas and policies (e.g., Majone 1980, 1989). While scientific research programs are sufficiently different from the process of setting risk management goals that we cannot borrow Lakatos's specific criteria directly, we have found it helpful to follow these earlier extensions of his work in thinking of "effective" goals as "progressive" goals. Below we enumerate some of the characteristics that might be associated with progressive goals for the management of global environmental risks, with illustrations taken from the empirical findings presented in more detail later in this chapter.

When It Promotes a Deeper Consideration of the Problem Effective management of complex social problems is seldom possible unless the problems generate consideration among social actors that becomes deeper over time. The initial consideration when the problem first enters the agenda is seldom sufficient to mobilize the appropriate amount of concern or to generate an adequate understanding of the risks and response options. Therefore, one way in which goals and strategies can be considered to be progressive is the extent to which they promote a deepening of consideration among social actors of the problem in question.

For example, the precautionary principle, a strategy employed in the climate change case, generated very different levels of social consideration than its counterpart, the no-regrets strategy. Evaluated strictly on their own technical terms, it is hard to consider one of these strategies as more effective or otherwise better than the other. But in examining the social effects they triggered, the differences are significant. The "no-regrets" strategy that was promoted by the U.S. government had very little resonance in the social debate over climate change in that country. It particular, it did little to deepen the social consideration of the climate change problem. On the other hand, in Germany, where the precautionary principle had the highest salience, it contributed directly to a deeper consideration of the climate change problem throughout German society.

When It Promotes Consideration of the Problem across Additional Countries and Additional Actors Complex social problems also commonly require broad-based participation, both transnationally and cross-sectorally. Therefore, goals and strategies that broaden the range of countries and social actors that are actively involved in the risk management effort can be considered progressive.

An example of a goal that had this effect is the Thirty Percent Club in the acid rain issue. Whereas Scandinavian countries and the Organization for Economic Cooperation and Development (OECD) had promulgated goals aiming at reduction of transboundary acidifying air pollution since the late 1960s, the goal as specified had little resonance outside a small circle of countries, scientists, and power-plant operators. But once a much starker goal was put forth in 1982—to reduce sulfur dioxide emissions by 30 percent—the pattern of spread was quite rapid. Serious debates about the acid rain problem spread to a much larger number of countries, and these debates involved a far broader range of social actors, including parliamentarians, nongovernmental organizations (NGOs), and the public.

When It Promotes Improvements in Other Risk-Management Functions Complex social problems, precisely because of their complexity, require effectiveness along more than one risk management function. A society that concentrated all its risk management efforts on formulating more sophisticated preventive strategies without seeking to improve risk assessments, monitoring activities, appraisal of response options, and so on would not be expected to manage risks well. Therefore, goals that promote improvement in other risk management functions can be considered progressive.

Examples of goals that had this effect include the Toronto 20 percent carbon dioxide (CO_2) reduction target, the goal of stabilizing chlorine concentrations in the stratosphere below 2 parts per billion, and the goal of reducing acid deposition so as not to exceed critical loads. In each of these cases the stated goals played a direct role in promoting useful option assessments. The critical-load goal also played a direct role in improving monitoring and risk assessment activities.

When It Promotes a Shift from Distributional Conflicts to Collective Problem Solving Complex social problems almost invariably give rise to distributional conflicts over who benefits and who loses in alternative social responses. Which CFC consumers must be forced to use substitutes? Do power-plant operators have to install expensive scrubbers, or must others endure damaged forests and lakes? Will society mandate technology that favors manufacturers of large cars or small cars? Such conflicts are endemic. When they dominate social consideration of a global environmental risk, they frequently generate self-serving proposals for change, heightened levels of mistrust, and deadlock. If these distributional conflicts can be reduced in salience, then it becomes possible to engage in collective problem solving in which levels of trust are raised, proposals can be evaluated according to collectively agreed criteria rather than self-serving ones, and collective movement can replace deadlock. Goals that encourage a shift toward framing of issues in collective terms can therefore, other things being equal, be considered as relatively progressive.

For example, the goal of eliminating chlorofluorocarbons (CFCs) in spray cans, while it served to rapidly eliminate a significant amount of CFC consumption in those countries that implemented it, helped promote a distributional deadlock thereafter. Between about 1978 and 1986, the central conflict between the European Community and the Toronto Group was over the question of whether to make a CFC ban on spray cans universal. The debate during this period is notable for its dominance by blatantly self-serving proposals (the European Community recommended that the rest of the world do what it had already done, and the Toronto Group urged the Community to copy what it had previously done). Once the shift in goals from spray cans to chlorine loading on the stratosphere occurred, it was possible to let these distributional conflicts fade in salience and pursue more creative and effective collective strategies.

18.1.4 Analytic Approach
This section describes our analytical approach to the study of goals and strategies in the management of global environmental risks. It begins with a review of the empirical materials we considered. We next outline the framework we used for describing and classifying the goals and strategies reflected in our empirical data.

We searched for goals and strategies across the entire range of sources employed in developing the arena histories reported in part II of this book. This included formal declarations from governments, private sector leaders, and other nongovernmental institutions, speeches and press releases, newspaper articles, public testimony before parliaments, and the like. Attempting to find every instance where a goal or strategy was mentioned would have been both pointless and impossible: an early survey

90

Marc Levy, Jeannine Cavender-Bares, and William C. Clark et al.

uncovered over a hundred statements from significant sources in a single country on a single risk over a two-year period. What proved both interesting and feasible, however, was to trace the emergence of goals that eventually achieved wide prominence in each arena and to discern patterns in such prominent goals and strategies across arenas and cases.

For each of the goals and strategies that eventually achieved relative prominence in a particular arena, we systematically documented for particular instances of declaration (such as a press release) the following:

• The date of the declaration,

• The actor group making it,

• The forum in which it was made (legislation, testimony, or interview),

• The "pedigree," if any, explicitly associated with the declaration (such as a reference to an assessment report or international conference from which the goal in question is drawn or adapted),

• The specific wording of the goal (such as promoting or opposing a particular form of action), and

• The substantive focus of the goal or strategy (see the discussion of classification below).

In our basic data collection we did not attempt to differentiate "honest" from "dishonest" or "strategic" from "tactical" goals. In particular instances where the motivations of those making goal statements seemed important for our understanding, we pursued the matter separately from our basic documentation of patterns in formal pronouncements.

The empirical material on goals and strategies that resulted from pooling our answers to these questions across arenas, cases, and years is necessarily incomplete and qualitative. We do believe, however, that it provides a sufficiently solid foundation for the tentative findings and propositions we advance here. As elsewhere in this book, our intention has been to sketch a broad-brush perspective from which subsequent more focused, quantitative, and critical studies can benefit.

Goal and strategy statements were classified using the same risk assessment framework applied elsewhere in this study. Goals, as statements of objectives, were classified to reflect the *target* of the changes they sought to promote or oppose: emissions, the environment itself, impacts on society, or more generic objectives usually involving improved understanding (see box 18.1 for details). Some goals were stated with no indication of the strategy that might be used to implement them. In fact, however, most goal statements were accompanied by at least an indication of the preferred *means*—the policy

instruments, programs, or other social mechanisms—through which they sought to attain their objectives. Such strategies were further distinguished according to whether, if implemented, their effect on the risk in question would be *direct* (through the means of incentives or commands) or *indirect* (through the means of increasing social capacity to deal with the risk later on). Again, details of the classification are provided in box 18.1.[3]

Combining the targets and means dimensions described above yields the classification framework shown in table 18.1. Into this framework we have mapped some of the most common examples of goals and strategies that we encountered in the course of our study. The nature of the specific entries and the relative frequency with which they appeared in our data set are discussed in the following sections.

18.2   Findings: Ozone Depletion

This section begins our summary of how goals and strategies for the management of each of our global environmental risks evolved over the study period. As elsewhere in this volume, we have found it most effective to convey the intricate dynamics of that evolution through reference to specific cases. For each of the environmental problems addressed in the study, we therefore begin with a broad account of goal evolution across multiple arenas. We then turn to a more detailed analysis of the emergence, spread, and efficacy of one illustrative goal for each case. While these selected stories by no means represent the full range of patterns that are possible, together they help show, in sharper relief than is possible in the aggregate, some of the important aspects of the goal-setting process that we have encountered in our study. General patterns and relationships that hold across specific cases are described and discussed in section 18.5.

18.2.1   The Big Picture

Goals and strategies for protecting stratospheric ozone from depletion through human activities first emerged during the late 1960s in the context of concerns for the possible impact of high-altitude engine emissions. These concerns were generally restricted to a small community of scholars and activists in the United States, the United Kingdom, USSR, and Canada, however, and never achieved wide prominence. We therefore leave this early history to the respective arena chapters in part II of this book. We turn here to developments following the publication of the article by Molina and Rowland (1974) postulating a CFC threat to the ozone layer.

The debate over goals and strategies for dealing with the CFC problem quickly became dominated by

Box 18.1

Classification of goals and strategies

Goals, as statements of objectives, are classified in this study to reflect the *targets* or *ends* of the changes they sought to promote or oppose. Strategies are classified in terms of the goals they address and by the *instruments* or *means* they specified for achieving those goals. Any goal or strategy statement can therefore be classified in terms of its targets and means. The finer-scale definitions used in our work are given below:

Goal Targets

Emissions Some goals target the human activity that is generating the environmental risk. In our three cases the shorthand label *emissions* suffices, though in other environmental risks land-use patterns, resource exploitation, and other forms of human activity might be relevant. If emissions (of greenhouse gases, ozone-depleting substances, or acidifying compounds) are believed by an actor to be causing an environmental problem, then one logical type of goal is to reduce emissions. For other actors, of course, the most important goal may be to oppose such emission reductions.

Environment A second type of goal aims directly at altering or restoring the state of the *environment*—a particular biophysical equilibrium or other condition. Seeking not to exceed certain levels of acidity in lakes or aiming to keep chlorine concentrations in the stratosphere below certain levels or keep climate change below a specified rate are examples of these kinds of goals.

Impacts A third type of goal focuses on the *impacts* of a risk on society and its economic performance. One can seek to reduce or eliminate the impact by pursuing goals of inducing people to stay out of the sun on days of high ultraviolet incidence, coating statuary with acid-resistant plastic, purchasing flood insurance, or increasing the availability of air conditioning.

Generic Some goals are specified in such a way that it is not possible to classify them as targeting any of the above categories. We refer to them as *generic* goals in this chapter. Sometimes goals do not permit specification according to target because they are specified too loosely: "Do something about this problem!" Other times, however, the goals may be very precise but not restrictive with respect to the kind of target whose achievement they might promote. "Mandate national climate change plans," for example, is a fairly specific goal, but it permits action that might reduce emissions, help cope with impacts, or aim at a particular state of the environment.

Strategic Means

The formulation of goals and strategies is a process that seeks to alter patterns of social action. It is possible, therefore, to classify strategies according to the goals or targets they address and the means or mechanism by which they aim to achieve such change. At the broadest level, such statements can signal an intention to alter what social actors *can* do (capacity), what they *want* to do (incentives), and what they are *compelled* to do (commands). In this study, we have found it helpful to distinguish the latter two strategies as *direct* interventions in the risk and the former strategies of capacity building as *indirect* interventions in the risk.

Unspecified strategies This category is used to handle those goal statements that propose no means for their implementation.

Indirect (capacity building) strategies We classify three kinds of indirect strategies in which the means of goal achievement is through changed social *capacity*—that is, what society *can* do:

• *Cognitive capacity* Strategies aimed at increasing *cognitive capacity* seek to improve the state of knowledge about an environmental risk (its causes, impacts, possible solutions, and so on).

• *Technical capacity* Finally, strategies may seek to improve the *technical capacity* of a society to manage a risk by seeking to enhance the technologies available for reducing emissions, coping with impacts, or maintaining desired environmental conditions.

• *Institutional capacity* Strategies that seek to raise *institutional capacity* try to improve the provision of such institutional functions as creating forums for bargaining, for collecting and evaluating information, for raising the salience of items on social actors' agendas, and for legitimizing participation in a collective enterprise.

Direct strategies We classify three kinds of strategies in which the means of goal achievement is through direct intervention in the risk through command and control or incentive measures:

• *Incentives* By *incentives* we mean measures that change what people *want* to do.

• *Informational* Incentives can be altered through changing the flow of *information* to social actors. Product-labeling schemes, air-quality and UV-level notices, and public-relations efforts all change the flow of information in such a way that help promote voluntary changes in behavior.

• *Market* Incentives are also commonly altered through the *market*—by changing the prices people pay. Taxes, subsidies, and financial penalties are all devices that have been used to promote changes in behavior through the price mechanism.

• *Commands* By *commands* we include strategies that simply require certain activities or changes in behavior. They change what people *must* do. These include the classic forms of government legislation (the U.S. Clean Air Act), as well as government programs carried out directly (German programs to lime forest soils). They also include commands issued by nongovernmental actors, such as NGOs and international organizations or treaty texts.

Table 18.1
A taxonomy of goals and strategies

Means	Targets	
	Emissions	Environment
Unspecified	Expand nuclear power and renewables. Stabilize carbon dioxide and greenhouse gases concentrations. Reduce carbon dioxide and greenhouse gases emissions by X percent. Reduce emissions in transport. Phase out CFCs. Save energy. Increase energy efficiency. Reduce deforestation.	Plant trees. Reduce ocean pollution to ensure carbon uptake.
Capacity: Cognitive		
Technical	Develop international technology transfer to reduce emissions.	Develop the ability to reflect sunlight with orbiting shields.
Institutional	Establish a working group to develop a strategy for 20 percent carbon dioxide reduction.	
Incentive: Information	Develop public-awareness campaigns on energy-use efficiency. Develop campaigns to reduce driving. Persuade the public to reduce carbon dioxide emissions.	Persuade people to plant trees.
Market	Taxes on energy, carbon, or carbon dioxide use. No taxes on energy, carbon, or carbon dioxide use. Financial transfers to reduce deforestation in tropics.	
Command		

discussions over the elimination of CFCs in spray cans. Spray-can CFC bans dominated the debate for about a decade, with some countries adopting bans, others pursuing bans partially, and still others rejecting bans outright. The emphasis on spray cans lessened a bit in 1980, when in the final days of the Carter administration the U.S. Environmental Protection Agency (EPA) issued a proposal to take the spray-can ban to the next logical step by regulating the production of CFCs on a sector-by-sector basis. This proposal called for parallel action at the international level and was also one of the first instances of a proposal to make use of marketable permits in a major environmental regulation. With the arrival of the Reagan administration in Washington, however, the effort to formulate broader ozone-protection goals was quickly submerged.

During the early 1980s two competing clusters of goals and strategies for addressing the ozone problems emerged.

On the one hand, a group of Nordic countries began promoting the adoption of sector-specific regulations similar to the earlier U.S. proposals. The coalition supporting such action, which came to be called the Toronto Group, later expanded to include Canada and to reengage the United States.

Opposed to the Toronto Group was the European Community, which argued that sector-specific goals and strategies were both unnecessarily burdensome as regulatory instruments and ultimately ineffective as safeguards to the ozone layer. The European Community instead proposed a cap on total CFC production, without regard to use. Because the actual target the Community proposed for capping production required no change in planned manufacturing, many actors, including U.S. negotiators, accused the Community of using its alternative as a smoke screen to cover its opposition to actual CFC reductions. The international agreement eventually reached at

Table 18.1 (continued)

A taxonomy of goals and strategies

Targets	
Impacts	Generic
Model the impact of climate change on agriculture.	Research climate change.
	Monitor climate change.
	Establish a national research program.
	Establish a national climate program.
	Assess the severity of the climate problem.
	Develop better monitoring technologies.
Establish an adaptation program.	Promote international cooperation.
	Establish the IPCC.
	Set up an advisory committee to government.
	Change the education system to address the risk of climate change.
Persuade people to accept impacts of changing climate.	Inform government of climate risk.
	Promote public awareness of climate risk.
	Include climate change in basic education.
	Subsidize developing country attendance at international meetings.
	Create an international climate convention.

Montreal, however, was seen by the Europeans as vindication of their approach.

That agreement was made possible when the stalemate over which type of goals and strategies were best suited to protect the ozone layer was broken in 1986 and 1987. The turning point came when Canadians and Americans began arguing for a focus on comprehensive control of CFC emissions (as opposed to production alone or consumption alone) in 1986. Once that view acquired assent, negotiations proceeded that resulted in the 1987 Montreal Protocol, which mandated a 50 percent reduction in production and consumption of CFCs (and a freeze in halons), with consumption defined as production plus imports minus exports.

The Protocol's 50 percent target and the freeze on halons were not grounded in an assessment of consequences for the ozone layer. Rather, they were chosen for reasons of political expediency. However, the debate had already started to shift by mid-1986 toward a search for reduction targets that could guarantee protection of the ozone layer. U.S. EPA officials were estimating reduction targets based on this objective in 1986, and simple dynamic models developed in 1988 made it possible to determine the consequences of different reduction schedules on total chlorine loading of the stratosphere.

As a focus on total chlorine loading displaced arbitrary reduction percentages in the goals debate, an implicit objective of achieving a total load of 2 parts per billion (ppb) had emerged as clear focal point of international discussions by 1989 (see, for example, UNEP 1989). It was generally accepted among those involved in the negotiations that the stratospheric-chlorine concentration prior to formation of the ozone hole had been in the neighborhood of 1.5 to 2.0 ppb. This provided a convenient benchmark for evaluating alternative phaseout schedules and rapidly became transformed into a goal

94

Marc Levy, Jeannine Cavender-Bares, and William C. Clark et al.

that, if reached, was presumed to be adequate to eliminate the ozone hole (see, for example, WMO 1991).

The effect of this shift in goal setting was twofold. First, because it was clear that the initial signatories of the Protocol could not alone achieve a reduction in CFCs use to keep total chlorine loading under 2 ppb, it led to heightened efforts to broaden participation in the Montreal Protocol. This in turn led naturally to the adoption of the strategy of creating a mechanism to transfer financial resources to developing countries in exchange for their participation in the Protocol. This financial transfer strategy emerged strongly in 1990 and was cemented in the creation of a Multilateral Fund at the 1990 London meeting of the Protocol signatories.

The second effect of the shift to goals focused on chlorine loading in the stratosphere was to broaden the set of compounds targeted for reduction. By 1988 there were goals being proposed to add methyl chloroform to the list of substances being regulated, and by 1989 carbon tetrachloride was also considered. These compounds were regulated under the London Amendments, when steeper reduction schedules for CFCs were adopted. By 1992, methyl bromide had also been added to the list of targeted compounds, effectively expanding the internationally shared goal to encompass loading of the stratosphere by ozone-depleting chemicals in general.

The focus above on debates over reduction in the emission of ozone-depleting gases accurately reflects the dominance that such goals had in the overall discourse about how to respond to the ozone problem. The primary exception concerns goals aimed at helping people reduce the health consequences of a thinning ozone layer (for example, by limiting exposure to sunlight or by using sunscreens, dark glasses, and so on). But among the arenas we studied, only Canada and the United States seriously debated "personal protection" as a potential goal for responding to the risk of ozone depletion. And even in those arenas, goals targeted on reducing social impacts were either ridiculed or neglected relative to those targeted on emissions reduction or the environment.

18.2.2 A Specific Example: Banning CFCs in Spray Cans

The story of how the goal of reducing CFC use in spray cans spread within and among arenas illustrates some interesting mechanisms by which goals move through society. It reveals insights into where new goals get considered and adopted readily and where they find resistance. And it reveals some of the dynamics that sometimes lead initial goals to serve counterproductive roles in the ongoing debate over how to respond to a global environmental risk.

Emergence of the Goal The goal of targeting the aerosol sector for immediate CFC reductions originated the United States in 1974, shortly after publication of the Molina-Rowland paper. It is easy to understand why this goal emerged so quickly among actors most concerned about the ozone problem. Spray cans accounted for a major portion of the CFC market (especially in the United States), they could be portrayed as "bad" technologies (both hazardous and frivolous) for which substitutes were easily available, and the cost of implementing the goal was minuscule relative to the potential risk from damaging the ozone layer.[4] The initial idea arose with Rowland and Cicerone, two atmospheric chemists who were involved in their professional capacities only in risk assessment but who faced few barriers to making public-policy recommendations in 1974. An NGO picked up the call, using existing statutes to petition the federal government for action. Cicerone pushed successfully for his home town to implement a voluntary ban on cans using CFCs. By 1975 a number of state and local governments had taken unilateral action to label, restrict, or ban CFC-containing aerosols.

The rapid emergence of the spray-can initiative in the United States was clearly facilitated by the intense environmentalism of the time, the home-grown character of the Molina-Roland paper, and the porous boundaries among actor groups in the United States. There was little presumption that action would have to await decisions by the national government, then in disrepute over Vietnam and Watergate, and little onus was attached to an actor or group "meddling" in the affairs of others.

In settings where the risk management process was more "orderly" and actor roles more bounded, such as in the United Kingdom, the idea of an aerosol ban never took root. Goals and strategies focused instead on regulating production capacity.

Spread of the Goal The idea of banning CFC use in aerosols spread rapidly to other actors and arenas through a variety of channels. Bans or boycotts were being called for by NGO and green party groups in the Netherlands by 1975[5] and in Germany and elsewhere by 1976. A number of national governments were debating action by then, with the first binding regulatory measures adopted by the United States and Sweden in 1977. Germany developed voluntary reduction agreements with industry at the same time.

Once the goal of an aerosol ban emerged onto the political agenda in particular arenas, the subsequent dynamics and ultimate resolution varied substantially. In Germany the idea seems to have taken hold at the national governmental level following the April 1977 intergovernmental meeting in Washington and then acquired an essentially internal dynamic. The salience of the goal would rise and fall with the perceived importance of the problem but was not dependent on external influence for its viability. Proponents in the Federal Environment Agency (UBA) and the Interior Ministry kept it alive.

In the United Kingdom, in contrast, the goal was never adopted by any domestic actor until 1987. Before then it entered the agenda only when foreign events forced it there, as in the initial reaction to the U.S. ban and the E.C. debate on a CFC directive in 1979 and 1980. Even when domestic actors in Britain finally picked up on the goal in 1987, foreign influence was central. U.S. NGOs[6] traveled to the United Kingdom beginning in late 1986, trying to get British NGOs off the acid rain crusade and on the ozone campaign. They finally succeeded with Friends of the Earth U.K., which decided to launch a consumer boycott of spray cans containing CFCs in August 1987.

The European Community also saw the goal enter its agenda only due to external influence—namely, German pressure. Once this pressure abated, the issue dropped from the Community's agenda. The innovation appears not to have spread to the Soviet Union at all and probably not at all to Japan. International institutions did not figure significantly in the spread of this early goal.

Different settings appear to have thrown up different roadblocks to the selection of this goal. The United States and Canada show evidence of the least resistance. Their regulatory machinery seems to have proceeded at a reasonably fast pace and to have selected the goal as authoritative within two years of its entering the public debate.

For Germany the central sticking point was provided by its membership in the European Community. The government determined early on that a unilateral German ban on aerosol production using CFCs would not achieve the desired outcome because retailers could import CFC-containing cans from neighboring countries. Germany could not ban imports without violating EC trade regulations. Therefore, the pace of Germany's movement on this goal was determined by what it could achieve at the Community-wide level, which never surpassed a 30 percent reduction. The movement to a full phaseout of CFC aerosols in 1987 occurred very differently in Germany than in the United States or Canada. In response to the growing unease about the ozone hole, a number of German aerosol producers began unilaterally phasing out their use of CFCs. Many feared the possibility of either a consumer boycott or a government-mandated ban. Industry was taking cues from the government and the public, therefore, but actually was ahead of both in selecting the goal authoritatively.

In the United Kingdom the goal was never adopted by government actors. Instead, the selection mechanism operated strictly at the societal level. Friends of the Earth (FoE) managed to launch its boycott in 1987 with two serendipitous boosts. First, they located a publication of the British Aerosol Manufacturers' Association that listed all brand names and their propellants. As a result, the campaign required few resources: a volunteer was able to use the book to draw up a list of "good" and "bad" spray cans. This was published in the newspapers and printed up in pamphlets. (Contrast this with the experience in Germany, where NGOs considered a boycott in the 1970s but couldn't find out what was in the different cans.) Second, one advantage of being so late to the issue was that FoE launched the campaign at the height of the public awareness of the ozone hole, in the fall of 1987 and early 1988. After some initial resistance, aerosol manufacturers began first labeling their CFC-free products as ozone friendly to escape the boycott and then switching propellants. By mid-1988 U.K. aerosols were CFC free, with the government taking a back seat in the whole affair.

By the end of 1987, the process was essentially over. The Montreal Protocol required reductions in production of CFCs, which obviated the need for sector-specific, first-step measures such as the aerosol ban. The goal of reducing aerosol use did continue to spread after 1988 but as part of a strategy to implement the Montreal Protocol and its amendments and not as an effort to take a precautionary first step.

Evaluating the Goal How can we evaluate the role of the aerosol ban in the broader process of responding to the risk of ozone depletion? The judgment is complex. In some settings, the goal served an expedient purpose but was otherwise essentially neutral with respect to the broader risk management process. Where the goal was adopted quickly and implemented completely (as in the United States and Canada), it served the function of reducing CFC consumption significantly and then disappearing from the public debate. In effect, these countries took advantage of an absence of barriers to goal adoption to take early action during a temporary period of high concern. On the other hand, once adopted this simple, straightforward goal often proved an obstacle to a wider, more inclusive conceptualization of the ozone-management issue, as was certainly the case in the United States.

In settings beset with tougher obstacles to goal selection, the goal was still being debated when the falling risk assessments in the early 1980s pushed the risk of ozone depletion off the agenda of national and international politics. In Germany scientists in the German Federal Environment Agency (UBA) and officials in the Interior Ministry who wanted to phase out CFCs in aerosols had to keep quiet until developments later in the decade

96

Marc Levy, Jeannine Cavender-Bares, and William C. Clark et al.

returned ozone depletion to center stage. In these latter settings the case could be made that the survival of the aerosol goal into the late 1980s was not a progressive result. When the overall debate had shifted toward competing goals for reducing total chlorine delivery to the stratosphere (by late 1987), active discussion of spray cans was at best a diversion and at worst a blinder.

18.3 Findings: Acid Rain

18.3.1 The Big Picture

In the case of acid rain, the goals and strategies adopted by most actors in most arenas were, by the end of the 1980s, remarkably similar. There was a clear convergence toward advocating the goal of emission reductions of both sulfur dioxide (SO_2) and nitrogen oxides (NO_x). In some countries the articulation of these basic goals, especially by governments, came earlier than others. In some countries the strategies embraced to achieve the basic goals changed over the decade. But across the countries the goals were very much the same by decade's end.

In contrast, the initial goals of national actors derived largely from the extent to which they perceived themselves as victims of transboundary air pollution or perceived themselves to be innocent of the blame attributed by others. From the beginning, Canada—like Sweden and Norway—saw acid rain damage more as the result of transboundary flows than as a domestic problem. All three consistently and vigorously called for cuts in the precursors of acid rain originating in "upwind" countries. The United States and the United Kingdom saw themselves as relatively unaffected by emissions of others. They never much worried about foreign emissions, remained for most of the 1980s officially unconvinced of the need for domestic controls beyond those already in place, and saw themselves as innocent until proven guilty of the accusations that they were responsible for environmental damage in other countries. Both stressed instead the need for more scientific research to determine the extent and causes of acid rain. The national governments of both countries came under increasing pressure from both domestic and external sources through the 1980s, however, and both eventually officially endorsed the goal of sulfur dioxide reductions.

In Germany tentative concerns about reducing emissions and about being seen as the source of acid rain in Scandinavia quickly gave way in 1981 and 1982 to overwhelming concerns about the impact of acid rain, both domestic and imported, on German forests. Germany was correspondingly the only case prior to the late 1980s of a dramatic shift in national goals, from opposing to promoting sulfur dioxide emission reductions. The shift came in part because of scientific evidence and in part because of domestic political factors (that is, the need of other parties to court the Green Party's agenda). It was also facilitated by the fact that Germany, unlike the United States and the United Kingdom but like Canada and the Scandinavian countries, was significantly affected by transboundary flows.

The Mexican case is also somewhat different in that the sulfur dioxide problem was seen as an urban air-pollution matter (largely in Mexico City) and became a long-range transport issue only after the United States pressed Mexico to put emissions controls on two large copper smelters located near the border with Texas. The overall pattern was the same, however, as Mexico accepted the need for controls and reductions.

The goals espoused by actors other than governments were even more consistent across the countries and less tied to the nature of transboundary flows. Scientists in all countries initially advocated research programs to investigate acid rain and soon were advocating emission-reduction programs. The reasons for these reductions varied but the content did not. (In Scandinavia and Canada, for example, the rationale was mainly to protect aquatic systems; in Germany, to protect forests.) NGOs similarly generally pushed reductions and, as a strategy, tried to broaden public awareness. In some countries at some points the NGOs were less active due to competing concerns; in Germany, for example, environmental groups were concerned that advocating acid rain controls not put them into the position of appearing to favor nuclear power, against which many had long fought. Industries, on the other hand, operating from their own conceptions of their own self-interests, tended to oppose emissions reductions. Those industries in countries that were significant importers of pollutants also tended to look for emission reductions on the part of other countries, which would ensure they did not bear all the brunt of the environmental action.

18.3.2 A Specific Example: Critical Loads

The critical-load concept in the acid rain debate involved the proposition that there exists a maximum amount of acidic deposition that can be tolerated by species or ecosystems without significant harmful effects. During the 1980s, the goal of keeping acid deposition below critical loads spread over many but not all of our arenas. The story of how this occurred illustrates the operation of a carefully planned, deliberate effort to propagate a goal. It reveals the dynamics involved when a debate over goals that is initially stuck on arbitrary reduction targets is liberated through a change in the conceptualization of goals. Finally, it provides evidence for the role of highly ambitious reference goals in prompting changes that reverberate across the risk management, option assessment, and monitoring functions.

Emergence of the Goal The idea of a scientific "threshold" concentration of pollutants above which adverse environmental effects would be manifest had been established in principle by the time of the Stockholm Conference and was widely applied in management of freshwater quality. By 1980 Canada and the United States had cast their Memorandum of Intent on acid rain in terms that put "appropriate deposition rates" at the center of the policy discussion. Swedish scientists began actively promoting the concept of an acid-deposition threshold in 1982 and formulated this in terms of critical loads beginning in 1985. Goals based on critical loads were put forward as superior to goals focused on flat-rate emission reductions on a number of grounds: they gave credit for prior reductions rather than penalizing them; they were based on scientific understanding and capable of revision as science evolved; and they permitted reductions to vary as appropriate with variations in sensitivity to acidification, instead of mandating equal protection for highly vulnerable and unthreatened regions.

Spread of the Goal Consensus among natural scientists regarding the merits of using critical loads as the basis for managing the acid rain problem spread rapidly if not uniformly among our study arenas in the latter half of the 1980s. A 1988 workshop generated a consensus on the various practical matters involved in defining and measuring critical loads. This strong scientific consensus resonated with and reinforced developments in the international policy realm. The critical-load goal surfaced in European political circles at the 1985 Helsinki Conference, was pushed thereafter by the Nordic Council of Ministers, and came under consideration by the Convention on Long-Range Transboundary Air Pollution (LRTAP) in 1986. In 1988 the LRTAP Convention's Nitrogen Oxides Protocol contained an explicit statement that subsequent NO_x regulations would be based on critical loads goals. Such a commitment was written into the 1991 Volatile Organic Compound (VOC) Protocol as well. By the time the second Sulfur Protocol was being negotiated beginning in 1991, critical-load thinking had come to dominate the formulation of international goals for the management of acid substances in Europe (Levy 1993; Staaf 1992; Ågren 1992).[7]

At the national level, the use of critical-load goals spread beyond Sweden but with substantial variation in the nature and extent of its adoption. The most thorough embrace took place in the Netherlands, which began setting deposition goals in terms of critical loads beginning in 1983. The United Kingdom adopted the strategic goal of not exceeding critical loads in 1990 but did not set specific target loads as was done in the Netherlands.

Two circumstances characterize the arenas in our study group for which critical-load goals failed to take hold outside of the international LRTAP framework. First, where acid rain was low on the domestic political agenda, the impetus necessary to introduce relatively radical critical-load goals simply failed to develop. This was the case in the former Soviet Union, Hungary, Mexico, and Japan. Second, even where acid rain was high on the political agenda, arenas with strongly developed policy styles emphasizing technology standards and emission limits found the move to critical-load goals uncompelling. This was the case in the European Community, Germany, and the United States.

Even where national emission regulations were not formulated in terms of critical loads, the goal of creating the capacity to assess acid rain options in terms of critical loads was widely shared. Acid rain researchers spent years preparing maps of critical loads, often requiring significant levels of effort devoted to monitoring and risk assessment. Detailed maps were prepared in many countries, including Germany, the Netherlands, and the United Kingdom. Critical-load data were incorporated into various national and international models, which then made it possible for negotiators to assess alternative reduction scenarios in terms of the degree to which they attained critical-load goals. Such assessments had become the dominant frame for European policy discussions of acid rain by 1991 (Amann 1992).

Evaluating the Goal The critical-load process has not lived up to the promise some held for it of making the policy-making process scientifically grounded and subject to politics. Politics proved quite capable of intruding itself into the process, and the scientific foundations of the concept remained contentious on technical grounds as well.

But judged by the criteria of whether progressive change in the broader management of acid rain came about, the critical-load approach was clearly a success. In the case of the United Kingdom, efforts to create critical-load maps led to discoveries of threatened ecosystems that completely shifted the debate over acid rain from one that focused on whether the United Kingdom deserved the "dirty man of Europe" label to one that focused on the degree of protection desirable for British ecosystems. This latter debate helped dislodge stale thinking about what levels of reductions would be possible or desirable. Whereas Britain fought very hard against a European Community directive on sulfur and nitrogen emissions from large combustion plants (eventually signed in 1988), in 1990 it committed itself in principle to adopting even stricter emission-reduction standards based on critical loads.

In the Netherlands the shift to critical loads played a role in broadening the focus of the acid rain debate to include agricultural sources of acidification. As critical loads were reassessed downward during the second half of the 1980s, in light of changing scientific measurements and understanding, pressure to seek more ambitious reduction targets increased. This orderly process of goal adjustment seems to have been at least in part responsible for bringing an increasing range of actors into the debate and for the development of a creative mix of strategies.

The contrast between the Dutch and British experience with goals targeted on environmental effects (that is, critical loads) and the United States' experience with goals targeted on emissions is striking. The U.S. debate became focused at an early stage on the goal of reducing national emissions of sulfur oxide by 10 million tons. Whatever scientific doubts persisted over the meaning of critical-load targets adopted in other countries, the U.S. 10 million ton reduction target for sulfur oxide had almost no scientific foundation at all. A serious debate on whether it was too strict or too lax for achieving environmental protection never materialized. Instead, a decade was spent arguing over whether the science was sufficiently "certain" (whatever that would mean) to justify action on the goal and who should pay for such action as was undertaken. Little change in the framing of the debate occurred over the decade with regard to either the actors engaged, the dangers posed, or the options considered.

18.4 Findings: Climate Change

18.4.1 The Big Picture

The goals and strategies invoked for managing the risk of climate change changed greatly over the period addressed in this study. The earliest sustained commitments to goals and strategies for managing the risk proposed research programs to produce better understanding. By the late 1970s and early 1980s, additional strategies began to emerge linking climate considerations to goals of energy development. Finally, starting in the late 1980s but only beginning to take hold internationally by the time of the Rio Conference, specific goals for the reduction of greenhouse gas emissions began to emerge. We expand on these general patterns immediately below and then turn to a more detailed look at the origins and fate of the "Toronto goal" for emission reduction.

The first formal goals for addressing the question of global climate change were advanced by small groups of natural scientists striving to establish government support for research and monitoring programs. Individuals such as Mikhail Budyko in the Soviet Union, Roger Revelle in the United States, Bert Bolin in Sweden, Hubert Lamb in the United Kingdom, and Hermann Flohn in Germany played central though largely independent roles in promoting research through the first decade of our study period. By the early 1970s, these and other scientists had enlisted international organizations in pursuit of their goals of better understanding of the climate change issue. The International Council of Scientific Unions' (ICSU) Scientific Committee on Problems of the Environment (SCOPE) established a formal goal of deepening international understanding of the global biogeochemical cycles in general and the carbon cycle in particular by the mid-1970s. The World Meteorological Organization's (WMO) commitment to what became the World Climate Program was already taking shape in the second half of the 1970s and was formalized at the first World Climate Conference in 1979. A few national governments adopted national goals for research into global climate change during the mid-1970s, notably the United States and the USSR. For most other arenas in this study, however, the seminal event in translating individual into national objectives for research on the climate issue was the First World Climate Conference. Largely in response to this meeting and the events surrounding it, all of the arenas in our study group had by the early 1980s articulated and begun to implement national goals for research and monitoring on climate change. Most of these programs expanded through the 1980s, with countries as different as Germany, Mexico, and the United States adding explicit public-education goals to their research programs by the early 1990s.

Connections between the risks of greenhouse gas emissions and energy policy were stressed by a range of actors including both nuclear energy and conservation advocates from the mid-1970s onward. In places and times where debates over comprehensive energy policies occupied a serious and prominent place on national political agendas, they occasionally carried climate issues to national prominence. In Sweden and Germany, for example, where serious energy debates were underway in the mid-1980s, there was a close interaction between proposed goals for handling the greenhouse problem and proposed goals for restructuring energy systems. If the result was most often cast as a primary goal of promoting some energy policy, justified in terms that included mitigating the risks of greenhouse gas emissions, the relationship nonetheless propelled climate concerns upward on the relevant national agendas. This situation contrasts sharply with the case of countries such as the United

States where high-level debates over national energy policy were more transient and less intense. In such circumstances, the climate debate remained unable through the early 1980s to generate much widespread enthusiasm for anything beyond goals of improved understanding. Small advocacy groups continued to push for a more serious consideration of national climate goals but to little immediate effect.

The growing attention to goals and strategies for sustainable development, culminating in the Brundtland report of 1987, provided another potential forum for the emergence of substantive goals for managing the risk of climate change. Such risks figured prominently in the report, which doubtless served to bring some knowledge of climate change to a broad development audience. Paradoxically, however, it was the climate issue rather than sustainable development that took center stage in the late 1980s and early 1990s, with the Framework Convention on Climate Change hijacking much of the attention at what should have been Rio's U.N. Conference on Environment and Development in 1992. How this came about is a complex story told in part in several chapters of this book. It is nonetheless a story in which the emergence and spread of a particular goal formulation for the reduction of greenhouse gas emissions played a central role. We sketch that story in the following section.

18.4.2 A Specific Example: The Toronto Goal

Prior to 1988, no government had announced a goal of specific, scheduled reductions in greenhouse gas emissions.[8] In the summer of that year, however, members of the NGO and expert communities inserted the goal of a 20 percent reduction in carbon dioxide emissions by the year 2005 into the final statement of the Toronto Conference on Our Changing Atmosphere. This goal statement lacked any supporting analysis suggesting what it would accomplish or what it would cost. It nonetheless precipitated an avalanche of governmental commitments to reduce the emissions of greenhouse gases. This was so complete that by the time of the Rio Conference in 1992, virtually every industrialized country was willing to endorse a goal of holding greenhouse gas emissions to 1990 levels by the year 2000. The Toronto goal thus marked a pivotal shift in the climate debate from its focus on scientific assessment and general discussions about the need to abate the risk to a focus on political debate about specific goals and strategies of emissions reduction.

The story of how the goal for greenhouse gas reductions developed at the 1988 Toronto Conference shows how varied and far reaching the repercussions of certain goals can be. It highlights the varied impact of an internationally, but not intergovernmentally, established goal. In

the case of Toronto, the goal sparked national response assessments in several countries and led to direct implementation in various municipalities around the world. In addition to the goal's widespread diffusion in various forms, it has also opened up the debate in several countries to consider options for other greenhouse gases and carbon dioxide sinks.

Formulation of the Goal For a goal with such far-reaching effects, however, the process by which Toronto's "20 percent reduction" was formulated was not well grounded in scientific calculations of the risk or assessments of the impact such a goal would have on achieving desired environmental conditions. Instead, as is so often the case in formulating goals, a compromise target was set that was both technically simple to specify and politically vague or innocuous enough to be widely acceptable.

The goal-formulation process itself was initiated by the NGO community at Toronto and continued by the other delegates: a mix of scientists, government officials, and industry representatives. At the start of the conference, after Norway's Prime Minister Gro Harlem Brundtland and Canada's Prime Minister Brian Mulroney both called for a global convention in their keynote addresses, the NGO community was alerted to the opportunity to push the conference toward setting a concrete goal. At the time, however, few technical studies had been done to calculate what changes would result if emissions were reduced by certain levels. There was a crude notion that a 50 percent carbon dioxide reduction would have some effect in stabilizing CO_2 concentrations in the atmosphere, but 50 percent was clearly too ambitious in political terms. Instead, the NGOs chose a pragmatic target for CO_2 reductions of 20 percent to be met by the year 2000. Additional options of including other greenhouse gases or any means of increasing CO_2 sinks were never considered. Subsequent negotiations with representatives of the energy community also present at the conference moved the target date to 2005 but otherwise stuck with the initial NGO position.[9]

Most saw the 20 percent target as only an initial step, and, indeed, the Conference Statement suggested that a 50 percent reduction of 1988 CO_2 emission levels would be required to stabilize atmospheric CO_2 concentrations. Nonetheless, it was the 20 percent figure that stuck in peoples' minds and that influenced subsequent developments.

Spread of the Goal Within Canada itself, the Toronto goal sparked an array of immediate responses, initiating a widespread debate on the appropriate action to take. A government Task Force decided that the 20 percent reduction goal would be too ambitious but suggested that

100

Marc Levy, Jeannine Cavender-Bares, and William C. Clark et al.

attention and effort should be devoted to conservation and energy-efficiency measures, renewable energy, alternative fuels, alternative energy, and examining CO_2 sinks.[10] Subsequent debate among Canadian domestic interests focused public attention on the issue, initiated response assessments, and led to policy proposals, the net result of which was Canada's CO_2 stabilization goal adopted in the Framework Convention on Climate Change in 1992.[11] The goal also found its way directly to municipal governments. Both Toronto and Vancouver, two of Canada's three largest municipalities, adopted the goal wholesale for their cities in 1990 (see box 18.2).

The Toronto goal's impact was not restricted to its country of origin. In Germany, the recently established Enquete Commission on Protection of the Earth's Atmosphere took the goal as a potentially reasonable starting point and asked the question, "Can we achieve this in Germany?" The Commission undertook a major option assessment to determine the feasibility and impact the goal could have. It determined that in fact a 30 percent carbon dioxide reduction was possible by 2005. The federal Cabinet subsequently agreed in 1990 on a compromise goal of 25 percent CO_2 reduction by 2005, the most ambitious national goal adopted at that time. An Interministerial Task Force was established to create a national strategy for meeting the goal. As was the case in other countries, local and regional governments adopted the goal, and it was at the subnational scale that real measures were implemented to meet the goal.

In the Netherlands, as in Germany, the Toronto goal stimulated a careful option assessment on whether the target would be feasible at the national level. Again, the assessment resulted in modifications of the goal, this time proposing a comprehensive approach that would include other greenhouse gases besides CO_2. The Dutch Minister of the Environment, who had announced the upcoming Nordwijk Conference at Toronto, was instrumental in carrying the momentum from Toronto to the first rounds of intergovernmental negotiations in Nordwijk. Although the Toronto goal was not the direct motivation for the Netherlands' eventual commitment in 1992 to reduce CO_2 emissions 3 to 5 percent annually by 2000, it did shift the political debate toward the development of a national policy. In addition, simultaneous local developments occurred along the lines of those in Canada and Germany.

In the United States, the Toronto goal served to initiate a debate about policy options on climate change. Entrepreneurs in the United States, including Senator Wirth of Colorado, who had attended the Toronto Conference, proposed the goal to Congress.[12] No national emission-reduction goal emerged in the United States as a consequence of Toronto. As elsewhere, the Toronto goal had its most tangible effects at the local level. For example, by 1991, the Los Angeles Department of Water and Power and Southern California Edison, two California electric utilities, had announced a goal of reducing carbon dioxide emissions by 20 percent from 1989 levels over the next two decades.

In the United Kingdom, the Toronto goal had little impact on the adoption of national goals. Official government policy remained, unsurprisingly and plausibly, that more scientific study was needed before specific reduction targets and timetables could be justified.[13] NGOs nonetheless used the goal, with limited success, to initiate or influence political debate. For example, in a formal report of Friends of the Earth in 1988, the British NGO explained warming trends and future impacts of climate change and offered clear policy recommendations for the United Kingdom and other governments. The main goal statement in its policy recommendations urged industrial nations to reduce carbon dioxide emissions by 20 percent of 1988 levels by 2005 and by 50 percent by 2015. The most that can be said for the Toronto goal's fate in the United Kingdom, however, is that it served indirectly as one of several international stimuli for the eventual development of a global environmental agenda in the Thatcher government.

In other arenas, the Toronto goal had even less effect. Although two Japanese academics attended the Toronto Conference, it is not clear whether the results of the Conference had any reverberations in Japan. In 1988 the Japanese began to take an interest in understanding the nature of the risk and its possible impacts, but it seems

Box 18.2
Local Responses to Global Goals

Ralph Torrie, who rapporteured the Energy Working Group at Toronto, subsequently initiated the International Council of Local Environmental Initiatives (ICLEI) in 1990. A major program of the ICLEI was the Urban Carbon Dioxide Reduction Project in which fourteen municipalities of the United States, Canada, Europe, and Turkey pledged to reduce energy intensity to CO_2 emissions. The Climate Alliance of the Cities was initiated by several European local governments and indigenous peoples of the rainforest. The cities committed themselves in 1990 to reducing CO_2 emissions by 50 percent by the year 2010, to stop using CFCs immediately, and to boycott the use of tropical wood. By 1994, 124 German municipalities, 117 cities in the Netherlands, and almost 130 other cities had joined (Klima Bündnis der Städte, ed., *Klima Bündnis zum Erhalt der Erdatmosphäre. Manifest europäischer Städte zum Bündnis mit den Indianervoelkern Amazoniens,* cited in Beuermann and Jäger 1996.

unlikely that this new interest was connected to Toronto. Representatives from Hungary, Mexico, and the Soviet Union attended the Conference, but attention paid to the resulting goal was short-lived in all of these cases. The Hungarians immediately agreed to the 20 percent goal but ultimately committed to a less ambitious stabilization goal that their economic recession had already achieved. The Mexicans also initially agreed to the goal in principle; not unexpectedly, however, the goal vanished from the political landscape shortly thereafter. By 1992, the only trace of the earlier enthusiasm was a commitment to planting trees. The Soviets were opposed to any notion of a carbon dioxide emission reduction from the start, although Soviet scientists continued to support strategies of afforestation and increasing carbon uptake in the oceans through ocean pollution control.

Evaluating the Goal The Toronto goal resonated deeply and widely through our study arenas. In evaluating its remarkable impact, however, three important external factors need to be kept in mind. These increased the potential for agreement on an international goal in Toronto and helped diffuse the goal once it was formulated. First, international public and political attention to the climate issue was already on the rise at the time of the Toronto Conference and had been growing for the previous two to three years. Second, in the week prior to the Toronto Conference, the 1988 G7 Economic Summit was held in Toronto. The result was that some 400 members of the international press who had come to Toronto to cover the G7 meeting stayed on to cover the Toronto Conference on Climate Change. This not only put pressure on the attendants of the Conference to work toward a policy goal, but it amplified global coverage of the goal itself giving it a life of its own. Finally, the North American hot summer of 1988 had been well publicized throughout the world. One week prior to the Toronto Conference, scientist James Hansen testified before the U.S. Congress with claims that the current warming trend was the result of the anthropogenic greenhouse effect. The heat wave, which persisted in Canada throughout October, made Hansen's testimony very believable. At least to an extent, this diverted international attention away from the uncertainties in the causes and impacts of climate change and focused them on the urgency of the risk. In summary, the Toronto Conference was held at a time—partially by design—when a wide range of relevant actors were ready for something tangible to do about the climate problem. To a first approximation, it may well be that just about any goal would have done as well as that actually adopted.

Following the Conference, in countries with a domestic predilection to take action on climate change—Germany, Canada, and the Netherlands, among our case studies—domestic advocates used the Toronto goal to promote and justify national responses. The goal also turned out to be very effective in initiating domestic debate, particularly at the local level where measures to meet the goal were actually implemented in numerous instances. In a wide range of countries, response assessments were undertaken as a direct result of the goal, which thrust the domestic debate forward and led to significant increases in political commitment. Because the original goal formulation was rather arbitrary, these domestic processes played an important role in legitimizing the goal and connecting it to scientific and economic assessments of outcome and feasibility. In some countries, however, the fact that the goal-formulation process was not seen as legitimate became an insurmountable barrier to further official action. This was clearly the case in Britain, where the Toronto goal was rejected outright for its arbitrary nature and because it was seen to have been formulated primarily by a group of NGOs.

Significantly, the Toronto goal seems to have avoided the "stickiness" pitfall that limited the effectiveness of such goals as the CFC-aerosol ban and the United States' 10 million ton goal for sulfur oxide reductions. Rather than limiting further discussion, the Toronto goal fostered assessments in a number of countries that actually broadened the debate and provided a stimulus for actors to consider options that they found more attractive for achieving some abatement of the climate change risk. Thus, tree planting and taking a comprehensive approach to count other greenhouse gases were not eliminated from the pool of legitimate goals. Specific reduction targets and years for carbon dioxide turned out also to be quite flexible. This flexibility may be partially a result of the continual series of international negotiations leading up to the Framework Convention in Rio. It may even reflect in part the nongovernmental character of the goal. Did this allow various state and nonstate actors to focus on discovering for themselves whether they could or should pursue some variant of such a goal rather than immediately pronounce reasons to adopt or reject a formal proposition?

Whatever the reasons behind the impacts of the Toronto goal, our summary evaluation is that the goal was good and "progressive" by almost any standards imaginable. It substantially widened the base of participation in deliberations on the climate problem, entraining not only scientists but much of the sustainable development community as well. It even, if temporarily, bridged the gulf between nuclear-energy advocates and "soft-path" NGOs. The Toronto goal clearly promoted a deeper consideration of the climate problem, particularly through the stimulation of a new generation of focused

102

Marc Levy, Jeannine Cavender-Bares, and William C. Clark et al.

and realistic option assessments. And if it did not quite reframe the policy debate in collective problem-solving terms, the very ambiguity of the goal's form and origins let it temporarily avoid casting climate policy as the distributional conflict that would resurface at the Rio Conference four years later.

18.5   Emergent Patterns and Possible Explanations

This section highlights the general patterns and relationships involving goal and strategy formulation that hold across a substantial fraction of the cases and arenas we have addressed in this study. We thus set them forth as tentative "models" or hypotheses about the role of goal and strategy formulation that we might expect to be applicable for other countries and issues as well. We begin with a broad sketch of overall patterns before turning to an account of goal and strategy dynamics: How and to what effect did patterns change over the course of the issue histories reviewed here? Finally, we then turn to a more detailed examination of several patterns revealed in those changes.

18.5.1   Overall Patterns

Most goal statements recorded in this research were proposed in association with, or as implicit afterthoughts to, specific strategies for their pursuit. In other words, most advocates of action were not indifferent to the means by which their goals were achieved. The fraction of goals associated with particular strategies was highest in the ozone case, where perhaps three-quarters of declared targets were coupled to preferred instruments (such as "reduce CFC production through taxation"). It was lowest for acid rain, where only half the goals were advanced in association with particular measures to attain them.

As suggested in box 18.1 and table 18.1, our study coded for goals targeted on emissions, environment, impacts on society, and generic efforts addressing the risk as a whole. Over the entire data set, goals were targeted on emission reduction about 60 percent of the time, generic targets (mostly involving nonspecific research and monitoring) appearing about half as often. Relatively few instances of goals targeted on the environment (such as lake liming) or social impacts (such as UV warnings) were encountered. The ozone-depletion case exhibited the greatest dominance by emission-reduction goals (on the order of three quarters), while the climate case showed the least (perhaps 40 percent).

The table and box also show the strategies or means for goal achievement encountered in the study. We coded for several indirect strategies that sought an effect through enhancing social capacity to deal with the risk

and several direct strategies that sought to alter events through incentives or regulatory commands. Indirect, capacity-building strategies dominated the overall picture, appearing in perhaps 60 percent of the statements coded. Incentives—market and informational—were next most popular, appearing in about a quarter of the strategy statements. Command-and-control strategies, though often implicit in goal formulations, appeared explicitly only half as often as incentive-based strategies. The highest frequency of capacity-building strategies occurred in the climate case (perhaps two-thirds); the highest frequency of command-and-control strategies, in ozone (perhaps one-third).

Finally, the goals and strategies that achieved the most play in the histories studied here had multiple origins. No particular country or arena and no particular actor group led or dominated the process. The often repeated premise that the formulation of "authoritative" goals is the exclusive purview of state actors finds no support at all in this study. On the contrary, as reflected in our overall database and the specific stories related above, NGOs, scientists, and management professionals may well have played a more influential role than state actors in the formulation of the goals and strategies that shaped the management of our global environmental risks. State actors appear less as leaders than as followers reduced to sanctioning social processes that had already taken on their own momentum.

18.5.2   Dynamics: From Generic Capacity Building to Direct Management of Emissions

This study revealed a temporal progression of management goals and strategies that was repeated in most arenas for most of the cases studied. In the early stages of issue development, most goals and strategies focused on generically targeted strategies to address the risk indirectly through enhancing understanding and management capacity. As the issues matured, such capacity-building measures were complemented by goals targeted on emission reduction, direct management strategies employing incentives, and, to a lesser extent, regulatory commands.

Early Capacity Building   In the early stages of development of the issues studied here, goal and strategy statements tended to be targeted generically on the problem as a whole and to emphasize the need for capacity-building measures rather than measures that would change the risk directly.[14] The most common strategy in these early years was to build the programs and institutions that would promote research and development on the problem. This initial emphasis is not particularly surprising, given the novelty of the risks we addressed. Moreover, a variety of actors had motivations to push toward the same general goals of enhanced understanding of the problem through more research and development (R&D): scientists wanted to discover what was happening, research managers wanted to build programs, technocrats wanted to develop solutions, activists wanted to promote concern, and politicians wanted to demonstrate responsible—but relatively inexpensive—behavior.

As the issues matured and their potential policy implications began to emerge, this broad coalition of R&D advocates almost inevitably became advocates for raising concern and attention within their own technical communities, among political leaders, and in the public at large. This advocacy of increasing concern was often focused on the need for better R&D resources and did not necessarily include overt advocacy of immediate policy action. But the pathway from increased concern to increased pressure for direct action was not difficult to imagine. Not surprisingly, then, the R&D advocates were joined in due course by others skeptical of the science or opposing the need for action or both. In particular, the first entry of industrial actors into the goal and strategy debate was often to put forth statements emphasizing the need for more R&D before any regulatory action could responsibly be taken. Those national governments inclined to give the views of industry a sympathetic hearing regularly echoed such goals, often resulting in even greater R&D support than was available under more activist governments.

This strategic advocacy of goals and strategies oriented toward R&D as an alternative to immediate direct action was ubiquitous in the cases examined in this study. Nonetheless, the evidence suggests that such strategies provided only modest delays even where they worked best. Explicit delay strategies were promulgated in the U.S. and U.K. acid rain cases and in most arenas in the ozone case. While actors that wanted to delay action on acid rain and ozone were able to do so, it is striking that even the most strident opponents of action all eventually shifted to emission-reduction goals that were essentially equivalent to those advocated by supporters of action. The time gained through R&D-justified delays in direct action was not of trivial significance to politicians and industrialists opposing immediate action on the risks studied here. In the United States, for example, the Reagan administration put off action on acid rain throughout its entire eight-year regime, relying in large measure on generous R&D support as a delaying tactic. But in the broader social context, taking a slightly longer view, the patterns in these cases reveal that (1) a decade is about the longest delay possible through "R&D first" strategies and (2) when the delay finally does end, the actions taken are unlikely to be much less severe than those

initially considered. In many cases, it appears that when R&D was carried out in part to delay action, the delay served inadvertently to increase the ambitiousness of the action goals considered when the delay stopped. This almost certainly occurred, for example, with U.K. research into domestic acid rain effects and in multilateral assessments of ozone depletion.

In addition to strategies aimed at increasing society's cognitive capacity to deal with risks through R&D, strategies of promoting increased institutional capacity also surfaced early in the issue histories studied here. These developments were linked, for in general the initial institutional focus was on coordinating relevant research and monitoring programs. This occurred both within countries and internationally. As related in part II of this book, many of the institutions for coordination of research and monitoring set up in response to these early goal declarations eventually grew in scope to include assessment and action as well (for example, the LRTAP apparatus). This pattern carried with it the potential risk that institutions built to handle R&D would eventually create obstacles to direct action. In most instances, however, the transition went surprisingly smoothly, with the R&D institutions either growing into their direct action roles or handing off such responsibility to other, more appropriately structured institutions as the need emerged.

By the mid-1980s, goal and strategy statements at both the domestic and international levels increasingly came to emphasize the need for enhanced institutional capacity to develop and implement international agreements on emissions reductions. Such international capacity-building strategies were particularly evident during the intensive institutional overhauls that many governmental environment programs undertook in the late 1980s and early 1990s. For example, the Canadian government's 1990 Green Plan was well populated with capacity-building measures, as was the legislation accompanying the overhaul of British environmental policy in the same year.

Later Direct Management Despite the early and sustained attention to capacity building, in each case examined in this study goals and strategies for capacity building eventually became less dominant. As the issues matured, the capacity-building measures were increasingly complemented by a proliferation of declarations aimed at directly managing the risk in question. The vast majority of the direct-management goals were targeted on emission reduction. Goals to protect or restore the environment or to lessen social impacts played little role. The patterns of, and possible explanations for, this dominance of the management debate by emission-reduction

104

Marc Levy, Jeannine Cavender-Bares, and William C. Clark et al.

goals are discussed in some detail later in this section. Suffice it to note here that the earliest goals targeted on emission reductions tended to focus on particular technologies: "bans" in the ozone case and exhortations for expanded use of less polluting alternatives in the acid rain and climate cases. Later—in all cases by the early 1980s—goals more generically targeted on "reducing emissions" began to appear. Finally, as issues emerged onto the broader public and political agenda, goal statements specifying targets and timetables for emission reductions became more common. This was occurring for acid rain (in Europe) by the mid-1980s and for ozone depletion a few years later, and it was still being debated for climate change at the Rio Conference in 1992.

The only serious challenge to the dominance of emission-targeted goals developed late in our study period as the political debate on each issue matured and a cluster of environment-targeted goals began to surface. In the acid rain case, as discussed earlier in this chapter, the late 1980s witnessed in some arenas a shift away from setting flat-rate reduction targets and toward goals framed in terms of "critical loads."[15] In the ozone case, sustaining suitably high ozone concentrations by keeping ozone-depleting substances in the stratosphere below a critical concentration became the motive behind the amendments to the Montreal Protocol. In the climate case, "tolerable rates of climate change" were proposed as a goal for management in the late 1980s but had not been widely adopted by the end of the study period. In no case, however, did goals targeted on environment fully overtake goals targeted on emissions in the overall social response to global environmental risks. Even in acid rain, where critical loads were employed most rigorously to guide policy, they did so by provoking a renewed examination of the adequacy of goals to reduce various emissions. A similar effect occurred with the chlorine-loading goal. It exerted a significant impact, but on a day-to-day basis, people active in developing a response to the ozone problem concerned themselves with what emissions to reduce, by how much, and by what year.

A Note on "Stickiness" and Convergence Two further patterns suggested by our data on goals and strategies deserve mention. First is what we have called *stickiness:* the tendency of specific goals and strategy formulations to persist longer than might be expected on grounds of relevance to the current debate or of helpfulness in shaping effective social responses. Examples discussed at length elsewhere in this chapter include the fixation of the ozone debate on reducing the use of CFC-containing aerosols, the persistence of the U.S. goal of a 10 million ton reduction in sulfur dioxide emissions, and, perhaps, the almost exclusive preoccupation of the climate debate with goals

of emission reduction. While such stickiness in goals and strategies may be less than optimally rational, however, it is not unexpected. In fact, the difficulty and rarity of changing goals to reflect new knowledge and experience are the central observations and predictions of the theories of learning reviewed in chapter 1. From that theoretical perspective, the stickiness we observed is precisely what a student of organizational, social, or policy learning from another field would have expected. We explore the implications of these observations in some detail later in this volume in chapter 20 (on evaluation), chapter 21 (on knowledge and action), and chapter 22 (on conclusions and implications for the future).

A second and not unrelated pattern in our goal and strategy data we call *convergence*. Although emission-reduction targets and capacity-enhancing measures dominated the overall picture, the specific goals and strategies advanced in individual arenas and actor groups initially differed substantially. Other things being equal, the stickiness noted above would have caused these differences to persist. In fact, however, all of the cases we studied were marked by episodes of rapid and widespread convergence across arenas on particular goal and even strategy formulations. These "globally" shared formulations tended to displace specific "local" ones. Moreover, these shared goals and strategies thereafter changed only slowly, if at all. Detailed patterns and possible explanations for this cross-arena convergence of goals and strategies are discussed later in this section. Implications for future goal and strategy formulation and management are dealt with in the final section of the chapter.

18.5.3   Why Do Goals Targeted on Emission Reductions Dominate?

As suggested in box 18.1 and table 18.1, all of the types of goals suggested by this study's theoretical typology actually appeared at some point in the histories of global environmental change that we studied. Those histories nonetheless reveal a clear dominance of goals to reduce emissions, at the expense of goals directly targeted at protecting particular aspects of the environment or society. This may not be surprising. But it is noteworthy since for none of the risks we studied can any plausible emission-reduction plan restore environmental conditions to their undisturbed state within policy-relevant periods or avoid the possibility of damaging impacts on society. And however inadequate a strategy of targeting social impacts or environment *instead* of targeting emissions might seem to some, it does not follow that targeting impacts or environment *along* with emissions is in principle ridiculous. Indeed, as the difficulties of achieving significant emission-reduction targets becomes increasingly clear, the neglect of goals targeted at protecting environment or social welfare emerges as a characteristic of societies' efforts to manage global environmental risks that is worth examining in more detail. We begin such an inquiry here with a look at the nature and sources of variation in the overall pattern of goal and strategy targets suggested above.

Patterns The acid rain case reflects the general pattern we observe in goal targets. Substantially more than half of the goal statements articulated over the study period were targeted on emission reductions. (The majority of these entailed no specific strategy but were rather of the character "Reduce emissions of sulfur dioxide.") The acid rain case is noteworthy in that it provides the widest array of goal statements directed toward environmental remediation or impact reduction. Most of the arenas we studied produced proposals to lime lakes, streams, and forests; breed acid-resistant fish and crops; protect buildings and other materials; and adjust drinking-water chemistry to prevent heavy-metal leaching. Such goals for environmental protection or restoration and reduction of social impacts were nonetheless overwhelmed in the debate and politicking by goals to reduce emissions of acidifying compounds. This occurred despite the opening provided by the widespread adoption of critical-load goals discussed earlier. The critical-load debate did indeed shift attention from limiting emissions to limiting environmental damage. But this was not generally matched by a commensurate broadening of the discussion of possible actions to include ones that would (for example) decrease the vulnerability of ecosystems to such acidifying emissions as did occur. The discussion of response options remained focused on emission reduction. The change was simply that reductions were now more directly related to their presumed effects than had been initial, more arbitrary reduction goals.

The ozone case provides an even more extreme example of the dominance of goals targeted on emission reduction. Throughout the long history of active debate on this issue, on the order of three-quarters of the goal statements advanced in our arenas were targeted on emissions. Next most popular, but far behind, perhaps a fifth of the goal statements on ozone were targeted on capacity building of various sorts (research, development, monitoring, and institutions). In contrast, despite the existence of technical assessments on the topic, there was virtually no serious discussion of goals or strategies to counter ozone depletion in the stratosphere. Goals of reducing the impact of a depleted ozone layer on society likewise received almost no sustained play. The goal of publicizing UV levels so that the public could protect itself is the most salient exception. But even this modest goal emerged only two

decades after the risk was identified and then only in a few forums, such as Canada's 1990 Green Plan (Parson, Fenech, and Homer-Dixon 1993). The only other instance of impact-reduction goals for ozone found in our study was the infamous "hats-and-sunglasses" plan that was floated by the United States government in 1987, only to die of ridicule in a matter of weeks.

Goals targeted on reversing environmental change or on reducing social impacts received somewhat more serious attention in the climate change case. Proposals to build dikes, develop new techniques of agriculture, and undertake other adaptive measures were sporadically advanced in various arenas from the early 1980s onward. Grand geoengineering schemes for changing the environment to counteract global climate change received very little attention in public goal discussions, despite their regular appearance in policy studies. More sustained attention was given to goals of reforestation. But while we coded reforestation goals as targeted on the environment, consistent with the fact that new forests will indeed withdraw carbon dioxide from the atmosphere, a closer reading of the actual goal statements shows reforestation goals to be less of an exception to our general pattern than they at first appear. Almost without exception, reforestation goals were tacked on without much comment to much more extensively discussed goals of halting deforestation—goals squarely targeted on emission reduction. In any event, discussion of environment and impact goals was overwhelmed in the broader discussion by emission goals. As suggested earlier, this became increasingly true following the 1988 Toronto Conference, when virtually all actors jumped on the "reduce carbon emissions" bandwagon. To cite just one quantitative example, a systematic catalog of goal statements in Germany over the period 1989 to 1992 records only one goal focused on impacts and five on reforestation out of almost 100 total goal statements by prominent scientists, industry, governments, and environmental organizations.

Possible Explanations Why were societies so reluctant to include goals targeted on the environment or on impacts along with goals for emission reduction in their discussions of what should be done about the risks of climate change, ozone depletion, and acid rain? We can shed a certain amount of light on several possible explanations for the patterns we observed.

Psychological Bias Do people prefer to think problems will be avoided? It may be that there is a systematic cognitive bias toward overestimating the ability of prevention strategies to succeed, thereby reducing the perceived virtue of considering alternative approaches. When a goal of preventing a risk is set, people are uncomfortable with the possibility that such action might be insufficient. So they tend to overlook that possibility and come to believe that the prevention measures will succeed no matter what. Once that idea is ensconced in peoples' heads, the idea of considering adaptation goals doesn't emerge because adaptations don't appear to be needed.

Availability   Are goals constrained by the focus of prior option assessments? Chapter 17 (on option assessment) documents a strong bias in the policy debate toward measures for emission reduction. An explanation for the observed absence of goals targeted on environment or social impacts might therefore lie with those option assessments and the factors that shaped them. There are, however, strong reasons to believe that the availability of supporting assessments does not explain the observed patterns of goal and strategy formulation. First, many actors involved in the goal-setting process do not routinely rely on formal option assessments to formulate their goals. The goal-setting process is one in which a variety of actors participate without particular sophistication regarding the state of scientific or policy debates. Why should they shy away from goals aimed at the environment or impacts? Many such goals are obvious and not dependent on in-depth response assessments. A "hats and sunglasses" goal comes to the mind as easily as a "ban spray cans" goal. Second, the bias against impact and environment goals is even more severe than the bias against impact and environment options revealed in the preceding chapter. Even where one finds serious analysis of interventions that could change the environment or impacts—such as the U.S. National Acid Precipitation Assessment Program report of 1990 (NAPAP 1990) or the Intergovernmental Panel on Climate Change (IPCC) Response Strategies report of 1990 (IPCC 1991)—one finds these possibilities disproportionately absent from discussion of goals. Third, as explored at length in chapter 21 (on knowledge and action), the risk management process observed in our study is far from a linear process in which scientists define problems, analysts then explore solutions, and politicians finally set goals for eventual policy implementation. In fact, in the histories we observed, goals were frequently discussed and promoted before the options for attaining them were formally assessed. It therefore appears unlikely that society shies away from goals targeted on social impacts and environment because of the timing or content of option assessments. Much more likely is that the preoccupation with emissions in option assessment and goal and strategy formulation share a common cause.

Ideological Bias   Is only emission reduction legitimate? Another explanation would focus on the way issues are framed in a broader social context. It could be that these frames contribute to a bias against goals targeted on social impacts or environment by making such goals appear to be illegitimate. Some goals don't enter actors' heads (or if they do, don't leave their lips) because they are not socially acceptable. Much of the uproar over the "hats and sunglasses" episode in the United States suggests that the issue had been framed in such a way that the only legitimate response was to prevent the risk. Destruction of the ozone layer had been framed as an absolute wrong that had to be rectified and not as an unfortunate blemish that could be dealt with through makeup.

Technical Feasibility   Are goals constrained by doubts over technical abilities to achieve them? As noted earlier, some policy literature suggests that choices about ends are constrained by knowledge of available means, as well as the reverse (e.g., Wildavsky 1979; Cohen, March, and Olsen 1972). *If* this were true in the case of global environmental risks, and *if* it were generally acknowledged that goals targeted on environmental remediation or impact reduction were technically infeasible, then this might explain the observed preoccupation with goals of (presumably feasible) emission reduction. In fact, however, common knowledge runs the other way. Air conditioning, irrigation, and glass greenhouses provide abundant and ubiquitous evidence of societies' abilities to alter local climates and their impact on people and economic activity. Programs of lake liming and statuary protection have a long history of combating the effects of acidification around the world. And the "hats and sunglasses" so scorned as a strategy for addressing the risks of ozone depletion have nonetheless long been recommended by health authorities to counter the adverse consequences of natural exposure to ultraviolet light. Indeed, some evidence collected in this study suggests that discussion of goals directed toward environmental protection or the reduction of social impacts was actively suppressed by groups favoring emission-targeted strategies. Why? It is because they feared that such goals, once considered at all, would in many cases appear sufficiently feasible and attractive to undercut the emission reduction efforts they preferred.

Political Economy Bias   Does the market favor prevention? Finally, it may be that there are political-economic factors that bias social goals in favor of the reduction of emissions. Emission reduction was often promoted by purveyors of technology that promised to reduce emissions. Such promoters are more greatly motivated to seek political influence than are social actors who might benefit from other forms of social response.[16] This was overtly true for the acid rain case where the competition among purveyors of "clean-coal" technologies was fierce and may have influenced some actors in the other cases as well.

As noted earlier, this study was designed to highlight patterns in the interactions between society and global environmental risks and not to explain those patterns. Nonetheless, on the matter of the dominance of the goal debate by declarations targeted on emission reduction, our evidence gives strongest support to a variety of interest-based explanations. Emission-reduction goals were backed by a wide variety of concentrated interests, many of whom also provided concentrated opposition to goals that might detract from emission reduction. Opponents of emission reduction, though powerful enough to slow the implementation of such measures during the period studied here, seldom tried and never managed to reframe the goals and strategies debate in ways that placed environmental remediation or social protection at center stage. Those who did try to promote environmental or social protection goals spoke for a diffuse audience and usually found themselves accused of a naive approach to environmental policy, at best, and more often of self-serving motivations. Whether these interpretations are valid and, if so, whether the pressures behind them serve the public interest, only further research and experience will tell.

18.5.4 Why So Much Convergence?

Why is it that across arenas our data show evolution from a variety of differing goals toward convergence on a small set of goals? Although goal convergence has been observed in other studies of risk management to the point of becoming almost conventional wisdom, explaining why convergence occurs has been elusive. Although we too are unable to provide a simple causal explanation for convergence, we can offer some reasons why it occurs and illustrate some of the mechanisms at work. (These issues are explored from a broader perspective in chapter 21 on knowledge and action.)

We find that the puzzle of goal convergence can be understood only by taking into account the social context in which the goal-setting process takes place. Although in some contexts it may make sense to conceive of the goal-setting process as one in which individual nations first formulate goals in a process dominated by governments and then set collective goals internationally through a bargaining process, in our case materials we observe a quite different context. What we observe is more like a social system in which multiple actors compete for influence over an issue whose general contours, while subject to change and competing interpretations, are generally known to all participants. As in most social settings, we

observe actors staking out differentiated (though certainly not fixed) roles. Some actors place utmost importance on the issue area in question and seek to influence others as much as possible. Others are more interested in balancing competing interests. Some specialize in creating or disseminating a particular type of knowledge; others in spreading certain forms of response. In spite of these differences, these actors, we observe, operate in a common social framework. They take cues from the same sources, and they frame their actions against the same benchmarks; this is so even when they value such cues and benchmarks differently.

In such a social setting, with common frames of reference and a highly pluralistic distribution of power and authority, we would expect to observe convergence for issues that rise high in salience. In the absence of convergence, actors will be mobilized to seek change. The high salience of the issue gives them the incentive to do so, the pluralistic political setting gives them the ability, and one would expect them to keep trying either until a degree of convergence is achieved or the issue evolves to diminish the actor's salience.

Conceived of in this way, the last thing one would expect about the process of convergence is that it would operate in the same way all the time. Pluralistic politics means that at any point almost any actor might be capable of seeking to influence any other particular actor so as to bring about a degree of convergence. Given a diffusion of relevant capacities, a variety of means of influence would be at their disposal. Consider some illustrations from our case material.

One area of convergence we observe is elimination of CFCs from spray cans. The British case is particularly instructive in revealing the nature of the process at work. British industry and government actors, for example, were devoutly opposed to this goal, and they resisted longer than any other group in our study. Initial, sustained pressure came from within the European Community, especially via the German government. Domestic pressure within Britain was almost nonexistent. A targeted campaign by U.S. NGOs helped spark interest among British counterparts, however. This in turn resulted in a consumer-led boycott of CFC-containing cans that was so effective that the ban became practically moot. The actors who put pressure on British society (German environmental officials and U.S. NGOs) were diverse, and yet they saw themselves as part of a more or less united social project (protecting the ozone layer); they took their cues as to appropriate models of behavior and benchmarks for action from the same sources. The targets within British society were similarly diverse (industry, government agencies, NGOs, consumers), and

108

Marc Levy, Jeannine Cavender-Bares, and William C. Clark et al.

they responded in distinct ways, without any monolithic orchestration from the state.

The "greening" of the Japanese government in the early 1990s also helps illustrate the social context in which global environmental goals are set. For most of the period under study, Japan consistently stood on the sidelines of global environmental discussions, sought to minimize obligations that would generate costs for Japanese industry, and for the most part did not engage in the social process of managing global environmental risks. When the government finally decided, for a variety of reasons outlined in the Japan case study in this book (chapter 8), to reevalute its role, it behaved as if responding to pressures of socialization much more than as if simply recalculating interests or striking new bargains. In particular, it incorporated new norms from the social setting from which it had previously excluded itself, and it charted new actions in reference to these norms and in reference to recognized group leaders rather than simply by calculating interests. In staking out a leadership position on the issue of Asian acid rain, for example, Japanese decision makers relied almost not at all on calculations of damage to Japanese ecosystems and much more on already established norms that acid rain was a social ill that required preventive action.

As these brief examples suggest, the process by which goals to manage global environmental risks converge resembles more closely a diffuse socialization model than it does a hierarchical interstate bargaining model. We think that for issues in which it can be demonstrated that a common social context exists, this social structure counts as the most efficient explanation for goal convergence. Clearly, for such an explanation to be useful one needs a fairly unambiguous sense of what constitutes a common social context:

• Common understanding of how the problem is framed (what the essential dynamic is and why it matters) and

• Common commitment to acting responsibly with regard to the problem in question (which means respecting the legitimacy of other members of the social group and respecting relevant group norms).

We now elaborate three distinct ways in which goals can converge.

Top-Down, Formal Efforts to Promote Convergence Goals may converge because actors deliberately try to get them to converge through campaigns orchestrated at high levels. For example, the international transition to the critical-load concept in the acid rain case was largely due to efforts by the Scandinavians (and later the Dutch), who held numerous workshops (and engaged in extensive diplomacy) to encourage scientific research that would facilitate the use of critical loads as elements of policy (Dietrich 1995). Much of this impetus can be traced to a single scientist, Jan Nilsson from Sweden. (The U.S. exception supports the rule. The United States, which never switched to critical loads, was never subjected to the Scandinavian and Dutch diplomatic initiative.) The inclusion of tolerable rates as an explicit goal in debates surrounding the Climate Change Convention was also attributable to the sustained effort of a group of exponents, as demonstrated in chapter 13 (on international institutions). Finally, the spread of acid rain goals in central and eastern Europe (see, for example, chapter 7 on Hungary) was attributable directly to Soviet government efforts to spread the goal there.

Bottom-Up, Informal Campaigns Goals can also converge through processes that percolate up from below. The spread of the Toronto target had elements of this. As noted earlier in this chapter, members of the NGO and expert communities inserted a goal of 20 percent reduction in carbon dioxide emissions by the year 2005 into the final statement of the 1988 Toronto Conference on Our Changing Atmosphere. In spite of its ad-hoc origins, the effort was highly successful; the goal diffused widely throughout the international community, sparking a series of governmental commitments to reduce greenhouse gases. This is also an interesting example of using forums that have no legal authority to help spread goals to forums that do have such authority. (Put another way, sometimes soft law is not just a second-best alternative to hard law but can be an instrumental step on the way to hard law.) When Britain was resisting the conducting of forest-health surveys in the early 1980s, it was eventually forced into accepting such surveys as a result of efforts by NGOs. Friends of the Earth carried out its own survey of British forest health and publicized its findings widely; this put pressure on the British government to embrace the goal to gain more control over the process.

Mimicry Goals may converge because of mimicry. One actor or arena fixes on a dominant goal, and others latch onto it in the absence of explicit efforts to promote spread. Mimicry appears to have been operative in Soviet ozone goals, Japanese ozone and climate goals, and to a more muted degree British climate goals. In all these cases mimicry derives its impetus from a desire to "join the club of responsible world citizens." There are other possible motives for mimicry, which may be important at other times. For example, mimicry can help function as part of a satisficing strategy to cope with information-processing constraints. The German NGO calls for aerosol boycotts, copying their North American counterparts, may have reflected this phenomenon. The various NGO and parliamentary cries for 20 percent reduction in carbon emissions following Toronto may have as well. Mimicry might also be part of an effort to promote harmonization for trade or other reasons.

Implications While our investigation into processes of convergence failed to uncover a causal explanation, it did generate findings with two clear implications. First, the process by which goals emerge and obtain support is so thoroughly decentralized and spreads in so discontinuous a pattern that it casts doubt on models that rely on states to occupy roles as leaders and laggards. We find no states that are consistently leaders, and more important we do not detect any dynamic in which states staking out leadership positions and cajoling, persuading, or pressuring laggards explains any significant portion of the change in goals or strategies over time. To be sure, some states tend to support environmental goals more than others, and states often stake out leadership positions and try to encourage their less enthusiastic counterparts. But in particular cases of goal diffusion, the more important dynamics tended to take place at the level of communities of interested actors, organized across national boundaries, extending influence into societies through a variety of means, only some of which traveled through conventional governmental politics.

Second, our findings are grounds for severe caution to analysts who would like to extrapolate causal models of goal and strategy formulation from the case, arena, or time period for which they were developed to broader conditions. We do, of course, attempt in this chapter to explain some broad patterns of goal adoption in a causal sense. Nonetheless, we found so many contextual and temporal contingencies at work and were so struck at the inadequacy of hypotheses developed to deal with one arena, case, or period for dealing with others, that we are left highly skeptical of extended theories or explanations whose range of applicability has not been empirically tested.

18.6 Toward a More Effective Evolution of Goals and Strategies

In this section we assemble some implications of this study for future research and action in the realm of goal and strategy formulation for the management of global environmental risks.

18.6.1 Trade-offs between Simple and Complex Goals

In the cases studied there appeared a clear dilemma that faces actors who wish to formulate goals that will generate the most effectiveness, defined in terms of promoting progress in risk management. In some circumstances, it appears that goals are most likely to be effective if they are simple, bold, fairly rigid in formulation, not tied to specific strategies, not necessarily consensually arrived at, and not necessarily scientifically sophisticated. The Toronto 20 percent reduction target is such a goal, as is cutting sulfur dioxide by 30 percent or CFCs by 50 percent. The advantage of such goals is that they do not take a great deal of time to formulate (a single meeting in a bar is sufficient), they accommodate a variety of preferences for particular strategies or means, they provide usable rallying cries for activists, and they help clarify which actors support taking action and which don't. The disadvantages of goals of this kind are equally obvious. They are often seen as illegitimate by those not involved in the initial formulation. They are often at odds with the best science. They fail to provide a strategy or program for their attainment. And they are not especially flexible with respect to differences in local conditions and changing scientific and technical knowledge. Because of these difficulties, in other circumstances simple, expedient goals are not effective, and more flexible, consensual, and scientifically sophisticated goals prove superior. Examples of these included critical-load targets in the acid rain case and the London and Copenhagen Amendments to the Montreal Protocol. Those interested in affecting social response to global environmental risks face a real dilemma in their choice of which sorts of goals and strategies to advocate.

18.6.2 Managing the Dynamics of Goals in Risk Management

What we do understand is that in the cases studied here, the answer to the dilemma is time dependent. Early on, before acceptance of the seriousness of the problem is widespread, sophisticated, complex, flexible goals appear to fall flat. Consider the ozone-depletion case. Canada offered as part of the Montreal Protocol negotiations a fairly nuanced and complicated formula for sharing the burden of reducing CFC consumption. It found, however, virtually no receptive ears: the far simpler 50 percent reduction was adopted instead. In the early 1980s in the acid rain case the stark goal of "reduce sulfur dioxide by 30 percent" was very effective at sparking reassessments of risks and options and at mobilizing concern across countries and actors. The goal helped break the stalemate that had emerged over the previous decade. Later in the decade the critical-load targets were able to move things even further, precisely because of their greater legitimacy, fidelity to science, and flexibility with respect to conditions and knowledge. Governments that fiercely opposed the 30 percent goal as unfair embraced critical loads (Britain). Governments that ignored the 30 percent goal as irrelevant because their domestic measures exceeded it came to take the international targets seriously and reevaluated domestic measures in light of them.

What are the necessary ingredients for moving from simple goals to more sophisticated ones? While it is difficult to generalize from a small universe of cases, some propositions can be formulated from the material we have examined here.

18.6.3 Rules of Thumb for Steering the Goal Setting Process Effectively

With some trepidation grounded in our earlier warnings concerning generalizability, we advance the following propositions as potential guides to the formulation of effective goals and strategies for the management of global environmental risks:

1. In the early stages of issue development, simple, rigid, unscientific goals are almost always more effective than more sophisticated goals.

2. Simple goals almost always outlive their usefulness because of the drawbacks outlined above. Remaining for a long time at the level of simple, rigid goals is prima facie evidence of failure.

3. Movement toward more sophisticated goals requires a process of socialization entailing the building up of a set of working relationships across arenas and social actors. This process is characterized by acceptance of a core set of common values, mutual recognition of the legitimacy of each actor, and a high degree of trust. In short, social learning requires a society or at least a community of actors interested in others' views, actions, and motivations. Such communities are not natural objects and must be created through the risk management process.

4. Success at the second stage of goal formulation often builds on contributions provided by actors who play the role of "opponents" at the first stage. In part this is because a classic ploy of opponents of action is perfectionism—arguing that the proposals on the table are not adequate. In the right circumstances perfectionist strategies can result in movement toward perfection as well as delay. (European arguments that regulation of CFC production was superior to specific consumption bans had some of this effect.) In part this is also because at the early stages proponents of swift action are concentrating their energy on mobilizing support for the early, simple goals and leaving it to the opponents to promote more sophisticated alternatives. In the acid rain case, in the early 1980s Scandinavia and Germany pushed for quick adoption of the 30 percent target, while Britain argued that international targets ought to take into account differential impacts of alternative reduction scenarios, a central component of the later critical-load approach.

18.6.4 Caveats and Future Directions

At the broadest level, it appears that by the time of the Rio Conference the acid rain and ozone depletion cases had successfully gone through the transition to sophisticated, flexible international goals. The climate change case remained dominated by simpler goals at that time for appropriate reasons, though nothing in the climate debate had yet suggested that more sophisticated goals would not eventually emerge. Does this mean that the process of evolution is automatic and unproblematic?

We don't think so and in support of our skepticism point to two sorts of evidence.

First, a note of caution is provided by the instances we discussed earlier of the goal-setting process getting stuck at an early, simple stage and failing to make the transition to more consensual, sophisticated, and flexible goals. In the United States, for example, the simple "reduce 10 million tons of sulfur" proved effective at generating attention and serving as a rallying cry but for a decade and more of changing science and politics resisted replacement by anything better. The deadening effect of the goal of a CFC aerosol ban on efforts to protect the ozone layer is another example discussed at some length earlier in this chapter.

Second, there are examples of other, older global environmental problems that have still today failed to manage the process of goal evolution effectively. Consider the example of pelagic whaling. There, international goals began as sophisticated but inflexible (for example, quotas set in terms of "blue-whale units"). Most analysts attribute the failure to prevent decline of stocks in part to the inability to correct the perverse incentives "blue-whale units" created (see, for example, McHugh 1977; Elliot 1979; Peterson 1992; Andresen 1993). By the mid-1970s goals became more flexible as well as more sophisticated, but these were seen as illegitimate by a number of important actors and rejected in favor of the far starker moratorium. The moratorium itself was seen as unfair by important parties, and as a result there was very little coordinated management of the international risk of depletion of whale stocks. It is hard to find the slightest evidence of progress (as defined in this chapter) in the global management of whale stocks, making it possible to conclude that not all global risks are bound to replicate the comparative success of the acid rain and ozone depletion examples.

We can conclude that

• Goal and strategy formulation exerts a substantial impact on the process of managing global environmental risks;

• It is intimately coupled through backward and forward linkages with other risk management functions;

• The innovation and diffusion of goals are not predominantly state functions; other actor groups have played significant roles in shaping perspectives on what ought to be done about global environmental risks;

• There exists today no deep and broadly applicable understanding of what makes some goals and strategies effective in promoting progressive risk management, while other formulations seem to bring its development to a halt; and

• There is nonetheless relevant history and simple rules of thumb that, if taken seriously, offer the prospect for improvement.

Appendix 18A.   Acronyms

CFC	chlorofluorocarbon
CO₂	carbon dioxide
EC	European Community
EPA	Environmental Protection Agency (U.S.)
FoE	Friends of the Earth
G7	Group of Seven (economically strong industrialized countries)
ICLEI	International Council of Local Environmental Initiatives
ICSU	International Council of Scientific Unions
IPCC	Intergovernmental Panel on Climate Change
LRTAP	(Convention on) Long-Range Transboundary Air Pollution
NGO	nongovernmental organization
NOₓ	nitrogen oxides
OECD	Organization for Economic Cooperation and Development
ppb	parts per billion
R&D	research and development
SCOPE	Scientific Committee on Problems of the Environment
SO₂	sulfur dioxide
UBA	Umweltbundesamt (German Federal Environment Agency)
UNECE	United Nations Economic Commission for Europe
UNEP	United Nations Environment Programme
UV	ultraviolet (radiation)
VOC	volatile organic compound
WMO	World Meteorological Organization

Notes

1. Many people contributed to the data collection and analysis on which this chapter is based. We thank those who participated in summer study working groups on the Goals and Strategies function, in particular Claudia Blumhuber, Sonja Boehmer-Christiansen, Renate Ell, Fen Hampson, Michael Hatch, Ida Koppen, Leiv Lunda, Donald Munton, Elena Nikitina, Edward Parson, and Vladimir Pisarev. For reviews of earlier drafts we are especially grateful to two anonymous reviewers.

2. To use Boothroyd's (1978) terminology, option assessment describes alternative "actions" and their consequences. Goal and strategy formulation presents "proposals" for which actions to implement.

3. One potentially confusing aspect of the coding should be noted. A common sort of statement treated as data in this study is "reduce emissions of sulfur dioxide by 30 percent by the year 1990." The *target* classification of the statement is clearly "emissions." The *means* classification, however, is somewhat problematical. One could (and we did in our preliminary coding) classify this statement as implying a "command-and-control" strategy or means, on the grounds that the speaker is stating a condition that he or she is (presumably) telling people they should bring about. On further reflection, however, the statement can be seen to include nothing about means: it would be compatible with a tax, a voluntary agreement, or a legislated regulation. For the analysis reported here, we therefore treated statements such as that reported above as a goal without any associated strategy and thus classified its means as "unspecified." Failure to make this distinction carefully would lead to a highly inflated estimate of the frequency of command-and-control measures advanced as strategies in the global environmental change debate.

4. It is true that the ban failed to suffice as a comprehensive strategy to cope with the problem of ozone depletion, a point often made by opponents. But all early proponents of the ban clearly saw the ban as first step while research continued to ascertain the relative merits of more difficult action. None of them claimed that an aerosol ban would solve the ozone-depletion problem: they claimed only that solving the problem would be made easier if society wiped out a major source of emissions immediately rather than wait.

5. Leaders were the Dutch Political Party Radicals, the Vereniging Milieudefensie, and the Stichting Natuur en Mileu. Their position is outlined in Vereniging Milieudefensie and De Kleine Aarde (1975). See Pleune (1997).

6. David Wirth of the Natural Resource Defense Council (NRDC) and Allen Miller and Irving Mintzer of World Resources Institute.

7. See also the following documents from the U.N. ECE LRTAP Secretariat: EB 5th Session, IO-ERD-87-064, WG Strategies 6/11/92; EB.AIR/WG.5/R.32. Staaf (1992) and Ågren (1992) both credited William Dickson of the Swedish Environmental Protection Agency with introducing the idea in 1980.

8. Prior to the Toronto Conference in 1988, goals to reduce carbon emissions, or greenhouse gas emissions in general, were simply not present at the national level of any country or in any policy arena. Some discussion had begun at Villach and Bellagio in the mid-1980s that emission reductions would be necessary to abate global warming, but no specific reduction measures received broad endorsement.

9. When the NGOs proposed their goal at a press conference midway through the week, they were met with considerable disagreement from other participants, particularly over the role that nuclear power could play. Partially as a result of the overwhelming media presence, however,

112

Marc Levy, Jeannine Cavender-Bares, and William C. Clark et al.

the 20 percent goal stuck, and options for achieving it were discussed within the Energy Working Group. The two chairs of the Working Group, Wolf Haefele of Germany and Jose Goldemberg of Brazil, represented diverging interests with regard to the role of nuclear power. But with the help of the rapporteur, Ralph Torrie, a Canadian NGO leader, they agreed that 10 percent could be met though decreased energy demand and 10 percent through changes in energy supply. Hence, increased energy efficiency and conservation as well as switching to alternative fuels, renewable energy, and nuclear energy were envisioned as means of achieving the reductions. The Energy Working Group also moved the target date to 2005, which would provide more time for nuclear power to play a role. Source: Personal communication, Jill Jäger.

10. The Canadian Council of Ministers for the Environment together with a parallel group of Energy Ministers initiated one of the major responses to the goal at the national level. In 1989 they established the Task Force on Energy and Environment, which investigated the feasibility of achieving the Toronto target. After taking economic growth and other factors into account, the Task Force reached the view noted in the text.

11. The Canadian position was redefined prior to the regional U.N. ECE conference in Bergen, Norway, where it moved toward a stabilization commitment on carbon dioxide emissions. Yet in 1990 the House of Commons Standing Committee on Environment persisted in recommending that the Canadian government adopt the Toronto Conference target. NGOs in Canada proposed their own versions of the goal, while representatives of various industry sectors contested it. Friends of the Earth adopted the 20 percent goal but moved up the target date to 2000 and encouraged Canada to become an international leader. Greenpeace offered the most ambitious version of the goal, promoting a 70 percent reduction of 1990 carbon dioxide levels. The Motor Vehicle Manufacturers' Association and other industry representatives, including the utility companies, made explicit arguments against the target.

12. Senator Wirth introduced a bill that would mandate a 20 percent reduction of carbon dioxide emissions from 1988 levels by 2000 and a 50 percent reduction by 2015 (S. 324 of 1989). Congresswoman Schneider proposed a 20 percent reduction of CO_2 emissions by 2000 (H.R. 1078 of 1989). Senators Leahy and Boschwitz proposed variations of the goal that emphasized emissions reductions from automobiles (S. 333 and S. 603 of 1989).

13. In December of 1988, the Secretary of State for the Environment was asked by a "back-bencher" in Parliament whether Britain would take measures to reduce carbon dioxide emissions by 20 percent by 2005. In response the Secretary announced that the government was not yet in a position, given present scientific understanding to, devise precise reduction targets that would be feasible or relevant to specific environmental goals. Instead, the first priority was to improve understanding of the problems of climatic changes and their impact on the environment.

14. It should be noted that strategies to increase society's capacity to cope with a hazard were not originally part of this project's research protocol. Therefore, as an analytic category it was not sought after in our research as exhaustively as strategies to directly manage risks. In spite of this, the picture that emerges from the data collected for this study is one in which strategies seeking to increase society's capacity for managing risks represent a substantial fraction of all the strategies ever enunciated. It is impossible to ascertain precise percentages, but capacity-building strategies may well have been as plentiful over the entire study period as those aiming to alter directly the risks in question. In an especially detailed analysis of goals in the German climate case, we found that even during the period 1989 to 1991—when the desire to "do something" about climate change was at its peak—a quarter of the

strategies enunciated nonetheless addressed capacity building rather than direct management.

15. There are exceptions to the general pattern. German and U.S. acid rain goals and strategies have nothing to do with critical loads but rather are framed in terms of straight technology standards, emission limits, or percentage reductions. Canadian climate policy goals and strategies are framed in terms of technical feasibility and costs. Many of these exceptions, on further inspection, nonetheless support the general argument. This is clearest in the acid rain case. German and U.S. emission and technology standards are so strict that there is little added protection (if any) that a critical-load approach could provide. That is, even though German and U.S. standards are not based on critical loads, the effect is very similar. Where technology and emission standards are more lax, we expect pressure to move toward a critical-load approach (as happened in Britain).

16. This hypothesis is consistent with the now common observation that political solutions that promise concentrated benefits will generate more intense support than those that will bring only diffuse benefits.

References

Ågren, C. 1992. Interview with William Dietrich. Gothenburg, Sweden, June 3.

Amann, M. 1992. Trading of emission reduction commitments for sulfur dioxide in Europe. Status Report 92-03. Laxenburg, Austria: IIASA.

Andresen, S. 1993. The effectiveness of the International Whaling Commission. *Arctic* 46: 108–115.

Argyris, C., and D. Schon. 1978. *Organizational Learning: A Theory of Action Perspective*. Reading, Mass.: Addison Wesley.

Beuermann, C., and J. Jäger. 1996. Climate change politics in the Germany: How long will any double dividend last? In Tim O'Riordan and Jill Jäger, eds., *Politics of Climate Change*. London: Routledge.

Boothroyd, H. 1978. *Articulate Intervention: The Interface of Science, Mathematics, and Administration*. London: Taylor and Francis.

Cavender-Bares, J. 1983. Goals and strategies for different societal actors in Germany. Social Learning Project Archives, BCSIA, Kennedy School of Government, Harvard University, Cambridge, Mass.

Cohen, M., J. March, and J. Olsen. 1972. A garbage-can model of organizational choice. *Administrative Science Quarterly* 17: 1–25.

Dietrich, W. 1995. The challenge of selecting goals: Case studies regarding the use of critical levels. CSIA Discussion Paper 95-05, Kennedy School of Government, Harvard University, Cambridge, Mass.

Elliot, G. 1979. The failure of the International Whaling Commission 1946–1966. *Marine Policy* 3: 149–155.

Hall, Peter. 1993. Policy paradigms, social learning and the state. *Comparative Politics* 25(3): 275–296.

Hirschman, A., and C. Lindblom. 1962. Economic development, research, and development policy making: Some converging views. *Behavioral Science* 7: 211–212.

Intergovernmental Panel on Climate Change (IPCC). 1991. *Climate Change: The IPCC Response Strategies*. Report from Working Group III. Washington: Island Press.

Kingdon, J. 1995: *Agendas, Alternatives, and Public Policies* (2nd ed.). New York: Harper Collins.

Lakatos, I. 1970. Falsification and the methodology of scientific research programmes. In I. Lakatos and A. Musgrave, eds., *Criticism and the Growth of Knowledge*. Cambridge: Cambridge University Press.

Lee, K. 1993. *Compass and Gyroscope: Integrating Science and Politics for the Environment*. Washington: Island Press.

Levy, M. 1993. European acid rain: The power of tote-board diplomacy. In R. Keohane, P. Haas, and M. Levy, eds., *Institutions for the Earth: Sources of Effective International Environmental Protection*. Cambridge: MIT Press.

Majone, G. 1980. Policies as Theories. *Omega* 8: 151–162.

———.1989. *Evidence, Argument, and Persuasion in the Policy Process*. New Haven: Yale University Press.

McHugh, J. 1977. Rise and fall of world whaling: The tragedy of the commons illustrated. *Journal of International Affairs* 31: 23–34.

Michaels, D. 1973. *Learning to Plan, and Planning to Learn*. San Francisco: Jossey-Bass.

Molina, M., and S. Rowland. 1974. Stratospheric sink for chlorofluoromethanes: Chlorine atomic catalyzed destruction of ozone. *Nature* 249: 810–812.

National Acid Precipitation Assessment Program (NAPAP). 1990. *Acidic Deposition: State of Science and Technology*. Vol. 4, *Control Technologies, Future Emissions, and Effects Valuation*. Washington: Superintendent of Documents.

Parson, E., A. Fenech, and J. Homer-Dixon. 1993. Stratospheric ozone depletion in Canada: A history of risk management. Version 1.1, Social Learning Project Archives, BCSIA, Kennedy School of Government, Harvard University, Cambridge, Mass.

Peterson, M. 1992. Whalers, cetologists, environmentalists, and the international management of whaling. *International Organization* 46: 147–186.

Pleune, R. 1997. *Strategies of Dutch Environmental Organizations, Ozone Depletion, Acidification, and Climate Change*. Utrecht: International Books.

Staaf, H. 1992. Interview with William Dietrich. Solna, Sweden, June 2.

United Nations Environment Programme (UNEP). 1989. *Assessment Synthesis Report*. UNEP/OzL.Pro.WG.II(1)/4*. Nairobi: United Nations Environment Programme.

Vereniging Milieudefensie and De Kleine Aarde. 1975. *Spuitbussen* 4(3/4): 42.

Vickers, G. 1965. *The Art of Judgment*. New York: Basic Books.

Wildavsky, A. 1979. *Speaking Truth to Power: The Art and Craft of Policy Analysis*. Boston: Little Brown.

World Meteorological Organization (WMO). 1991. *Scientific Assessment of Ozone Depletion: 1991*. WMO Global Ozone Research and Monitoring Project, Report No. 25. Geneva: WMO.

19
Implementation in the Management of Global Environmental Risks

Rodney Dobell with Justin Longo, Jeannine Cavender-Bares,
William C. Clark, Nancy M. Dickson, Gerda Dinkelman,
Adam Fenech, Peter M. Haas, Jill Jäger, Ruud Pleune, Ferenc L. Tóth,
Miranda A. Schreurs, and Josee van Eijndhoven[1]

19.1 Introduction

Implementation, the subject of this chapter, is action to realize a strategy or fulfill a purpose. It is concerned with giving faithful effect to declarations, putting decisions into motion, and selecting and carrying through a course of action intended to achieve identified (not necessarily announced) objectives.[2] For present purposes it does not extend to considering what strategy or purpose to embrace, which was the subject of chapter 17 (on option assessment) and chapter 18 (on goal and strategy formulation) or to examining whether actions taken actually achieved their intended impact or ultimate purpose, which are the subjects of chapter 16 (on monitoring) and chapter 20 (on evaluation).

This discussion envisages risk management activities for a national society or arena as an entity, tracing a course of action in response to a perceived risk. Implementation is one component in that overall response. But there are many individual and organizational actors involved in this ongoing social interaction, and each has an implementation component in its own activity.[3]

In general, implementation is not an event: it is a chain of events, a process. This process is not precisely structured, unidirectional, or divorced from the process of policy formulation; on the contrary, it is open, evolutionary, adaptive, and disorderly (Majone and Wildavsky 1979; Palumbo 1987) and inextricably integrated with policy development. But for expository purposes the process can be viewed as starting from identification of a strategy or a purpose and a decision to pursue it. This process may then be imagined to move on to development of more closely defined policies, themselves perhaps conceived of as packages of more specific governing instruments to be exercised, and ultimately to the deployment of concrete operational measures and the exercise of day-to-day discretion by managers and "street-level" agents (Lipsky 1978; Calista 1994).[4]

The crucial observation is that a policy is not fully defined, or a strategy realized, until this sequence of events works its way out to discretionary action on the ground, where autonomous individual actors alter their behavior to comply with the strategy adopted. And this discretionary action must continue through an endless sequence of individual decisions in changing circumstances: commitment to adapt behavior continuously to comply with the spirit and the intent of the strategy is ultimately what counts in determining whether implementation has occurred.

We cannot hope to observe all this. In studying implementation at the level of generality of this project, all we can do is note illustratively the commitment of resources, the passage of legislation or regulations, or the initiation of other action undertaken to begin the chain of events that it is hoped will lead to realization of some intent, down the road.

Three main themes emerge as the most significant conclusions to be inferred from the outline sketch of implementation actions in this chapter.

The first—perhaps most striking—result is that a plausible pattern of emergent order can be observed in the record of social action initiated, even though there may be no neat "policy cycle" anywhere. This order entails creation of capacity at the international level for debate about frameworks and standards and devolution of responsibility for action. Such supranational capacity to debate ideas and pursue agreement on commitments to principles has to be matched by social capacity to enlist commitment to changing behavior on the ground. Action thus evolves toward a focus on compliance, which also reflects, in a fascinating way, an orderly view of social evolution.

Indeed, the key idea with which we leave is that the arrival of the "congested global village" (Dobell 1997) opens another chapter in the story of the "civilizing process" (Elias 1982). In this story, increasing density and interdependence in human communities force implementation efforts increasingly toward norms, conventions, and self-constraint as the ultimate disciplines on the forceful pursuit of self-interest in a global commons where rights and purposes conflict. These provide the ultimate instruments to bring about the behavioral change essential to any effective social response to global environmental risks.

The second theme is the growing emphasis on creation of institutional capacity at the international level and the growing significance of international agreements as a spur—indeed, in many cases a prerequisite—to domestic action.[5] This chapter reveals—as a result, not as an imposed structure—a history in which the early parts of the story are domestic or national (though they may reflect cross-border influences), but later developments stress collective international efforts in response to risks increasingly perceived as inherently transboundary or global in nature. These may lead on to normative declarations at international level. But thereafter, the capacity to ratify such agreements and make truly authoritative declarations rests again with sovereign nations or entities (Putnam 1988; Chayes and Chayes 1995; Keohane and Ostrom 1995). The story of rule making, monitoring, and enforcement thus returns to the domestic setting.[6] Commitments to reporting and surveillance maintain the link with international institutions, however, and the development of effective international regimes for those purposes is a key part of current implementation activity. International negotiations may also be necessary to provide assurance that nations initiating domestic measures to comply with agreements will not be left at a competitive disadvantage relative to nations that do not take comparable action.

A third theme underlines the way in which this institutional emphasis parallels the academic evolution toward understanding implementation as an organic, uncertain process of interactive negotiation and feedback. This theme emphasizes understanding of compliance as an emergent phenomenon reflecting discretionary adaptation of local action in order to pursue agreed goals in diverse and changing circumstances. It suggests diminishing appeal to coercion to eliminate an "enforcement deficit" and increasing appeal to cost-effective, negotiated, differentiated action in place of uniform directives coupled with uniform enforcement. It also suggests growing emphasis on "reporting and review," therefore, rather than "targets and timetables." In this evolution an essential development is the growing role of NGOs and other informal networks and their growing capacity to provide audit and scrutiny of the actions necessary to ensure compliance with commitments undertaken.

In this chapter we review (section 19.2) very briefly and selectively the record of implementation action in individual arenas and at the international level, illustrating the range of governing instruments employed and noting patterns in action across arenas and over time. Section 19.3 sets out seven stories selected to convey, in aggregate and impressionistically, the changing emphasis and evolution in implementation action over the period covered by this study. Section 19.4 suggests that performance in implementation efforts is improving and

examines what might explain the improvements claimed. In that section and the subsequent conclusion (section 19.5), we return to the above themes as part of the social response to an increasingly interdependent "full world" (Daly 1991) pressing against the limits of its atmospheric commons.

19.2 The Empirical Record and Some General Patterns

19.2.1 The Data

What are we actually examining? In this study and even more in this chapter, the pool of items examined can be only a very small sample from the very large universe of instances of implementation. By their nature, as just noted, these implementation episodes are extended processes, of which we have only highly abridged accounts. We do not have products analogous to an assessment report, nor do we have primary materials in the same way. We do have products in the sense, for example, of a Clean Air Act as a tangible object whose passage represents one step in an implementation process. But it is only one step. Ideally, we should look at the whole sequence of such steps in an implementation process as an entity in itself.

Nevertheless, having recognized this problem, we are forced, in a comparative study of this size, to deal with only a highly selective shorthand sampling of the isolated implementation events featured in the various arena accounts. The discussion here will highlight just two of the many features identified as significant in the theoretical literature relevant to implementation questions—namely, international organization or international environmental negotiations and the choice of governing instrument at the national level. In effect, we add to the traditional spectrum of governing instruments the class of new measures specifically directed at capacity creation in the form of international institutions,[7] but we lose most of the detail on action by individual agents either inside or outside government.

This chapter therefore adopts a descriptive and empirical approach to implementation issues, focusing on the selection, in specific examples, of governing instruments or other executive actions as sketched in the arena chapters of this book. Emphasis is on trends in the nature of the instruments actually deployed in various arenas and on the evolution of implementation processes themselves rather than on any particular outcomes.

For each of the three risks or cases, one can examine highlights of action undertaken, in each arena, at the level of major legislation, authorization of major program, or notable action by other actors. Each such account provides a thumbnail chronology of implementation action in a single arena, in response to—or bearing on—a particular risk. For purposes of analysis, it is these capsule histories or sequences of implementation action that are of most interest. The phenomenon we wish to examine and perhaps interpret as a learning process is thus a time path of implementation actions and not a single implementation event. Again, however, it has to be emphasized that the sheer magnitude of the record here precludes more than a subjective synthesis of selected arena stories, as set out below.

19.2.2 Selection of Instruments

Table 19.1 summarizes what was done, classifying action according to the instruments and targets tables developed in the two preceding chapters. This record of action actually taken, as laid out in these tables and the arena chapters themselves, suggests the following general observations.

First, it is notable that implementation efforts tended to cluster around regulatory action to reduce emissions, with substantial further commitment of resources to build cognitive capacity through research (by scientists, government, nongovernmental organizations, and industry), institutional capability—both domestically (especially NGOs in the later years and especially with respect to climate risks) and internationally (especially governments and international organizations themselves)—and consciousness-raising (especially NGOs). Since both option assessments and declarations of goals have been found in this study to be dominated by emission-reductions options, it is hardly surprising that implementation actions turned out to be similarly focused.

Government action tended to be in the regulatory domain, through either directives or incentives, though it was also noted in the commitment of resources to scientific research programs, coordination capacity, or creation of institutional capacity at domestic or international levels. Action by industry (and to some extent local authorities with operational responsibilities) tended to be concentrated in the commitment of resources to create or improve technological capacity, though actions to build institutional and consultative capacity are also noted in the arena histories, as are actions to create cooperative programs aimed at awareness or information exchange.

A crucial limitation of the analysis here, of course, is the impossibility of achieving any comprehensive inventory of the myriad actions undertaken on a daily or hourly basis by operational authorities. The holes in the tables when it comes to action by industry actors or individual economic agents reflect the level of aggregation in our discussion rather than the absence of discretionary activity.

One suspects, for example, that the few examples noted of implementation efforts directed toward adaptation (such as the work of Dutch local authorities to maintain and extend coastal defenses, the introduction of information advisories like the UV-B index, or the selection of acid-resistant varieties of seedlings in forest renewal operations) are illustrative of a vast universe of activities to discharge continuing operational responsibilities, both public and private, to adapt effectively to a changing environment for local works. Similarly, action at the local level to offset acidifying deposition through liming (of the Boston water supply, for example, or Canadian lakes) is illustrative of use of technological capacity through actions that cannot be captured in this abridged account.

Action taken by individual scientists, by analysts within NGOs, and in government scientific initiatives focused on instruments directed toward building cognitive capacity. Action taken by NGOs also deployed instruments designed to build institutional capacity—to participate effectively in consultative activity, to engage persuasively in public debate, and so on.

Deployment of traditional "governing instruments" by governments rested on devolution of responsibility for the subsequent exercise of more specific operational instruments by other actors, investing in the creation of institutional and technological capacity, modifying ongoing operations, and continuing economic activities to respond to regulatory requirements or to increased awareness of risks to be dealt with. Much operational activity may have been in the nature of action to offset or adapt to those risks, but the selection of instruments at the more aggregate level involving governments and major actors tended initially to concentrate significantly in the area of directives aimed at emission reductions and subsequently to fall back more substantially on less coercive arrangements negotiated with industry.[8]

Contrary to first impressions, then, we do not see simply a continuous increase of scale in implementation activities, or solely an escalation to global institutions. More accurately, perhaps, what was occurring was a bifurcation, a passage to a global arena and global market (and an attempt, as discussed below, to build international institutions of governance around these global markets), but with concrete implementation responsibilities and discretionary action increasingly devolved to subsidiary units, ultimately to the individual economic agent acting within the market (and social) framework set globally.

19.2.3 The Three Cases Compared

The story of action taken to deal with global atmospheric risks started with an integrated approach to "air issues" as

Table 19.1
Actions and instruments

Means	Targets	
	Emissions	Environment
A. Acid rain		
Capacity:		
Cognitive	EMEP established by OECD (1972/II); continued through the LRTAP Protocol (1984/II)[a] Thirty Percent Club (1984/II)	UNCHE (1972/II): acid rain issues raised by Scandinavian countries Critical-load concept
Institutional	OECD Guidelines to reduce emissions (1974/II) Federal/Provincial Emissions Reduction Accord (1983/Canada)	Liming of aquatic and terrestrial systems proposed by United States in negotiations with Canada (1982)
Technical	Transportation Efficiency Act (1991/U.S.) Fuel switching; nuclear CEGB proposes to refit three power stations with Germany by 1997 (1986/U.K.) Catalytic converters	Sudbury "Superstack" erected as best solution to local air pollution (1971/Canada) Experimental liming of lakes rejected as inadequate (1970s and 1980s/Canada)
Regulations:		
Directives	LCPD Agreement (1988/E.C.) Sulfur Dioxide Protocol (1985/II) MITI regulations for oil companies to control sulphur content (1992/Japan)	
Incentives	Sulfur oxide emissions charges (1973/Japan) CAA Amendments: tradeable emissions quotas (1990/U.S.) Clean Air Acts Covenants with electricity producers (1990/Netherlands)	
Education/information	"Un dia sin auto" (1987/Mexico)	State of the Environment Reports
B. Ozone depletion		
Capacity:		
Cognitive	UNEP establishes CCOL (1977/II)	Increase ozone artificially
Institutional	MOF established to fund technology transfer (1990/II) Vienna Convention (1985/II) Montreal Protocol (1987/II)	
Technical	CAA: CFC substitutes, replacements, and recycling (1990/U.S.)	
Regulations:		
Directives	Congress stops SST funding (1971/U.S.) Aerosol bans (various/1970s) Retrofitting Production freezes	
Incentives	CFC exise taxes (1989/U.S.) Tradable production quotas	
Education/information	Blue Angel ecolabeling program (1979/Germany)	

a. II is international institutions.

Table 19.1 (continued)
Actions and instruments

Means	Targets	
	Emissions	Environment
A. Acid rain		
Parliamentary Hearings on air-pollution impacts (1971/Germany)		IIASA RAINS model developed (1982/II) National Acid Precipitation Assessment Program (1980/U.S.)
Acidification Fund		LRTAP (1979/II) U.S./Canada Air Quality Agreement (1991)
Acid-resistant trees and crops		
Forest tenures		Clean Air Acts: U.K. (1962), U.S. (1963), Japan (1968), Canada (1971), Mexico (1971), Netherlands (1972), Hungary (1972), Germany (1974), EC (1976), international institutions (LRTAP-1979), former Soviet Union (1981) National/local round tables
B. Ozone depletion		
Health and ecosystem effects of ozone depletion		NASA/WMO/UNEP ozone assessment (1986/II) World Plan of Action on the Ozone Layer: coordinated research (1977/II) Ozone Action Program established (1991/II)
Protective devices UV-resistant crops		
		NRDC sues to have CFC regulations enforced (1984/U.S.)
Ozone Watch and UV Index Programs (1992/Canada) Greenpeace: health and ozone-layer campaign (1992/U.K.)		NGO information campaigns on health and ecosystem impacts NGOs promote CFC alternatives (1992/U.K.)

(*continued*)

Table 19.1

Actions and instruments

Means	Targets Emissions	Environment
C. Climate change		
Capacity:		
Cognitive	Climate Impact Assessment Program (1971/U.S.) Canadian Climate Program (1978/Canada) UKMO $2 \times CO_2$ modeling (1979/U.K.)	
Institutional	U.N. FCCC (1992/II)	Joint implementation (tree planting) : AES/WRI "trees for power plants" (1988/U.S.) Debt for Nature Swaps
Technical	National Energy Strategy (1991/U.S.) National Energy Efficiency Act (1991/Canada)	Carbon sequestration
Regulations:		
Directives	Transportation Efficiency Act (1991/U.S.)	Massive reforestation program (1990/Mexico)
Incentives	Carbon taxes and energy taxes Convenants (Netherlands, U.K., Canada) "Pledge and Review" (1990/Japan)	Emissions offsets: tree planting and greenbelts
Education/information	Labeling of energy-efficient appliances	Green Plan: tree planting programs (1991/Canada)

an undifferentiated concern (an approach to which it began to turn again at the end of the period we study). The record beginning from the 1950s and 1960s built on a long tradition of earlier measures to deal with the "nuisance" of pollution but showed an upswing in concern and initial widespread appeal by governments to hardcore "command-and-control" measures embodied in national legislation. These measures, however, were initially a response to growing concerns about risks to human health arising out of local pollution from industrial activities rather than a direct response to concerns about acid rain or other specific atmospheric risks. Indeed, in most of the arenas studied, growing awareness of problems of discharges of pollutants into the surrounding environment and the collective-action problems involved in dealing with such threats locally led to formal action to establish responsibility and organizational capacity at the national level for concerted action to address concerns about protection of human health.

As awareness of problems of long-range transport and transboundary issues grew, the acid rain story assumed a life of its own. Early commitments of resources to research and coordination of research were followed by international negotiations and ultimately international agreements on national commitments to action. Growing experience with enforcement problems coincided with

growing commitments of NGOs to active campaigns to raise public awareness, coupled subsequently with appreciation of the need for still stronger action to realize national commitments. Regulatory approaches were developed that offered greater discretion in choice of technological means to meet imposed standards (integrated permits, bubbles, regional offsets). Experimentation with fiscal instruments or tradable emissions schemes broadened still further the discretion offered to industry (principally) in selecting actions to realize national targets for emission reductions. One innovation in the development of more flexible arrangements is the so-called flexible permit based on a capability to ensure continuous compliance with each individual emission cap.[9] As another illustration, the 1991 Pollution Prevention Act in the U.S. state of New Jersey gave the state Pollution Prevention Office authority to negotiate with selected firms an integrated permit focused on the production process as a whole; a pilot program was launched in 1992. Thus Dutch, German, and Japanese models of negotiated covenants (see section 19.3.6) were emulated in implementation measures at the state level in the United States.[10]

Action taken to deal with the perceived risks of ozone depletion pursued a different course. In a first round of activity, campaigns and consumer boycotts were led by ad hoc local voluntary action groups (for example, in the

Table 19.1 (continued)
Actions and instruments

Targets	
Impacts	Generic

C. Climate Change

Capacity:

Living with Climate Change meetings: food security (1975/Canada)	Dahlem Conference (1976/Germany)
Human impacts meeting (1976/Mexico)	Enquete Commission on global climate change (1987/Germany)
Association of Small Island States (1990/II)	World Climate Conferences (1979 & 1990/II)
	GEF (1990/II)
	U.N. Framework Convention on Climate Change (1992/II)
Preventive measures against rising sea levels (Netherlands)	
Friends of the Earth campaign: Impacts of Climate Change (1988/U.K.)	

Netherlands, Germany and the United States). Later, government bans on aerosols and spray cans were announced, first at local, then at state, and then at federal levels in the United States, subsequently in other jurisdictions. At the same time, resources were committed by producer interests to initiate research programs designed to identify substitutes for CFCs. Substantial resources were committed also by industry to campaigns designed to emphasize the uncertainty of available scientific knowledge. In most arenas the issue disappeared from view through the late 1970s and early 1980s; the most notable implementation actions involved scaling back or terminating the scientific search for substitutes, until concern was revived by dramatic new scientific findings. In the meantime, however, action was pursued through commitments of international institutions, particularly the United Nations Environment Programme (UNEP), to the international exchange of scientific information, to the coordination of discussion of response options, and ultimately to the animation of international cooperation through the Vienna Convention, which created the capacity on which to build the Montreal Protocol.

Implementation actions in the climate case focused substantially on initiating research programs (either for substantive reasons or perhaps as a ploy to forestall regulatory action, or both), organizing conferences, and establishing institutional capacity for risk and option assessment. NGO advocacy and participation in consultative and negotiating processes assumed a degree of intensity unprecedented in either of the other two cases. International negotiations leading to the signing of the United Nations Framework Convention on Climate Change (FCCC) followed the framework and protocol model used in previous settings but developed a graduated approach of reporting, scrutiny, and suasion to a considerably greater degree. In several arenas, government action embraced substantial programs of afforestation to sequester and offset increasing emissions of carbon dioxide (CO_2), including a number of carbon sequestration efforts worked out as joint initiatives across nations, and also established research efforts to explore adaptation possibilities. Almost all arenas shrank back from any substantial regulatory interventions, either coercive or incentive-based.[11]

Exceptions to this last observation include modest energy- or fuel-tax initiatives (generally undertaken for reasons unrelated to commitments to reduce greenhouse gas emissions) and some regulatory actions by local governments embracing the 1988 "Toronto targets" discussed in chapter 18 (on goal and strategy formulation).

In all three cases, a plausible pattern can be seen in the observed sequence of activities and division of effort.

Initial commitments of resources were uncoordinated, reflected in the activities of individual scientists and isolated commentators attempting, respectively, to identify and characterize possible risks and bring them to public attention. A period of contested science followed, in which implementation efforts were concentrated generally in attempts by NGOs, industry and industry associations, and government scientific establishments to shape the image of the character and severity of the risk in the public mind. To a substantial extent the same actors moved on into social debate and interaction around formulation of goals and assessment of possible responses to the emerging awareness of the risk (though not necessarily to the risk itself). Governments offered authoritative pronouncements about national targets for containing the risk, while NGOs and industry groups mounted campaigns, respectively, to strengthen or contain pending government action to conform to these pronouncements.

19.2.4 The Record Overall

In the implementation record, the acid rain story began before substantial action was taken in relation to ozone-depletion concerns, and the climate change story was only beginning by the end of the period we study. Over this whole sequence of activity, then, one can discern an attempted evolution toward greater discretion in technological decisions in action taken to meet national targets, ultimately through appeal (even though still fragmentary in fact by 1992) to market-based incentives. Greater discretion was also sought in determining what emission levels individual firms must in fact attain, as well as by what means. Evolution toward greater screening to constrain regulatory initiatives through regulatory impact assessments and testing according to cost-benefit threshold requirements is also evident in North America, Japan, and perhaps the United Kingdom.

Over the period, some NGOs created significant capacity to review ongoing scientific research and produce persuasive commentary as part of risk and option-assessment processes and, more strikingly, began to direct their own implementation actions toward participation in regulation writing processes as well as the monitoring and audit of compliance with regulations and standards. Individual scientific efforts initially identifying possible risks were overtaken by government action to initiate substantial research programs. Initiatives to create deliberative capacity, such as the Royal Commission on Environmental Problems in the United Kingdom or the Enquete Commissions in Germany, appear to correspond well to the conventional wisdom that the European tradition is inclined to embed its scientific disputes in discrete authoritative bodies addressing polit-

ical choices, while the American way is to fight the political battles more openly on the backs of contending scientific positions.

Thus, in the record of national action on implementation measures up to 1992, the focus was clearly on the visible middle of the causal chain, where perhaps the most obvious technological opportunities lie—namely, at the point of discharges into the atmosphere. The arenas studied here achieved some success in arriving at action to reduce emissions in the two cases where doing so did not demand going back behind the emissions and the production activities themselves to the underlying human activities and human behavior driving those activities. No arenas and no implementation efforts pushed back beyond that threshold to significant impacts on behavior or lifestyles or aspirations toward affluence. Nor was there will to push forward to implement substantial action to adapt to impacts or consequences.[12]

Organizational initiatives must not be overlooked as important examples of implementation action with symbolic value as well as operational impact on information flows and patterns of decision making. We see this in the creation of national agencies to deal (as discussed below) with problems of transport across regional frontiers and a shift in perspective from air pollution as a local problem to a national concern with human health; with a shift of responsibilities among agencies, as, for example, in the Netherlands to reflect a further new framing of the issue as one of ecological impacts rather than human health (Dinkelman 1995); and with the creation of institutional support structures for the international regimes formed to address atmospheric risks as global concerns. The parallel institutional evolution in the formation of scientific NGOs and epistemic communities; in the evolution of environmental and industrial NGOs; in the development of informal "virtual communities" and networks; and ultimately in the appearance of "transborder citizens" with channels of access (if not influence) into diplomatic activities at many levels should also be noted.

19.3 Seven Stories

In this section we illustrate, with a highly selective array of examples, the range of instruments employed in implementation of action to realize the goals, strategies, and normative declarations described in the other chapters of this book.

With respect to air pollution and specifically precursors of acidifying precipitation, we see (in section 19.3.1 below) a confluence across arenas both in organizational initiatives in the creation of national agencies and in the development of legislative capacity at national level to deploy directives or regulatory (command-and-control) instruments to limit emissions.

With respect to risks associated with stratospheric ozone depletion, an interesting two-part story is sketched in section 19.3.2, with an opening act relying on consumer boycotts and local action against consumer products and a still unfinished second act centered on international cooperation, employing a variety of market-oriented instruments as well as an overall regulatory framework to control industrial production and use of ozone-depleting substances.

With respect to climate change, action directed toward offsets and adaptation was more evident than mitigation efforts, though neither was much in evidence by 1992. Section 19.3.3 very briefly mentions attempts to offset industrial emissions of greenhouse gases through forest-related initiatives aimed at carbon sequestration, as one example illustrating the general difficulties of implementing cost-effective indirect measures targeted toward offsetting or adapting to emissions.

The story of economic instruments illustrates a significant trend. An underlying theme in the literature on choice of instruments is the tension between the theoretical argument for fees and charges and the political predisposition toward standards. In action at national level we can identify some evolution toward experiments with instruments that offer more discretion in the manner in which compliance might be achieved. "Bubbles" or other offset provisions represent a more generalized, less direct instrument than regulation of discharges at all individual sources and are preferred on theoretical grounds as more effective, less costly means to achieve target reductions in emissions. Chronologies of individual national action reveal some recent examples of appeal to such more generalized instruments (defined in some cases over geographically more extensive regions) and some appeal as well to market instruments, including tradable emission permits, charges, taxes, and fiscal incentives such as accelerated depreciation allowances designed to encourage introduction of emission reduction technologies. Despite their popularity in pronouncements, however, few examples of fully implemented incentive schemes or "bubbles" were found in place up to 1992.

The use of fiscal instruments to alter incentives in production and consumption is illustrated in section 19.3.4, and tradable permits, or "markets for pollution," in section 19.3.5. These latter are seen to blend in a variety of ways into nonmarket initiatives to achieve more flexible permits and regulation.

And these in turn are seen to lead directly to "regulatory negotiation" or voluntary agreements in domestic arenas (section 19.3.6) or, more ambitiously, to agreements at the international level, where no central authority exists and

sanctions can be mandated only by consent (section 19.3.7). Thus the spectrum of instruments for addressing global atmospheric risks appears in the end to circle back to flexible realization of regulatory standards or, even more fundamentally, back to efforts to build cognitive capacity and awareness of consequences, rather than on to the promise of "free-market environmentalism" deploying economic instruments in a trading environment free of regulation.

19.3.1 Capacity to Address Air Pollution and Acidifying Deposition: Clean Air Acts

Awareness of problems of transboundary transport and deposition of atmospheric pollutants was predated by widespread concerns about ambient air quality and local impacts of emissions of pollutants from industrial activity. A tradition of local legislation to deal with this "nuisance" goes back more than a century. The story of concerted national or international action to deal with transnational atmospheric risks, and specifically acid deposition, itself was anticipated, by the time of the 1972 United Nations Conference on the Human Environment (the Stockholm Conference), by a flurry of Clean Air Acts or similar legislation in several countries. These moved concerns about adverse impacts on human health arising from smog and local pollution toward a capacity to deal with industrial emissions as a national issue. Measures to create national agencies with the capacity to address perceived threats to human health arising out of air pollution were announced in several arenas.

This burst of national legislation in response to what were initially perceived as local or at least domestic threats to human health later provided a legislative basis within arenas for regulations to control sulfur dioxide (SO_2) emissions, and indeed nitrogen oxides (NO_x) and volatile organic compounds (VOCs), in response to emerging concerns about acid rain and photochemical smog. Subsequent rounds of legislation to address concerns about long-range transport and transboundary fluxes were undertaken in the context of international agreements negotiated over a long period.

Highlights in implementation action (or nonaction) include the following (for a more substantial comparative account of national legislation, see IUAPPA 1995):

• In the United Kingdom, the "killer fog" of 1952 was followed by the Beaver Report (United Kingdom 1954) and the Clean Air Act of 1956.

• In Germany, emissions limits were established in 1964 in the first German Technical Directive for Air Quality (Technische Anleitung zur Reinhaltung der Luft), which was issued as an administrative directive under the General Trade Regulations but reflected standards proposed by the Commission for Clean Air of the German Association of Engineers founded in 1957. The Federal Air Quality Protection Act of 1974 provided general enabling legislation under which the federal administration may issue regulations and directives and provided that emissions from industrial plants requiring authorization should be reduced to the extent achievable by use of economically feasible state-of-the-art technologies. While not specifically directed toward concerns with acidifying deposition, this legislation provided the legal foundation for government action to deal with acid rain.[13]

• In the Netherlands, an Act on Air Pollution developed by the Ministry of Health was promulgated in 1970. In 1971 a new Ministry of Health and Environmental Hygiene was created.

• The U.S. Clean Air Act of 1963 assigned to the federal Department of Health, Education, and Welfare the responsibility for overseeing state implementation plans to ensure attainment of ambient-air-quality standards. With the perceived failure of that initiative and the Air Quality Act of 1967 to achieve reasonable compliance, Congress subsequently, under the Clean Air Act Amendments of 1970, passed to the Environmental Protection Agency (newly established by Congress to reorganize and consolidate various environmental responsibilities) the responsibility to ensure state compliance with new emission limits.

• In Japan, national-level action with respect to air pollution stemmed from the Basic Law for Environmental Pollution Control passed in 1967, requiring that air-emission standards and ambient-air-quality standards be established. In 1968, as a result, a new Air Pollution Control Law set for the first time a national emission standard for sulfur dioxide discharges. The influence of the Ministry of International Trade and Industry (MITI) ensured that the basic law was written with full consideration for economic interests and industrial competitiveness and indeed with a "harmony requirement" that specified that environmental standard setting must proceed in harmony with industrial objectives. This "harmony" clause was abolished by parliamentary amendment in 1970. In 1972, the Environment Agency (established in 1971) formally issued nitrogen oxide standards for 1976 equivalent to the standards established for the United States in the 1970 Clean Air Act Amendments. Standard setting after 1980 once again introduced an explicit concern for balance with economic considerations, in what has been characterized as "a more rational approach" (IUAPPA 1988).

• Creation of the Canadian Department of the Environment in 1971, as authorized in the Government Reorganization Act of 1970, accompanied passage of a Clean Air Act without significant control or enforcement features.

• Constitutional amendments in 1971 established federal government responsibility for environmental protection in Mexico, with a federal law for the prevention and control of environmental pollution passed in the same year. A local transborder (El Paso–Juarez) agreement on air quality was developed the following year.

Thus, in the run-up to the 1972 Stockholm meeting, growing awareness of local pollution issues and possible health consequences led to the formal creation of national agencies to assume responsibility for protection from industrial emissions.[14] Legislation reflected a conventional command-and-control approach to "end-of-pipe" measures for the reduction of emissions, although industry generally argued a case for "solution by dilution" through the use of tall stacks. In all the arenas studied here, this institutional and regulatory response reflected a general recognition that local action would run into unresolvable problems of free riders and capital mobility. But also a general concern for economic costs militated against effective implementation. In general, emission reductions were mandated, and failures in monitoring and enforcement were subsequently identified as leading to general failure in implementation (Downing and Hanf 1983).

19.3.2 Banning Chlorofluorocarbons

Early action in banning chlorofluorocarbons (CFCs) in North America and Northern Europe followed activist campaigns that highlighted the newly perceived threat to the ozone layer. Chapter 18 (on goal and strategy formulation) describes the way in which the goal of reducing CFC use in aerosol cans spread from arena to arena. How those desired reductions were achieved is also an interesting story. The European version of this story shows bottom-up NGO-led aerosol boycotts culminating in government negotiation of agreements with industry to reduce CFC use in spray cans and aerosol products; in this account, grassroots commitment and enthusiasm are harnessed to achieve direct action. By comparison, the process witnessed in North America in the later part of the 1970s, though not without its conscientious-consumer element, was heavily influenced (especially in the United States) by a new breed of scientist—the policy activist urging regulatory action.

There followed a pronounced lull, fed by the popular illusion that the problem had been fully addressed. When this lull ended, this unusual process mixing science and activism was later developed in a second round, in the story of the Montreal Protocol, an international agreement that attempted both to create a framework of adaptive management to exploit in a continuing fashion advances in scientific understanding and also to build a capacity for monitoring impacts. Challenging questions of implementation through the hierarchical structures of federal states and subsidiary organisms became central.

Contrasting the European and North American versions of the implementation of early CFC bans highlights the evolution of ideas on the need for consultative mechanisms to build commitment and harness enthusiasm at the later stages of top-down processes that begin from "sounder science" and diplomatic negotiations, and ideas on the concern for continuous monitoring and evaluation provisions built into international agreements and/or domestic implementing legislation.

Banning CFCs in Aerosols Following the seminal Molina and Rowland article in *Nature* in 1974 and sparked by the personal involvement of committed scientists, environmental groups in 1974 began calling for a ban on CFCs and urged Americans to boycott CFC-based aerosol spray cans. Sales of CFC-propelled aerosol products dropped 25 percent in the first half of 1975.

Late in 1974, the U.S. Natural Resources Defense Council (NRDC) petitioned the Consumer Product Safety Commission (CPSC) to ban nonessential uses of aerosols containing CFCs. The CPSC ultimately rejected the petition in July 1975, but the NRDC suit further raised awareness of the issue.

The first response from industry to the connection between aerosols and the ozone layer was to launch a massive public relations campaign, spearheaded by the U.S. Council on Atmospheric Sciences (COAS) (an industry-funded public relations effort), the Aerosol Education Bureau, and DuPont. This solidarity was lost when the Johnson Wax Company announced it would discontinue all use of CFCs in spray cans (which, in any event, amounted to only 5 percent of its total use of propellants), setting off a flurry of activity within the whole industry to distance their products from CFCs by converting to alternative propellants and packaging.

The earliest governmental policy action directed toward CFC-based aerosols identified in this study was a voluntary ban passed by the Ann Arbor, Michigan, City Council in 1974. This action, of no force or effect except in raising awareness of the issue, can be attributed to the presence of a University of Michigan ozone-layer research team—in particular, team member Ralph Cicerone. The mayor of Ann Arbor then persuaded a local congressman to cointroduce a bill to amend the Clean Air Act (Roan 1989, 30).

In June 1975, Oregon banned the sale of spray cans containing CFCs (the act went into effect in March 1977), New York passed a law effective April 1977 that required all spray cans containing CFCs to carry a label that warned that they contained a propellant "which may harm the environment," and Ohio passed a limited ban in 1977. Many other states and municipalities considered the possibilities, with much resulting controversy.

Subcommittee hearings in the U.S. Congress in 1975 and 1976 pointed toward a federal ban on CFC-based spray cans being adopted. In October 1976, the Federal Drug Administration, Environmental Protection Agency (EPA), and CPSC formed an interagency working group chaired by EPA and announced their intention to propose regulations concerning nonessential (cosmetic and household) uses of CFCs in spray cans. These regulations—which at the time covered over 50 percent of CFC consumption in the United States—were generally viewed as a definitive response to the ozone-layer threat.

As efforts to extend action to cover industrial uses and production developed, industry advocates in the United States argued that in the absence of international cooperation to establish guidelines on CFC production and use, domestic regulations would handicap U.S. industry while doing little to address the global nature of the problem of stratospheric ozone depletion. Although this argument led to congressional rejection of a 1980 U.S. EPA proposal to freeze CFC production, U.S. regulatory agencies continued to press for domestic regulations to show U.S. leadership on this issue.

In Canada, the link between science and policy in the early regulation of CFCs was more direct, with little organized involvement from the environmental community but with close consultation between government and major industry players. The Atmospheric Environment Service (AES) of Environment Canada was the primary vehicle by which information on the effects of CFCs was delivered to federal officials and legislators.

An AES Working Group on CFCs recommended (on scientific grounds only) immediate regulations to achieve a significant reduction in CFC releases. Immediately on receipt of the AES report, the Environment Minister announced that Canada would regulate nonessential aerosol uses of CFC-11 and CFC-12 (Environment Canada 1976). Following extensive consultations with industry, regulations banning the import or manufacture of CFC propellants in the three largest aerosol uses—hair sprays, deodorants, and antiperspirants—took effect May 1, 1980, over three years after the Environment Minister's initial announcement.

Early attention in Mexico to the CFC and ozone-layer issue is attributed to the activities of Mario Molina, a Mexican-born chemist, who circulated the Molina and Rowland findings to colleagues and government officials in Mexico in the 1970s. At a trilateral workshop in 1976, the issue of ozone depletion and the risks to human health were highlighted, without any subsequent implementation action identified.

In general, the potential threat from CFCs was not addressed in a governmental regulatory response in Europe until some time after its initial emergence.[15]

In Germany, the federal government commissioned a study in 1975 on the scientific and economic aspects of the CFC issue, and by the end of the year the government was supporting some twenty projects studying the earth's atmosphere. The following year, the German Association for Environment and Nature Protection (BUND) and the Working Group of the Society of Consumer Associations both called for a boycott on aerosol cans—marking the beginning of extensive NGO involvement in the issue. The boycott was brief, however, due in part to the inability of consumers to differentiate between CFC and non-CFC aerosol products. In 1977, a government-industry agreement was negotiated in which industry voluntarily undertook to reduce the volume of CFCs in aerosols by 25 to 30 percent from their 1976 levels by 1979. When the Federal Environment Agency instituted its environmental labeling program (blue angels) in 1979, CFC-free spray products were the first to receive the labels.

In the Netherlands, the Political Party Radicals (PPR) and environmental groups launched action in 1974 against spray cans. In February 1975 a full consumer boycott was launched, leading to a substantial shift in the spray-can industry toward non-CFC products and subsequently to labeling requirements. In 1980 the government issued a request to industry to reduce use of CFCs in spray cans by one-half, relative to 1976 levels, by 1980. Although the boycott resulted in a reduction in domestic spray-can sales, there was a smaller effect on actual CFC production, where exports continued to represent a large share of sales.

At the level of the European Union, a 1978 Council Resolution on Fluorocarbons in the Environment, based on a 1977 Commission proposal, led to regulations two years later aiming for a 30 percent reduction in use of CFCs in aerosols from 1976 levels by 1981. As in dealing with other risks, concerns about the trade consequences of national regulatory initiatives in interaction with measures proposed at the European Union (E.U.) level can be seen as contributing to a delay in implementing either timely unilateral measures or a consistent policy overall.

The only activity with regard to spray-can bans in Japan came from the consumer group Citizens' Alliance for Saving the Atmosphere and the Earth, which called for a consumer boycott of spray cans. However, at a 1980 Organization for Economic Cooperation and Development (OECD) environment committee meeting—responding more to a sense of international obligation than to scientific conviction—Japan announced that it would freeze production capacity of CFC-11 and CFC-12 at 1980 levels through administrative guidance and seek ways to decrease the use of these substances.

There appears to have been little social concern for the CFC and ozone-layer issue in Hungary. Similarly in the Soviet Union, a long-standing background of scientific interest, contribution, and awareness failed to generate public interest or action until after the Vienna Convention and Montreal Protocol were endorsed. A generally comprehensive 1980 Law on Atmospheric Pollution contained no explicit provisions regarding CFCs or other ozone-depleting substances.

Reducing Industrial Uses of CFCs   While there was a focus on consumer end-uses of CFCs during the mid-1970s—and those efforts were successful in achieving reductions in the use of aerosols—the achievements that resulted from that campaign were soon overshadowed by growth in the use of CFCs in other sectors. In many ways, the targeting of consumer use of CFCs had been relatively easy: substitutes were generally available, and those applications for which CFCs were used were often trivial. However, one problem with the focus on consumer applications was that it gave the public the impression that banning spray cans had solved the CFC and ozone-depletion problem. The more important target—a reduction in the amount of CFCs produced—demanded an international agreement.

After 1985, interpretations of the scientific evidence on ozone depletion and the impact of CFCs on stratospheric ozone began to converge. Coupled with the (possibly inadvertent) admission by DuPont that CFC substitutes could be made available, the Natural Resources Defense Council called for a phaseout schedule that would culminate in a worldwide ban of CFCs and halons by 1998.

Interest in the CFC and ozone-layer issue was revived in Germany following the discovery of the Antarctic ozone hole. Most actors were still focused on spray cans as a major source, though other CFC uses such as foam packaging were also targeted. In 1987 the Federal Council passed resolutions calling on the federal government to prohibit the production and sale of CFCs, and a voluntary agreement with industry was negotiated to provide for a reduction of CFC use in aerosol cans by 75 percent by 1989 and by 90 percent by 1990. By the end of 1987, one month after adoption of the Montreal Protocol, the German parliament created an Enquete Commission on Preventive Measures for Protection of the Earth's Atmosphere.

In 1987 a group of NGOs in the Netherlands initiated a further consumer boycott, with a substantial impact on sales. In January 1988 a covenant between the Ministry of Environment and the organization of spray-can producers was signed.

In Britain, the Stratospheric Ozone Review Group (SORG) appointed by then Prime Minister Thatcher dismissed, as late as 1987, the CFC and ozone issue. Antiaerosol campaigns by NGOs, particularly a boycott led by Friends of the Earth (FOE) U.K., were launched but foundered. In February 1988, Prince Charles announced that CFC aerosols would be banned from the Royal household—an action that had a significant impact on public opinion—and FOE relaunched its campaign. In 1988 the SORG endorsed the conclusions of the Ozone Trends Panel. This report contributed to a dramatic environmental turnaround on the part of the Prime Minister and to her offer to host the 1990 London Meeting of the Parties to the Montreal Protocol.

A significant evolution in NGO action should be noted here, in the commitment of resources not to raising public awareness of the risk, on which there was already substantial acceptance of the science, but toward influencing the design and procurement decisions of industries highly dependent on the chemical giant Imperial Chemical Industries (ICI) the major British producer of CFCs.

While negotiating in the international arena, the European Community (EC) negotiated also with user industries of the Community member states to establish voluntary agreements on the use of CFCs, drafting three separate Codes of Conduct for CFC users.

Social concern about CFCs seems to have remained generally dormant in Hungary, despite highly regarded contributions to international scientific understanding of ozone depletion. Hungary did not attend the Vienna Convention but acceded to it in May 1988 and to the Montreal Protocol on April 20, 1989, several months after its entry into force for others (on January 1, 1989) and then as a result of international, not domestic, pressure.

Thus, the story of action up to 1992 to respond to risks of stratospheric ozone depletion falls into two distinct parts. A first, bottom-up, implementation round was substantially led by local groups and individuals, with boycotts or bans stemming from subnational or national regulation focused on consumer products. The second round led to more comprehensive international cooperation in regulation of industrial uses and production of CFCs themselves. In this latter stage, one can see the escalation of the issue to an international scope, with a concern for environmental consequences winning out over worries about competitive position, and the focus of standard-setting on possible environmental impacts

and health risks rather than direct regulation of specific products.

This internationally coordinated action, recognized explicitly as a first transitional stage, was undertaken subject to regular review of standards in light of evolving scientific evidence. These considerations, however, take the story into the account of international negotiations, which is the subject of section 19.3.7.

19.3.3 Offsetting Emissions: Carbon Sequestration
Arresting deforestation can provide an important contribution to an optimal climate policy. Important policy elements in this area include the conservation and protection of forested areas, changes in land tenure systems (including the development of private property rights) and the removal of logging subsidies, and a move to an ecologically based pricing of forest resources.

There are many reservations about the effectiveness of these instruments. Carbon-sink inventories are highly uncertain, and in several cases sink-enhancement policies may be contrary to biodiversity protection needs, where such enhancements rely on plantations of exotic and/or monoculture species. Nevertheless, tree-planting programs appeared rather prominently in announcements of action to offset industrial emissions. For example, the Tree Plan Canada Program, announced in 1990 in the Canadian government's Green Plan, was designed as a six-year program to plant 325 million trees; the 1990 America the Beautiful Act encouraged tree-planting projects in part to reduce global carbon dioxide levels; in Mexico the 1990 World Environment Day speech of President Salinas committed the government to planting 32 million trees as a response to global warming and other ecological problems.

Sinks and reservoirs of greenhouse gases figure explicitly in the U.N. Framework Convention on Climate Change (FCCC). The Convention on Biodiversity, the nonbinding Statement of Forest Principles, and several sections dealing with forestry and carbon sequestration agreed to in Agenda 21, all negotiated separately, also deal with the role which forests might play in offsetting industrial emissions of carbon dioxide.[16]

Early experience with the general idea of joint implementation (JI)[17] focused on forests as an offset to industrial emissions. Examples include a series of projects in which the World Resources Institute (WRI) worked with Applied Energy Services, an independent electric power company based in Virginia, to identify least-cost strategies to reduce carbon dioxide emissions. In 1991, AES donated U.S. $2 million to the Nature Conservancy, which covered the full purchase price of a 585 square kilometer section of forest in Paraguay. This preservation was expected to more than offset the 13.1 million tons of carbon dioxide emitted over the lifetime of a new AES coal-fired plant to be constructed in Hawaii.

There is a similar story for the Joint Electricity Producers (SEP) in the Netherlands, whose proposals to invest in afforestation projects outside the Netherlands generated substantial controversy, in part because of concern for the possible consequences for people living around the areas to be forested. In the end, SEP elected to arrange with a Foundation to undertake the tree-planting scheme on its behalf, but independently.

"Debt for nature" swaps as a means to protect forest areas in developing countries gained some prominence in the 1980s and were endorsed at the 1988 Toronto Conference on the Changing Atmosphere as a means to halt deforestation and deal with global warming. By the time of the Rio meeting more than two dozen such swaps had been initiated by NGOs, along with a number of debt-reduction programs undertaken by Western governments, including three under the U.S. Enterprise for the Americas initiative (Jakobeit 1996). Ultimately, this momentum was not maintained, however, and no major implementation of the idea came out of Rio.

Thus, while a variety of mechanisms to encourage carbon sequestration and other offsets to emissions of greenhouse gases have been envisaged, actions of this sort had made, up to 1992, no significant contribution to efforts to deal with global atmospheric risks. Symbolically, however, they played a greater role in buttressing industry arguments for cost-effective action to offset, rather than reduce, industrial emissions.

19.3.4 Economic Instruments: Taxes for Environmental Purposes

Tax measures of various kinds have been mentioned in connection with the accounts of national action to deal with both acid rain and ozone depletion risks. Such measures have been enthusiastically advocated for decades by economists but just as stoutly resisted by politicians and legislators. Up to the time of the Rio Conference implementation was confined to a very few examples. This section outlines some of the experience with emissions charges and carbon taxes.

Emissions Charges Japan has long levied a pollution tax on both stationary and mobile sources of air pollution for certain environmental costs and for the purpose of financing compensation payments to pollution victims. Under the 1973 Law for the Compensation of Pollution-Related Health Injuries, a pollution levy was to accrue to a fund administered by the Pollution-Related Health Damage Compensation and Prevention Association (directed by industry representatives). The Association also admin-

istered a ¥50 billion fund designed to finance various air-pollution-prevention initiatives. Polluting industries contributed 80 percent of this amount, and government 20 percent. Sulfur oxide charges were based on actual emissions, with rates differing substantially between regions.

Attempts to create economic incentives for purchase of more fuel-efficient automobiles with reduced emissions were found in several arenas. In the United States, excise taxes have been applied since 1980 on cars that do not meet federal fuel-efficiency standards. These tax rates were increased in 1991. Product charges identified in the Canadian arena included federal sales tax differentiation for vehicle weight, federal taxes on automobile air conditioners, and some provincial taxes on cars that are fuel-inefficient. Federal and provincial excise taxes resulted in differentiated taxes for leaded versus unleaded gasoline.

The 1990 integrated program to control air pollution in Mexico City included a fuel-pricing policy that taxed gasoline at above 50 percent of world market prices. The United Kingdom and Germany applied differentiated taxes for leaded versus unleaded gasoline. Germany taxed cars not equipped with catalytic converters at a higher rate and applied annual vehicle taxes on cars based on an emissions index covering major pollutants. The Netherlands introduced a tax on cars not meeting emissions standards as well as differentiated tax rates for leaded versus unleaded gasoline.

Title V of the Clean Air Act (as amended in 1990) in the United States authorized states to levy emission charges as a condition for permits. An excise tax was introduced on all ozone-depleting substances, equal to a base amount multiplied by an ozone-depleting factor for the chemical. This measure was designed primarily as a windfall profits tax, however, to account for the windfall profits created under the production limits imposed by the complementary system of tradable production quotas (see Section 19.3.5 on emissions permit trading), rather than as a means to discourage production or use directly.

The primary environmental enforcement tool in Hungary was a system of noncompliance fines, which, while not technically economic instruments, provide clear economic incentives to change polluting behavior (Klarer 1994, 68). The fines apply to ozone-depleting substances (ODS) controlled under the Montreal Protocol and to general air-pollution controls. Product charges on all types of fuel were introduced in 1992.

Similarly, in the former Soviet Union the instrument of choice for dealing with atmospheric-pollution problems was emission charges.

Thus one can point to a history of implementation efforts using tax instruments to deal with industrial discharges into the air and emissions from automobile use and other transport. But the efforts were scattered and frequently too modest to have substantial impact. For many reasons, the vigor of implementation did not match the enthusiasm of economic theorists for Pigovian taxes. Despite this experience of somewhat limited implementation success, however, recent theoretical argument has been particularly insistent on the merits of tax instruments to deal with threats of greenhouse warming. The case of carbon taxes is particularly interesting in illuminating implementation difficulties arising in the legislative setting.

Experience in Implementing (and Not Implementing) Carbon Taxes In February 1990, the government of the Netherlands introduced a carbon tax, but more as a means to raise revenue than to influence consumer behavior on carbon dioxide emissions. According to the Netherlands' own Environment Ministry deputies, the tax was too small to make real changes in the use of fossil fuels (GECR 1990). The tax rate was raised 300 percent on July 1, 1992.

Also in February 1990, the European Community began investigating the potential implications of a Community-wide carbon and energy tax to be implemented as an instrument to assist the European Community countries to meet their carbon dioxide emission reduction targets. It immediately ran into opposition from within and outside the Community administration. After considerable debate, the proposed carbon tax was rejected.[18]

In the United States and Canada there were also debates about carbon taxes but no implementation. The record up to 1992 thus suggests that when it comes to the crunch, the general preference in theory and in rhetoric for market instruments faces many practical barriers. Formidable reluctance was displayed not only by consumers and the public but by industry as well.

19.3.5 Economic Instruments: Emissions Permit Trading

Overall, there have been many attempts to introduce economic instruments into the arsenal of government measures designed to promote environmental objectives. While these generally took the form of taxes and charges in Europe and Japan, the United States led in the use of an alternative tool, a property-rights-based measure as contrasted with a fiscal instrument.

Title IV of the United States 1990 Clean Air Act Amendments has been described as a "major policy innovation: the first broad-scale use of a market for pollution" (Hausker 1992, 553). Its introduction of tradable sulfur dioxide allowances can be seen as an important test of the concept of "market solutions" to environmental problems.

But in application the purpose is in fact more modest. Tradable-allowance schemes operate as complements to regulatory structures to provide greater flexibility to firms in meeting the requirements of imposed legislation or in achieving standards stipulated by regulation. They are intended simply to provide a cost-effective means to meet prescribed emissions targets. It is therefore not surprising that most appraisals of implementation experience to date conclude that such instruments are successful in achieving significant cost savings but have little effect on environmental quality.

When finally the time came for implementation of action to deal with the issue of acid rain in North America, the lessons learned from almost two decades of experience with the trading of emission-reduction credits under the Clean Air Act Amendments of 1970 were available to carry into the design of Title IV of the 1990 Clean Air Act.

Tradable sulfur dioxide allowances were created in Title IV to facilitate achievement (and, it is argued, acceptance) of a permanent U.S. national cap on aggregate sulfur dioxide emissions at 8.9 million tons annually, a reduction in annual emissions of 10 million tons. Under this trading scheme, EPA issues annually (in perpetuity, it is expected) to the electricity-generating plants involved allowances equal to the permitted emissions of the source. At the end of the year, the source must deliver to EPA allowances equal to the actual emissions recorded (by audited continuous emission-monitoring equipment). Since allowances are issued annually, firms have the possibility of accommodating fluctuating requirements by buying allowances (in advance or on spot markets) to meet changing needs as forecasted with, presumably, lower costs as a consequence.[19]

In Europe, on the other hand, no such market structures had emerged by 1992. But flexibility in realization of imposed standards at lower cost was pursued through similar though nonmarket means.

In Germany, the 1974 technical guidelines permitted one firm to locate a plant in a nonattainment area if it replaced a similar plant, whether of the same firm or different firm, in the same area. In 1983 this provision was extended to recognize the offset offered by renovation, rather than outright closure, of the existing plant.

Covenants in the Netherlands offered similar flexibility. In 1990, the electricity producers' association SEP and four individual electricity producers negotiated with the government a covenant that provided for an aggregate target for reduction in acidifying pollution that went beyond the standards and targets existing at the time. Existing plants were required to meet existing standards, with the newest plants having to meet more stringent standards negotiated in the covenant. But targets for further reduction were expressed in aggregate form and not on a plant-by-plant or firm-by-firm basis. Substantial cost savings were anticipated by this move to an overall ceiling. Existing policies for oil refineries allowed a similar averaging of emission-reduction targets over individual plants within the same refinery. In the same way, under Canada's acid rain–control program, mandatory targets for Ontario Hydro, for example, offered the possibility of switching among plants within their overall grid, provided aggregate targets for the corporation were met.

At the European Community level, the bubble approach was incorporated eventually in the Large Combustion Plant Directives directed toward acid rain precursors. A proposed 1984 directive for which the initiative came from Germany was rejected by the Council of Ministers. In 1986 the Netherlands advanced a proposal that envisaged a regulatory bubble with an allocation to each country of a target reduction of emissions to be achieved as a percentage of an overall Community target. This proposal also failed. In 1988 a Large Combustion Plant Directive finally established agreed emission-reduction targets for sulfur dioxide and nitrogen oxides (NO_x) with different targets for each country—thus retaining, in effect, the bubble approach.

Thus the concepts of bubbles over plants, regions, and nations had been built into implementation decisions dealing with acid rain problems by the time that national governments had to face the task of meeting commitments established under the Montreal Protocol (and, indeed, by the time that the idea of a global bubble was being carried by former Dutch Environment Minister Winsemius into the 1988 Toronto Conference, where it emerged as a call for a World Atmosphere Fund) (see Dobell and Parson 1988).

The innovation was carried over into implementation activities dealing with ozone depletion. In 1988 the United States EPA proposed a system of fully tradable CFC production permits as part of the U.S. program to phase out CFC production. These permits, allocated annually, were to decline in number with the phaseout schedule but ultimately were set aside in favor of the excise tax described above. Similarly Canada's ozone-depleting substances regulations to implement commitments under the Montreal Protocol allocated permits for production or import of CFCs and provided for trades of production permits within firms, among firms (via the power of the Minister to reallocate unused permits among firms) and among the five controlled forms of CFCs, as weighted by ozone-depleting potential. Extensions to cover methyl bromide and hydrochlorofluorocarbons (HCFCs) were later put in place.

In Mexico, SEDESOL and the CFC industry agreed on a phaseout schedule for CFCs, and SEDESOL adopted a tradable-permit scheme to deal with the supply side of the phaseout targets.

In Europe, the provisions in the Montreal Protocol that permit the European Union member countries to implement jointly their allocated consumption quota created, in effect, a regional bubble. Canada and the United States had also made use, by 1992, of the "industrialization" provision of the Montreal Protocol, which allows transfers between countries of CFC production rights.

Thus, in the process of learning about means to pursue flexibility in regulation of emissions, European countries moved toward creating broader bubbles to permit transfers among plants or countries, while the United States moved more toward creating markets for the same purpose. In both cases there is strong evidence of transfer of experience over three decades, involving all three issues.

19.3.6 Voluntary Initiatives and "Challenge" Arrangements

Economic instruments, whether exercised through pricing mechanisms or property rights, offer ways to achieve environmental targets more cost-effectively than rigid directives. But the standard-setting exercise itself cannot be so readily delegated or decentralized. Political machinery (value-based judgment) is required at some stage to reconcile economic concerns about the survival of individuals and groups with ecological concerns about the survival of other species or the biosphere itself. A number of streams in actions undertaken in the social response to global atmospheric risks converge around negotiated arrangements—possibly undertaken on the basis of voluntary initiatives put forward by firms or industrial associations, perhaps within a framework of general legislation. Within such negotiated agreements, the possibility of an informal or implicit trading of commitments to emission reductions is obvious.[20]

Concern about such "bureaucratic agreements" has been expressed by de Beus (1991), but in a variety of forms they seem to be widespread in the portfolio of action taken to date to deal with atmospheric risks. Covenants, or voluntary pollution-control agreements, are a preferred approach not just in the Netherlands but in Japan also, where they emerged historically prior to development of any national environmental policy framework and were used by local prefectural governments as flexible measures in addressing local air-pollution problems. In Japan the pattern was set early; by 1970 there were over 800 arrangements negotiated by local governments with industry groups to go beyond the standards specified in national legislation governing emissions into the air. By 1991 the number exceeded 31,000.

Other voluntary initiatives (such as proposals by Unocal Corporation in California to finance, in exchange for emission-reductions credits, the withdrawal and scrapping of older automobiles) are worth noting in this same connection. These amount to broadening the scope for recognized offsets or joint implementation: they permit, through flexible administrative arrangements, a broader range of more imaginative ways, beyond the scope of regulations to envisage exhaustively, to meet commitments to achieve target reductions in emissions. Moreover they permit the building-in not just of continuous emission monitoring but also of a philosophy of continuous improvement that ratchets down emissions levels through negotiation rather than through a sustained reduction in aggregate numbers of emission permits, for example, or through the actions of other actors in buying and retiring such permits.

In a changing and uncertain world, implementation can in the end be achieved only through continuous adaptation and refinement of policy and instruction. Ultimately, the problem of implementation, as is noted in the conclusion to this chapter, reduces to the question of whether these necessary adaptations are being made by actors committed to the purposes of the legislated goals and regulatory targets or by those committed, through self-interest or ideology, to thwarting the process and negating the targets. Scrutiny matters here, to ensure that flexibility and discretion are not granted without accountability for the outcomes. Liability provisions or environmental bonding may matter crucially here also, in spelling out financial accountability for failure to achieve agreed standards. (The prompt response of industry in Japan to emission-reduction requirements can surely be linked to the establishment of a Pollution Victims Compensation Law that put a price on industrial pollution, for example.) Thus the trend toward negotiated arrangements becomes coupled closely with assurances about the adequacy of data, reporting, and access to information, as well as broad participation in regulation-writing, monitoring, and enforcement processes.

19.3.7 International Agreements: Covenants without Sanctions by a Central Authority

The previous stories illustrated briefly the use of a variety of governing instruments and national measures to deal with global atmospheric risks, principally through efforts to reduce emissions. This section examines efforts to orchestrate internationally some commitments to such measures by agreement on national targets for emission reductions. Such exercises may be considered analogous

in some fashion to the voluntary covenants favored in many national programs, except of course that they lack the threat of direct enforcement in the case of default or noncompliance. Sanctions cannot flow from any central international authority but only by agreement among sovereign nations collectively committed.[21] The development of dispute settlement and noncompliance provisions within international regimes reflects an important element of learning in implementation, as noted below.

In the case of acid rain, the Convention on Long-Range Transboundary Air Pollution (LRTAP) provided the framework within which successive protocols dealt on a gas-by-gas basis with sulfur dioxide, nitrogen oxides, and VOCs in agreements concluded in 1985, 1988, and 1991, respectively.

Hungary and the former Soviet Union signed the LRTAP Convention early, as a signal of their desire to be more fully a part of an international community. As the Convention evolved, however, the arrangement moved toward a more systematic treatment of acidifying deposition, until by 1992 it was able to design national commitments on emission reductions with the aid of an elaborate decision support system.

Once initiated, international management emerged more quickly in dealing with the risks of stratospheric ozone depletion, but again with the familiar framework and protocol structure (and, as Parson 1993 emphasizes, only after a lapse of eight years from early international discussions initiated and orchestrated by UNEP). The Vienna Convention signed in 1985 (which came into force in 1988) laid the groundwork that made possible significant action on the Montreal Protocol (1987) and its subsequent strengthening in London (1990) and Copenhagen (1992). Parson (1993) links this framework convention and protocol structure not directly to the earlier experience with the problem of acid rain and the LRTAP Convention of 1979, but (via the 1981 UNEP Montevideo meeting of experts in international law) to the perceived success of UNEP's Regional Seas Programme and the Mediterranean Plan (Haas 1990), where the 1976 Barcelona Convention and its protocols became the model for eight subsequent framework treaties and their protocols in other marine regions (Sand 1988).

The chronology of action at the international level to deal with stratospheric ozone depletion, viewed over two decades, reveals a substantial evolution that could be characterized, as was the U.S. Clean Air Act of 1970 in a different context, as "rational, patterned and explainable" (Ingram 1978). The account given by Haas (1992) of the commitment of resources by UNEP to coordinate, first, scientific research and exchange, then diplomatic preparations, leading to the Vienna Convention and onward to the Montreal Protocol and the subsequent strengthening of commitments through amendments at succeeding conferences of the parties, suggests overall a strikingly orderly evolution. Further, it can be noted that the times required to negotiate and ratify the Vienna Convention were significantly longer than those for the Montreal Protocol and subsequent amendments. For those countries ratifying the Protocol and amendments, the time required was generally under one and a half years.[22]

Effective provision for successive measures to strengthen and adapt commitments in response to accumulating scientific evidence and understanding is another feature of the evolving implementation process illustrated by the experience with risks of stratospheric ozone depletion. This capacity, achieved through institutional structures that channel the necessary decisions through technical working groups rather than political or diplomatic bodies, stands in interesting contrast to the blunter instruments of direct action and consumer boycotts with which action to deal with ozone risks began (Sands 1995, 114):

The provisions of the 1985 Vienna Convention and the 1987 Montreal Protocol illustrate new techniques [for updating targets and commitments in light of new scientific evidence]. . . . The 1987 Montreal Protocol also provides an alternative to formal amendment by the adoption of "adjustments and reductions" by the parties . . . adopted by a two-thirds majority of the parties present and voting which represent at least 50% of the total consumption of the controlled substances, and these are binding on all parties without possibility of objection.

As Palmer (1992a) notes (in the climate change context), this move away from insistence on unanimity represents a significant break with tradition in the international arena.

As noted in the discussion of individual country action, the instruments adopted by parties to the agreement to meet their commitments under it evolved toward more general measures, including quantitative permits to produce or trade ODS, with the permits themselves in some cases being tradable inside individual countries and fungible across substances at exchange rates reflecting their estimated ozone-depletion potentials.

Experience with the Framework Convention on Climate Change reveals how much more difficult it was (and is) than even the very challenging cases of acid rain and ozone. Much more fundamental changes in behavior and lifestyle are at issue, as has been widely noted. And neither dramatic photos of forest dieback and dead lakes nor colorful scientific charts of ozone "holes" were available as evidence to warrant wrenching and costly changes. As a result, national measures likely to achieve committed reductions in emissions of greenhouse gases were not in prospect as the Convention was negotiated over the months leading up to Rio.

The movement, rather, was even further in the direction of a new and "softer" (Palmer 1992a) approach, with the hope that an initial emphasis on goals, reporting, and scrutiny might later harden into more formal targets and timetables for emission reductions.[23] Indeed, the Convention on Climate Change illustrates the extent to which the reporting requirement is becoming more onerous and emerging itself as a central technique for ensuring implementation. The agreement required parties to publish national inventories of emissions and removals and communicate information regarding implementation efforts. It also extended the provisions of the 1990 London Amendments to the 1987 Montreal Protocol for meeting some costs to ensure the participation and compliance of developing countries. In the United Nations Framework Convention on Climate Change (FCCC), developed-country (Annex II) parties undertook to provide new and additional resources to meet agreed full costs incurred by developing-country parties in complying with reporting obligations and agreed full incremental costs of implementing measures related to substantive commitments (Sands 1995). This provision links directly with the observations below about the greater role to be played, necessarily, by nonstate actors to provide adequate ongoing scrutiny and audit capacity.

Thus, by 1992, action at the international level to deal with air issues was moving toward a new phase. With the LRTAP Convention this entailed a second round of protocols, with analytically supported differential obligations. With respect to ozone, international action enjoyed great success in reducing the production of CFCs but with the real challenge of compliance issues still outstanding. And with climate change, while it may be viewed as a triumph of a new softer approach to making international law, the task of establishing targets and timetables or even a realistically precise process of "commit, review, and report" remained outstanding with the passage of the FCCC.

In all three cases, international action followed the pattern of a framework convention coupled with subsequent, more detailed protocols. The initial framework conventions did little more than establish agreement on the character of the problem and create the capacity to talk about it in successive meetings of an increasingly detailed nature. Serious reservations about the possible cost of such an approach to treaty making in the area of international environmental risks are expressed—and some suggested alternatives explored—by Susskind (1994). On the other hand, a variety of observers (e.g., Palmer 1992a, 1992b;

Johnston 1997; Széll 1995) appear to hold the view that few alternatives to such a staged approach exist.

The important pattern or lesson to note from these very brief accounts is probably the fusion, in the pursuit of international measures to address global atmospheric risks, of continuing diplomacy with ongoing administration. Continuing negotiation, previously viewed as part of treaty preparation only, had become by 1992 an accepted part of treaty implementation, viewed as an ongoing process of adaptation to ensure continued commitment and compliance. What also can be noted, more particularly, is the consistent pattern of transition in implementation action first from general diplomatic and political maneuver to more concrete institutional capacity-building and collective commitments, and then onward to operational tasks of monitoring country compliance with their agreed undertakings (just as countries setting standards for themselves have to move on to monitor the compliance of individual economic agents, industry associations, or industrial sectors with new national rules). As Parson and Greene (1995) emphasize, it is this transition to the concrete tasks of implementation that poses serious operational challenges on which there cannot yet be any confidence as to ultimately successful outcomes. Though commitments have been negotiated and put in place in an international setting and successively refined and made more precise, it remains still to be seen whether these will be fully reflected in national action adequate to deal with the risk, openly and fully reported to the international institutions supporting the agreements, and also in the subsequent continuing operational measures necessary to meet national commitments.[24]

In each of the three cases studied, signatory countries charged international institutions with establishing technical advisory panels to assist states with compliance. In the case of European acid rain, workshops and working groups were established to explore new technologies and processes and to identify combustion methods and industrial designs that emitted relatively lower amounts of acidifying compounds.[25] In the ozone case the various panels have been described by Parson (1993), with the effect being that laggard states and companies were made aware of scarce technical information that they lacked and that warranted continued international cooperation and also were assisted in meeting compliance with past commitments. Finally, similar expert panels were established under the FCCC to identify and appraise technology that may help countries to reach their commitments under the treaty. Thus, to the extent that implementation includes the capacity to disseminate information on "best practices" and to support transfer of knowledge and technology, it reflects the interest of all parties in helping weaker or laggard members of the international community to do better and not just information dissemination for the purpose of scrutiny and shaming to force compliance.

The successive protocols also display a process of refinement by which each builds on the experience of the previous elements of the series. This process entails both increasing precision in some of the language and increasing flexibility or discretion built into the specification of means to meet commitments, as well as the adjustment of commitments to reflect accumulating scientific evidence and apparent convergence in the interpretation of that evidence.

The growing focus on compliance rather than enforcement is a significant feature of the story at the international level: "Our analysis leads away from the search for better enforcement measures—'a treaty with teeth'—to better management of compliance problems" (Chayes, in Lang 1995, 89).

So also is the growing emphasis on provisions to encourage participation—a "global bargain" in which technology transfers, resource transfers, or financial transfers now are seen as less directly related to development aid and more as measures to induce participation of nonparties in treaties that otherwise they may see as not sufficiently in their interest, at least in the short run. The trade provisions of the Montreal Protocol, it is argued, are not sanctions to deter noncompliance by parties but are measures to get around the "least common denominator" problem where noncompliance by nonparties can confer a competitive advantage (Campbell, in Lang 1995, 233).

Along with the voluntary character of the undertakings assumed by sovereign national authorities goes the requirement for scrutiny and capacity to report on—if not enforce—compliance. An interesting feature in the story of international action is the emergence of nonstate actors in the process of scrutiny and, indeed, in more formal audit roles.[26] So also is the emergence of industry-initiated voluntary certification programs that, at a higher level of abstraction, certify the existence and adequacy of systems that firms must have in place to be able to report appropriately on their compliance with regulatory standards established nationally to meet commitments under international agreements (Rozniak 1994; Schmidheiny 1992).

More generally, the important and growing role of nonstate actors, including NGOs, is increasingly recognized by many observers: "The end of the century will be characterized by increasing NGO influence" (Bekhechi, in Lang 1995, 161). Sands (1995) also notes this feature as one of the four general conclusions emerging from his review of changing patterns in international environmental law. A key development here is the part played by new communications technologies. An anonymous reviewer of this volume cites, as "something of growing significance in terms of learning," U.N. and NGO use of the Internet and EcoNet in dissemination activities (for example, the use of EcoNet since 1990 to carry all documentation submitted to preparatory meetings for the Earth Summit, a practice now being replicated for climate and other issues) and notes that "often the local NGO community will start focusing on issues in newly issued U.N. documents before the government receives the formal copy and puts it into circulation."

These developments prompt more general questions about the possible nature of an overall international community to which sovereign states may (by virtue of emerging customary rules) owe obligations in respect of their actions in the global commons, beyond more conventional notions of liability to other sovereign states harmed by those actions or other legal obligations to other states that are party to agreements relating to such actions.

Finally, the record just reviewed also serves as a reminder that action may initially be taken more with the intent simply of demonstrating commitment to a cause, underlining the importance attached to a problem, or demonstrating a symbolic sympathy with others concerned about an issue than as the result of any firm conviction that the action itself will have objective impact on a substantive target. Further, it seems evident that in the early stages of some of these negotiations national authorities were entering into substantial commitments internationally in advance of any clear knowledge as to exactly how these commitments would be met when the stage of concrete implementation through national rule making arrived. Nevertheless, all three cases examined here suggest that there has been significant learning about—or at least evolution in—action at the international level.

19.4 Did the Performance of the Implementation Function Improve?

19.4.1 Criteria to Assess Improvement

We conclude that societies are getting better in undertaking action to address global atmospheric risks. This section suggests some reasons behind these claimed improvements in implementation efforts, flowing largely from more successful attempts at international cooperation. But one caveat must be stressed at the outset. One dimension of success may reflect compliance with agreements, or treaty effectiveness. A second may reflect actual changes in behavior of actors, or substantive effectiveness. The really significant test hinges on target effectiveness—actual improvement in the state of the environment or ecosystem

health.[27] These are all questions for a deeper evaluation than this chapter can undertake. Here we talk about "improvement" in a more mechanical sense of better realization of intent: better linkage from idea to outcome, not better idea or better outcome.

On the basis of criteria proposed by Clark and Majone (1985), in a somewhat different context, we argue that implementation has improved, in at least three respects.

First, the range of governing instruments actually used was broadened. The theoretical arguments for "least-cost implementation" have been discussed for a long time and have led to advocacy on behalf of more generalized, flexible regulatory measures (permitting greater discretion in the means adopted to achieve broadened standards) and economic instruments (permitting discretion as to how stringent a standard will be attained). An upswing in support, on general philosophical grounds, for market instruments is obvious. And in more recent times, arguments for specific economic instruments like green taxes have been advanced, sometimes with the revenue motivation dominant and the environmental consequences adduced merely as a favorable side benefit. All these theoretical considerations were strongly advanced throughout the period covered by this study but encountered significant practical objections on grounds both of feasibility and, more particularly, political acceptance. Nevertheless, on the basis of practical small experiments, the record on implementation does show attempts to include a variety of economic instruments and broader, more flexible regulatory processes, even though action up to 1992 was fragmentary.

Second, objectives expressed in the implementation process—the commitments to be achieved—began to be phrased in terms of more fundamental measures that were more logically related to the underlying character of the risk and the responses to it. In all three cases, commitments with respect to national action were by 1992 being calculated in light of anticipated impacts (such as critical loads or target loads) or potential contribution to the risk in question (such as ozone-depletion potential or global warming potential). Environmental harm, not mere discharge, became more often the focus for regulation. In this respect, implementation action became more coherently related to fundamental concerns with human or ecosystem health outcomes or consequences of the risks in question.

Third, the implicit policy making frequently buried in the implementation process was dramatically opened up through the growing involvement of NGOs and other outside groups. While it risks slowing down and diffusing the process, the growing involvement of national and international environmental nongovernmental organizations (ENGOs) in the development of national action plans to meet international commitments does mean a more democratic, and presumably therefore more legitimate, process of implementation. Greater acceptance of regulatory or fiscal action can be expected.

To date we do not have the informationally efficient and environmentally appropriate global price signals to guide the corporations and economic agents at the center of all the little operational decisions that add up to implemented social response (Hawken 1993). This might be said to be the big implementation failure so far. The alternative to such pricing mechanisms is to express changing (consumer) preferences through purchasing alliances (boycotts) or shareholder (more broadly, stakeholder) activism. The growth of environmental auditing and reporting, and the emergence of voluntary standard-setting activities are major developments for this purpose.

When it comes to dealing with the global character of the risks, the implementation record can be seen as directed in part toward increasing concern and commitment (public education campaigns and so forth) and in part toward an improved contracting environment (some attempt at ecological pricing and ecological taxes) but mostly at increasing capacity at the national and international levels. Using the categories proposed (Levy, Keohane, and Haas 1994) for assessing international negotiations, then, we can also argue, in addition to the above improvements, that we have certainly seen executive action taken and resources committed to

• Increase concern and commitment (such as through NGO and government campaigns to heighten awareness of risks);

• Increase capacity (for example, to audit and monitor follow-through on implementation through Friends of the Earth and the Climate Action Network), to exchange information, to assess risks and possible responses through a long array of research programs initiated and continuing, and to undertake transfers of resources to encourage participation in agreements that would not otherwise be sufficiently in the short-term interest of nations that should be involved); and

• Improve the contracting environment (such as pilot projects with regulatory innovation, bubbles, or economic instruments; multilateral funds; and initial efforts to improve prices and signals through "green taxes").

19.4.2 Explanations for Improvements

These procedural improvements represent an improved capacity to carry through implementation processes appropriate to the challenge of responding to global atmospheric risks. They stem in part from substantial learning, since 1957, about the social and institutional

systems within which implementation takes place, as well as accumulating scientific understanding about the nature of the risks themselves. In particular, we argue that by 1992 negotiations and decisions reflected better understanding of the

• Concept of sustainability and the significance of social and natural capital in processes of global environmental change (O'Riordan and Jäger 1995), the implications of the precautionary principle for rules of evidence and the burden of proof (Shrader-Frechette 1991), and consequences for participation, representation, and principles of standing (Stone 1993);

• Dynamics of common pool resources and problems in management of the global commons (Hardin 1968; Ostrom 1992; Brown Weiss 1995);

• Issues of collective action and enlightened self-interest (Olson 1965; Axelrod 1984; Dawkins 1989; Frank 1988) (see also Parson 1992);

• Dynamics of implementation processes as evolutionary, interactive bargaining processes (Majone 1983; Downing and Hanf 1983); the separate issues of institutional design, instrument choice, and organizational innovation (Calista 1994); and the consequent need for negotiated agreements on voluntary action coupled with specific obligations for reporting and review.

This improvement in understanding the nature of implementation processes led to some reduction in earlier expectations about dramatic central action to cure implementation failures or reduce enforcement deficits directly. It recognizes the likely wisdom of decentralized decisions in achieving faithful implementation of legislative or strategic intent in diverse and changing local circumstances and focuses on the need to secure the commitment of myriad independent agents in the process. This challenge leads on to the question of whether such commitment can be legislated through formal regulation or explicit accountability frameworks, whether it can be bought through incentives and economic instruments, or whether it must be inculcated in and flow from education and changing beliefs. Greater awareness of these questions has led to a more promising choice of governing instruments for implementation purposes.

So also has greater capacity in recognizing the constraints imposed by, as well as the potential benefits of, using markets for purposes of implementation of measures to achieve environmental goals. In particular, the need for full information, widely available, to enable market mechanisms and market disciplines to work effectively has been clearly identified.

More pluralistic, democratic, and open processes of international agreement and national rule making encouraged greater "buy-in" by essential participants and hence greater legitimacy in implementation processes.

As noted earlier, the growing role of NGOs as actors at the international level has emerged as an important theme in this review. In the acid rain story in Europe, along with Friends of the Earth and Greenpeace and the International Institute for Applied Systems Analysis Regional Acidification Information and Simulation (IIASA RAINS) model, influential public-information and lobbying campaigns were undertaken by a limited number of groups such as the Worldwide Fund for Nature (WWF), World Information Services on Energy (WISE), and a minuscule Swedish and Norwegian NGO Secretariat on Acid Rain. In the Canada–United States story also, the NGO presence was largely limited to the special-purpose Canadian Coalition on Acid Rain. But in both the formal and informal negotiating processes leading to Rio and the FCCC, the NGO presence was substantial and apparently influential. In the aftermath of the United Nations Conference on Environment and Development (UNCED) there is formal provision for NGO participation in the activities of the United Nations Commission on Sustainable Development (CSD) and in other institutions for development of international environmental law.

An important—indeed central—explanatory factor underlying this increased involvement and consequent improvement in the legitimacy and acceptability of implementation efforts is the revolution in information dissemination and computer-communications technologies. The growing use of email and specific networks such as EcoNet has been mentioned above.[28] The use of this capacity by NGOs and others (including the United Nations and other international institutions themselves) to disseminate information and build pressure on governments had become, by the time of Rio, a key new feature in processes of implementation, particularly in the implications for private-sector and nongovernment capacity to comment on, if not monitor, implementation and compliance internationally and domestically.[29] In addition, the establishment and maturing of epistemic communities has meant the growing experience of individuals in a position to carry the lessons from one stage of an implementation process to a later or from one case to another.[30] Lesson drawing among international institutions in respect of measures for implementation of international agreements may in some cases emerge in a manner seemingly analogous to lesson drawing among national bureaucracies in respect of implementation of national policies.[31]

Széll (1995) spells out details of the way in which implementation efforts following the Montreal Protocol attempted to build a supportive and nonconfrontational noncompliance regime to bridge the gap between routine reporting requirements and confrontational dispute-settlement provisions. Again the parallels between these developments in building international regimes for implementation of international agreements (in the sense of effective monitoring of national compliance) and the approach taken to voluntary covenants in seeking realization of national policies is interesting. Just as the sword of potential future coercive regulation continues to be held over the performance of industries entering into negotiated agreements on meeting improved national environmental standards, so Széll argues (1995, 107) that the presence of dispute-settlement procedures, even though possibly never utilized, helps to give force to a nonconfrontational and supportive noncompliance regime.

Finally, international scrutiny of compliance by individual countries with international agreements and commitments, and even audit of domestic legislation and policy measures for achieving compliance, may be becoming more generally acceptable and regarded as legitimate. The concept of "intrusive sunshine" as an important instrument for enforcement may lead to systematic provisions for dealing with citizen complaints, thereby multiplying vastly the capacity for ongoing monitoring and audit of compliance at the local level.

19.5 Conclusions

19.5.1 Evolution in the Implementation Function Is Evident

Inferences about learning or changes in patterns in the process of implementation are hard to draw out from the background of many ongoing social developments. Nevertheless, examination of the (highly selective) empirical record set out in this chapter suggests the following inferences about the way in which the process of implementation has evolved in the arenas featured in this study over the thirty-five years leading up to the 1992 Rio meeting.

1. National societies have responded to increasing awareness of the transnational character of the risks studied by building institutional capacity at the international level, moving the starting points for implementation efforts to the international arena. That is to say, the record of implementation measures taken shows an evolution toward attempts to deal with transnational risks by creating international institutions to negotiate national commitments and monitor national action to meet those commitments. These institutions provide a forum for international cooperation to address problems where the short-term self-

interest of individual nations would otherwise drive each toward action inconsistent with the longer-term collective interest.

2. In the design of these international institutions, emphasis shifted from instruments targeted simply on emissions (action to achieve a reduction of discharges on a pollutant-by-pollutant basis) toward standards intended to bring about targeted levels of environmental health or quality (taking into account differing contributions to environmental harm, possibly through measures differentiated across nations).

3. Domestic action to deal with problems of acid rain reveals some tendency to establish hard symbolic targets and then back away in light of implementation difficulties (Majone 1983). The particular ideological orientation of the Reagan administration in the midst of the deregulation wave of the late 1970s and early 1980s made it difficult for Canadian-U.S. negotiations to build on the substantial progress made in European regional arrangements for purposes of developing a regional agreement in North America. While international science bodies flourished in Europe to deal with this issue, there was an almost total failure to create comparable bodies in North America.

4. The framework convention and protocol approach pioneered in the UNEP Regional Seas program led in the case of ozone depletion to successful implementation of an agreement that is flexible and responsive to new scientific results.

5. In a more complex setting, with many more actors concerned, the FCCC moved further toward a philosophy of "soft law" with its focus on reporting, transparency, and commitment to general objectives rather than fixed targets. With its explicit provisions for scrutiny and justification, the convention opens the way for a more central role for activist NGOs.

6. In federal systems, a key barrier to implementation can arise in the ways in which subnational units are involved (or not) in the development of international agreements and in cooperative approaches to implementation of commitments undertaken. Similar issues arise within the supranational structure of the European Union.

7. In the review of national experience, the most striking feature of the implementation record is the growing appeal to voluntary arrangements or covenants, whereby governments negotiate with industry associations or sectors a long-term program of measures to be closely audited for results. The increasing sophistication of monitoring equipment and specifically the feasibility of continuous emission monitoring are often crucial prerequisites to the credibility of such arrangements.

8. Economic arguments for the effectiveness of economic instruments are strong in all arenas but face strong political barriers to adoption. Among the countries studied here, only the Netherlands (in company with Scandinavian countries) had implemented a carbon tax by 1992, and that was not significant in terms of behavioral changes to achieve environmental objectives. While the United States pioneered the introduction of tradable emission permits, few other countries drew positive lessons from that experience, at least up to 1992. For the most part, it seems, economic instruments were adopted only where they were easy, mostly for reasons other than pursuit of environmental objectives (though Japan may be a noteworthy exception).

9. The most fundamental structural feature in the evolution of the implementation function is the move of NGOs into central roles in the implementation process. Their increasing technical capacity to focus on scrutiny, audit, and disclosure of activities throughout the process of negotiating regulations and enforcement provisions had the effect of moving the implementation activity from an elite negotiation around technical concerns to a more open accommodation involving value-based differences. This evolution, in effect, carried the ongoing value-based differences in perspectives over from the debates around policy formulation into the continuing processes of standard setting, inspection, enforcement, and compliance assurance. In the international arena, business entities mobilized international associations to intervene earlier in the processes of policy formulation. NGOs learned that they must stay involved longer in the (domestic) process of implementation, carrying their activities through into the minutiae of definitions, standards, and monitoring equipment to ensure that the purposes embraced in strategies are not eviscerated in their realization and in particular to ensure that international commitments are reflected in subsequent organizational action at the subnational level.

10. In the traditional literature on policy analysis, emphasis was placed on the formulation and design of clear policy mandates without (until recently) much conscious concern for implementability. Our examination of the implementation record shows that in the broader context of institutional design and selection of supporting instruments, emphasis did shift from mandate concerns to issues of organizational acceptance and individual compliance; design was redirected toward self-regulation and incentives for compliance at the "pointed end" of the system. And skepticism about ultimate outcomes drove much of the debate.

11. In the end, the view of implementation as unending interactive negotiation comes down to the issue of

commitment to a process of constructive adaptation designed to ensure that individual outcomes in a changing environment will achieve effective compliance with the spirit and goals of covenants embraced. It seeks a relationship in which the environmental responsibility and social purposes of citizens are not thwarted by reluctant agents, either corporate or civil, throwing "sound science" and "cost-effectiveness" in the way of efforts to achieve responsible adaptation of general policy mandates to diverse and changing circumstances.

12. Although a central feature of the implementation record is the creation of capacity at the international level to deal collectively with global risks, there is not a uniform escalation of implementation to an ever-larger scale. Central also is the trend toward increasing "subsidiarity" and devolution of responsibility through appeal to more flexible directives—economic instruments, incentive-based regulation and performance-oriented bureaucratic covenants, designed to achieve designated (indeed, improving) standards at least economic cost. Increasingly accepted, moreover, is the need to provide specifically for monitoring, reporting, and disclosing measures bearing on the degree of compliance with declared targets or international commitments.

Thus we see implementation action at all levels but with very different emphasis. At the international level, action is directed toward building cognitive capacity, crafting contracts and commitments, and working from esoteric knowledge and epistemic communities. Locally, by contrast, action is directed toward meeting ongoing operational responsibilities through discretionary adaptation to changing circumstances, drawing on traditional knowledge and cultural communities. Nationally the focus is on designing legislation and regulatory frameworks, crafting mandates, and increasingly, perhaps, simply on mediating between the demands of international commitments and the local actions that industry and individuals need to take to meet these demands.

19.5.2 An Orderly Flow Emerges
The overall impression that emerges from the record of action independently taken by myriad individual actors over a few decades to deal with each of the three risks studied here is one of an unexpectedly orderly aggregate flow and progression emerging from the interplay of uncoordinated intentions. A review of the story sketched in this chapter suggests the following:

• A reasonable sequencing of resource commitments to implementation action by various agencies, with social resources flowing first to initiation of research programs to identify and characterize risks, then to the commissioning

of work to appraise possible policy responses, then to the creation of capacity (particularly at the international level) to negotiate a collective response, and subsequently to the creation of capacity to monitor action and refine targets within an increasingly structured and predictable administrative framework;

• A reasonable division of labor among social actors with governments or independent scientists committing the resources to launch research programs to explore the nature of the risks identified, governments and non-government organizations committing resources to the debate as to possible responses, and governments negotiating (through regulatory mechanisms or voluntary agreements) with industry actors on action necessary to achieve compliance with the measures agreed nationally or internationally as appropriate responses;

• Attempts by industry associations, just beginning at the international level by the time of the Rio meeting, to encourage industry to show itself socially responsible and "green"—to respond to both government and consumer pressures by adoption of environmental audit and reporting practices.[32] (Schmidheiny 1992);

• Consistent with a general trend in social attitudes toward participatory decisions and the essential accountability of government, a strong pressure from nongovernment organizations to be involved in the regulatory or administrative measures necessary to implementation and in attempts to ensure scrutiny and audit of industry's compliance.

19.5.3 Convergence toward Adaptive Learning and Focus on Commitment Can Be Claimed
Thus, given time, the process of implementation does seem to work its way toward action to meet the declared goals emerging from deliberative processes. Given the case in favor of incremental approaches (Collingridge 1992) or adaptive management (Lee 1993), the record reveals an appropriate approach to implementation, not just in the small actions but in the pursuit of fundamental measures to deal with global atmospheric risks. The growing capacity of emerging international institutions to support orderly international action has been noted above. Moreover, change gradually worked its way downward from declarations and initiation of executive action at the top, into the standard operating procedures, the prices, and the values of subsidiary component units in the economic (and political) system.

Experience with the implementation process does not show many examples of successful attempts to engineer, through authoritative executive action, massive leaps to new and less ecologically demanding lifestyles on the planet. But it does show some gradual adaptation toward a new awareness and new values in human activities, new community interactions, and new institutional arrangements. What would seem to be crucial in this more gradual and adaptive implementation process, then, is simply continued vigilance, audit, scrutiny, and pressure on all the myriad operational decisions at the grassroots level that must mirror constantly tightening overall goals established collectively in international institutions.

The growing appeal to covenants and soft law thus is, in this argument, the key transition, the greatest social lesson to be learned from the reading of the implementation record. At the core, there must first be commitment. This must then be backed by capacity and accountability if significant and sustained reductions in environmental risks are to be achieved. Action will not result solely from the exercise of executive authority. The lesson from implementation experience is that implementation cannot be decreed centrally. In an uncertain, plural, changing world it will be blocked by individual indifference or resistance unless there is commitment at every stage in the chain of events.

And as always there are two schools on how the necessary commitment is to be achieved. The first argues the need to respect self-interest and ensure that incentives are built through the appeal to market structures. This school is individual, transactions-based, and contract-oriented (Coase 1960; OECD 1994). The second rests on shared goals and sensed mutual obligation, on responsibilities and duties rather than market opportunities, on building structures of cooperation (Lichbach 1996). This school is networked and oriented to sustained cooperation in sustained relationships (Axelrod 1984) and trust where explicit monitoring may not be plausible (Frank 1988).

The necessary relationship of these—the need for an ethical setting within which to place the incentives of market structures—is clear. Market mechanisms permit decentralized action in the vast bulk of human activity, and formal or informal institutions of group ownership serve to organize access to local open-access resources in an orderly, efficient manner (Ostrom 1992). But in dealing with the global commons, the market and the invisible hand cannot be sufficient. Only a more conscious cooperative structure of international governance, based on some explicit framework of ethics or relationship to the land, will do.

This chapter has traced this evolution from action to create national agencies reflecting the scientific concerns about long-range transport, through the U.N. Conference on the Human Environment (UNCHE) meeting in Stockholm in 1972 (which focused those concerns on transfrontier political issues and the need for international cooperation), to the implementation efforts focused on

the creation of monitoring and administrative capacity as well as negotiation processes, and now onward to the next phase of national rule making to realize commitments undertaken internationally and to coordinate operational measures in subnational and nongovernmental organizations, both corporate and civil.

Whether this coordination was pursued through regulatory instruments or market mechanisms, it appears that the action on the ground converged toward similar sets of measures. On the one hand, regulatory interventions widened and softened through creation of bubbles and regulatory efficiency measures, leading ultimately to voluntary agreements and covenants. On the other hand, market instruments extended information and price signals to serve the "green consumer," thus forcing voluntary measures on the "green corporation" through the civil society (market) equivalent of the "bureaucratic covenant." Possibly to forestall either citizen action through bureaucratic agreement or consumer action through market pressures, international business associations such as the Business Council on Sustainable Development (BCSD), now the World Business Council on Sustainable Development (WBCSD), or International Chambers of Commerce (ICC) urge corporations to undertake voluntary action as a reflection of their own social responsibility.

In either case, these developments amount to a general renegotiation of property rights in light of new knowledge about the global consequences of individual action and new concepts of social responsibility, in a new world grown small. In the language proposed by Haas and McCabe (chapter 13), one might then argue that at an aggregate level this evolution should be characterized as "learning" and not simply "adaptation."[33]

Over the thirty-five years from the International Geophysical Year (IGY) in 1957 to Rio in 1992, concern with atmospheric risks moved from local nuisance to apprehension about a global commons. The starting point from which implementation is initiated—the framing of commitments and mandates—moved upward from nations to international level and substantially outward from the formal nation-state system to the broader and denser network of global civil society. At UNCED these commitments were cast comprehensively and ambitiously.

Cognitive and institutional capacity aimed at meeting those commitments certainly has been built. Experience with the limits of governing instruments and the organizational challenges of implementation has grown. But, as always, implementation must in the end come down to discretionary action by individuals sufficiently committed to the intent and the principles to pursue them through changing and uncertain circumstances. Frameworks of institutions, rules, and incentives influence such individual action. But finally it is personal preferences, belief systems, and moral codes that determine whether the implementation process is successfully completed. Much machinery has been put in place. Whether it is enough remains to be seen.

Appendix 19A.   Acronyms

AES	Applied Energy Services
AES	Atmospheric Environment Service (Canada)
BCSD	Business Council on Sustainable Development
BUND	Bund für Umwelt und Naturschutz Deutschland (German Association for Environment and Nature Protection)
CAA	Clean Air Act
CAN	Climate Action Network
CCOL	Coordinating Committee on the Ozone Layer
CEGB	Central Electricity Generating Board (U.K.)
CFC	chlorofluorocarbon
CIAP	Climate Impacts Assessment Program
CO_2	carbon dioxide
COAS	Council on Atmospheric Sciences (U.S.)
CPSC	Consumer Product Safety Commission (U.S.)
CSD	Commission on Sustainable Development (U.N.)
EC	European Community
EMEP	Cooperative Programme for the Monitoring and Evaluation of the Long-Range Transmission of Air Pollutants in Europe
ENGO	environmental nongovernmental organization
EPA	Environmental Protection Agency (U.S.)
EU	European Union
FCCC	Framework Convention on Climate Change (U.N.)
FGD	flue-gas desulphurization
FOE	Friends of the Earth
GECR	Global Environmental Change Report
GEF	Global Environmental Facility
HCFC	hydrochlorofluorocarbon
ICC	International Chambers of Commerce
ICI	Imperial Chemical Industries
IGY	International Geophysical Year
IIASA	International Institute for Applied Systems Analysis
IUAPPA	International Union of Air Pollution Prevention Associations
JI	joint implementation
LCPD	Large Combustion Plant Directive
LRTAP	(Convention on) Long-Range Transboundary Air Pollution
MITI	Ministry of International Trade and Industry (Japan)
NAPAP	National Acid Precipitation Assessment Program (U.S.)
NASA	National Aeronautics and Space Administration (U.S.)
NGO	nongovernmental organization
NO_x	nitrogen oxides
NRDC	Natural Resources Defense Council (U.S.)
ODS	ozone-depleting substance
OECD	Organization for Economic Cooperation and Development
RAINS	Regional Acidification Information and Simulation (model)
SEDESOL	Secretaría de Desarrollo Social (Mexico)
SEP	Joint Electricity Producers (Netherlands)
SO_2	sulfur dioxide
SORG	Stratospheric Ozone Review Group (U.K.)
SO_x	sulfur oxides
SST	supersonic transportation
TA Luft	Technische Anleitung zur Reinhaltung der Luft. (German Technical Directive for Air Quality)
UKMO	United Kingdom Meteorological Organization
U.N.	United Nations
UNCED	United Nations Conference on Environment and Development
UNCHE	United Nations Conference on the Human Environment
UNECE	United Nations Economic Commission for Europe

UNEP	United Nations Environment Programme
UV	ultraviolet
VOC	volatile organic compound
WBCSD	World Business Council on Sustainable Development
WMO	World Meteorological Organization
WRI	World Resources Institute
WWF	Worldwide Fund for Nature

Notes

1. Many people contributed to the research on which this chapter is based, and we are particularly grateful to participants in the discussions on implementation at the summer meetings of the project. In addition, this chapter has benefited from the comments of anonymous reviewers and from the editors of the book.

2. More fully, *implementation* as used here refers to actions taken by any actors in purposive pursuit of their own (articulated or undeclared, immediate or delegated) goals related to environmental risks; to actions taken in pursuit of other goals but impinging significantly on social capacity to manage the specific environmental risks studied here or the achievement of social objectives with respect to them (such as an energy tax that incidentally helps to reduce greenhouse gas emissions); or to preemptive measures undertaken to avoid other action considered by the actor in question less desirable than those embraced (such as a research program financed to delay regulatory decisions). It is not usually as easy to be sure of the motivations for action as the simple definition in the text suggests.

3. Within each component of the overall social system, some actor is taking action. Individual scientists commit time and resources to investigate ozone reactions. The journal *Nature* elects to publish articles on the subject. The Natural Resources Defense Council elects to launch legal action. The industrial giant ICI elects to pursue research on HCFCs as substitutes for other substances. The United Nations Environment Programme (UNEP) establishes a coordinating committee to exchange information on the ozone layer. National governments sign a Vienna Convention. National regulatory agencies promulgate rules. Friends of the Earth (FOE) International undertakes a report card on compliance with those rules. Government departments publish an ultraviolet (UV) index. Individuals buy sun hats. Academics initiate a five-year study of social learning. All these actions represent implementation decisions for the organizations and individuals concerned, even though they occur within quite different components of any overall social process.

4. For a single actor, this sequence might be viewed as simply a plausible, commonsense approach to applied problem solving. What this chapter claims as a central result is evidence that an overall social process, aggregated over all the independent actors and interdependent chains of ongoing implementation activity, can be discerned and itself displays an orderly sequence of a similar character.

5. In an informative precursor to this chapter (Downing and Hanf 1983), the editors express their hope that their project, initiated in 1979, can be seen as "the first progress report on a . . . growing effort to understand implementation issues through cross-jurisdictional comparisons." In the years between their work and this, however, the fram-

ing of implementation questions may be seen as having evolved considerably:

• Their project studied domestic pollution laws and implementation of national action to control discharges. Now the issue would be framed in terms of the impact of human activities on ecosystem health and the possibility of action to mitigate those impacts through changes in the human behavior underlying those activities themselves.

• Just as the Convention on Long-Range Transboundary Air Pollution (LRTAP) was being signed, their project studied national differences in pollution-control outcomes, possibly attributable to institutional and cultural differences across nations. Now in this book the emphasis has turned out to be on differences in national action to implement international agreements reached in global institutions and the capacities of these institutions themselves to influence implementation outcomes.

• A third key difference is also hinted at in their text, which emphasizes statutory measures and authoritative regulation to close an "enforcement deficit" (see also Mayntz 1976). Yet as implementation is increasingly recognized to be a continuing interactive process of negotiation to adapt central statutory designs to changing realities in the field, the notion of "policy outcome" fades, the appraisal of effectiveness of implementation becomes ambiguous, and "compliance" becomes commitment to the intent or spirit of a text or "bureaucratic covenant" (de Beus 1991) rather than observance of the letter of a law.

Thus, in the years since the LRTAP Convention and the start of the Downing-Hanf comparative study, the focus of questions about implementation has shifted considerably toward the capacity and effectiveness of global institutions that did not exist in 1979 and (to some extent) away from the functioning of national agencies created in most countries in the decade prior to the LRTAP Convention. It has also shifted away from domestic pollution-control legislation and toward a body of international environmental law and principle that is largely a product of the last two decades (Sands 1995) and that aims to a considerable extent at avoiding potentially harmful human activities (although planned adaptation to the anticipated consequences of those activities also remains a central topic for debate).

6. This evolution leads attention toward the lags in negotiating cooperative outcomes in the face of commitment problems, achieving ratification of these outcomes to the extent necessary for agreements to come formally into force, and ensuring effective action at national level to meet the commitments undertaken. It also raises, however, an important question. Such appeal to the international arena may be necessary to achieve abstract commitment to agreement on rational undertakings that serve an enlightened self-interest and to put pressure on domestic processes to honor and pursue those commitments. But does the very fact of escalating the process to international discussion also risk losing the personal commitment "on the ground" that is essential to implementation—that is, necessary to carry the idea into conduct?

7. By focusing on a discussion at this level, we are effectively leaving aside most of what is usually central in theoretical work on implementation—namely, the body of materials in political science, public administration, organizational dynamics, and organizational psychology dealing with program implementation, the exercise of administrative discretion, and organizational barriers to realization of mandates or the management of innovation. (For reviews of these literatures, see the surveys of Palumbo and Calista 1990, Howlett 1991, and Calista 1994 and the references there cited.) Some attempt is made in this chapter, however, to identify some changing characteristics of the implementation process itself, with respect to actors outside government as well as agencies of government at various levels.

The study of action taken to address global environmental risks forces us to plunge deeper into the domestic activities of many actors in a post-ratification period than do, for example, Susskind (1994), Spector and Korula (1993), or Sands (1995). It also forces more explicit attention to the international arena in establishing the initiating purposes of implementation action than does the conventional political science or public administration literature on implementation. National governments remain, therefore, the fulcrum of the implementation stories told here, but only one part of a dynamic, interactive, not consciously coordinated sequence of actions involving many actors at both subnational and supranational levels.

8. The approach said to prevail in the Netherlands is representative of a general orientation toward voluntary arrangements, which was becoming widespread by 1992 (Netherlands 1995, 9):

Since the drafting of the first environmental policy plan [in 1988], the government has offered to hold discussions with the various sectors to decide what contributions should be made by specific industries to meet the national environmental targets. The government is making centralized agreements (covenants) with almost every sector to cover the whole of a particular industry. These agreements give highly detailed quantified targets for the years 2000 and 2010. If the companies abide by these agreements, they can count on a good relationship with the government. If they don't, the government will reimpose its mandatory environmental licensing requirements.

9. Rabe (1995) describes the example of 3M in Minnesota, in which a flexible permit for all volatile organic compounds (VOC) emissions from a single facility was proposed by the company and approved by the Pollution Control Agency, conditional on 3M's acceptance of enforceable limits on all VOC emissions at less than half the aggregate level previously permitted and development of an approved procedure for continuous emission monitoring and reporting. Flexibility for the company to change product lines and processes—hence emissions— was ensured, provided the overall cap was not exceeded. The innovation in this flexible or integrated permit lies in the possibility of extending a single facility permit to cover a wide range of emissions and media and thus ultimately to deal with interaction among air issues.

10. Canada's proposed Regulatory Efficiency Act also pursued flexibility through administrative agreements and alternatives to designated regulations, but was strongly opposed by environmental groups, and ultimately abandoned.

11. Several years after it came into force, efforts to ensure national compliance with the commitments undertaken within the FCCC still rest to a remarkable extent on covenants and "voluntary-challenge" programs to encourage industry to undertake cost-effective "no-regrets" measures to reduce emissions of greenhouse gases through investment in new technologies or production practices.

12. Exceptions include the following:

• UV-B watch, sun hats, and sun blocks in the case of ozone depletion represent efforts at adaptation; U.S. EPA concern about the need to stockpile CFCs against demands for servicing of automobile air conditioners illustrates that regulatory measures did have some effect on lifestyles and human behavior;

• Liming of water supplies or lakes in the case of acidifying precipitation represents implementation action directed toward restoration of environmental quality;

• Dutch coastal defenses, building design criteria, and zoning changes around vulnerable coastal areas illustrate adaptation measures in the climate case.

13. In the second such Act, in 1986, provisions were strengthened to permit regulations requiring the introduction of emission-reductions technologies where benefits warrant the costs. A third such Act was passed in 1990, providing that all new plants must use state-of-the-art technology and that all old plants must be cleaned up within five years to meet new plant standards or be shut down.

14. The spread of this type of legislation internationally is documented by Sand (1990).

15. About the same time that regulations to ban CFC-propelled aerosols were being developed in the United States, two European importers of CFCs—Sweden and Norway—banned nonessential uses of aerosols, cutting CFC aerosol use dramatically. It has been argued that differences in the speed with which individual countries reacted to the scientific research on CFCs and the ozone layer and took action to regulate CFCs can be explained to some degree by the extent of the epistemic community's influence on various governments and the difficulty of nongovernmental actors in gaining access to the decision-making process (Haas 1992, 215).

16. In the negotiations leading up to the final text of the Framework Convention on Climate Change, the term "net emissions" was explicitly rejected. Despite this, Australia, Canada, New Zealand, the United States, and others have made plans to trade off increased carbon sinks against increased carbon dioxide emissions (Climate Action Network 1995).

17. Joint implementation (JI) refers to the approach in which countries invest in emission-reduction projects in other countries where such projects would be more cost effective.

18. In 1994, the United Kingdom imposed a uniform sales tax on domestic energy supplies that was justified, in part, as an environmental measure designed to help meet carbon-reduction commitments. But it was criticized as having a "disproportional impact on poor households" without containing "any significant incentives for energy conservation, nor have its revenues been earmarked for environmental investment" (Wright 1995, 78). It was also an initiative not of the Department of the Environment but rather of the Treasury Department, leading to the conclusion that the stronger motivation was financial rather than environmental.

Within the European Community the issue of economic instruments for environmental protection was raised again in late 1994 with a major policy communication from the European Commission to the European Parliament and the Council of Ministers. The document, *Economic Growth and the Environment,* was seen as a "significant shift in Commission thinking, away from regulation and towards encouragement" (Wright 1995, 78). The document made a strong case for the use of market instruments generally.

19. An interesting sidelight arises because explicit language in the Act ensures that these allowances do not represent property rights and hence "could be limited, revoked, or otherwise modified in future without compensation." Subsequent appraisals have examined the trading and auction activities involving these allowances and have concluded that fewer trades than expected are occurring but that a working market is slowly evolving, with potential for considerable cost savings to result.

20. In the case of climate change, such voluntary arrangements or "challenge" schemes, together with reporting requirements, represent almost the only concrete action taken at the governmental level to bring about emission reductions. They appear to be dominant options in the United Kingdom, Canada, and the United States, among other arenas.

21. The process of negotiating an international agreement on a framework convention and subsequent more concrete protocols may be lengthy, but it is of course not the end of an implementation story. A further process of postagreement political activity is necessary to achieve ratification of the agreement by each participating national government (Spector and Korula 1993). In some cases this ratification process within each arena may demand extensive further negotiation with domestic interests concerned and may itself be lengthy. But although ratification may be considered to be a rather precise statement of purpose by a national government, it is still only the starting point from which a long, complex, chain of implementation action starts. This chain is intended to lead on to changed behavior on the part of subnational governments and agencies, street-level bureaucrats, firms, industries, economic agents, and citizens. In many jurisdictions, this postratification process of negotiation among domestic stakeholders as to the nature of the rule making and other action necessary to ensure realization of the international commitments undertaken may be a much more significant political challenge than the formality of ratification. Certainly in federal states the question of effectiveness in the implementation of rules or regulations established at the national level is a significant question in itself (Mayntz 1976; Wood 1992; Kux 1994).

22. Soroos (1991) and others suggested that this experience and that of acid rain held useful lessons to be drawn in shaping efforts to deal with the problem of climate change. At the same time, Haas (1992), Skjaerseth (1992), and many others cautioned that special features (ready availability of a technical fix, relatively few major actors directly affected, and so on) that distinguished the case of ozone from many others made it unreliable as a basis for dealing with the issue of climate change or environmental risks more generally.

23. However, one might note that the initial Conference of the Parties, in Berlin in 1995, endorsed no specific targets or timetables for reductions in emissions of greenhouse gases and agreed only on the necessity for a more detailed protocol covering carbon dioxide emissions to be concluded by 1997. Some progress toward criteria to govern pilot projects in "action undertaken jointly" was achieved, however, opening the way to possible implementation of offset measures in the development of carbon sequestration and carbon sinks. These topics were of course pursued in the Kyoto Protocol concluded in December, 1997.

24. "Whilst UNCED and its follow-up do not provide a clear sense of direction as to likely future developments, one feature emerges as international environmental law moves into its next phase. . . . [It is] no longer exclusively concerned with adoption of normative standards to guide behavior, but increasingly addresses techniques of implementation which are practical, effective, equitable and acceptable to most members of the international community" (Sands 1995, 61). The importance of the Hague Declaration, which envisaged the possibility of international action even in the absence of unanimity, provides an interesting marker in this respect. More specifically, in the summary of conclusions from another discussion of international environmental negotiations it was noted that "The speaker [Széll] stressed that it [the Montreal Protocol] has already had a useful impact, but has still to embark on the difficult task of determining which individual parties have not met their substantive obligations under the convention" (Lang 1995, 296).

25. The "critical-load" concept is one example of results emerging from this work. An unusual feature of this UNECE structure is the "lead-country" system, whereby each program network is self-financed around permanent research institutions in a host country, with other countries funding participation of their respective national institutions on a voluntary basis.

26. Though beyond the period of this study, the examples of Friends of the Earth International (which encouraged national FOE groups to prepare "report cards" on the implementation of national action to conform to commitments under the Montreal Protocol) or of the Climate Action Network (which undertook to assess the national reports prepared by signatories to the Framework Convention on Climate Change) are also notable in this respect.

27. This question of target effectiveness remains as the urgent, unspoken, but formidable challenge to be faced. "The adequacy of international law will be judged by its ability to protect and preserve the environment. [One must question] whether that objective is being met, and whether existing legal and institutional structures can meet that objective" (Sands 1995, 3–4). On the other hand, there are also much more optimistic readings (e.g., Easterbrook 1995; Simon 1995).

28. Ronfeldt and Thorup (1994) emphasize this burgeoning capacity in their examination of the growing importance of citizen networks and virtual communities as new institutional forms. More intensive examination of the importance of connectivity as a determinant of democracy is contained in Anderson et al. (1995).

29. A question of importance for the future will be the possibility of an evolving role in formal scrutiny and audit and, in particular, possible provisions for direct NGO participation in secretariats or implementation committees at the international level.

30. The case of Patrick Széll provides a good illustration of the role individuals can play. He chaired the drafting group set up for implementation procedures under the Montreal Protocol and worked on the corresponding rules for subsequent protocols under LRTAP. He then became involved with preparing similar procedures for the Climate Convention.

31. Accumulating experience across cases (which here means among distinct international institutions or regimes) thus helps in the slow confidence-building process that may be seen as enabling sovereign nations to sign on to increasingly stringent commitments and to accept explicit monitoring of their performance in carrying out action to realize those commitments. As increasing recognition of interactions among atmospheric risks creates pressure for greater institutional integration of the regimes constructed separately to deal with particular risks and ultimately perhaps for some institutional structure that might advance a more comprehensive "law of the atmosphere," these common elements in the implementation process may make such operational integration feasible.

Problems of data quality, secrecy, and absence of external audit may still remain serious concerns. Particularly in a world with enterprises pressing strongly their concerns for cost control and international competitiveness on national authorities, any barriers to open scrutiny of the progress made by national governments toward attaining their agreed targets for phaseouts and emission controls must be seen as serious challenges to successful implementation. International depositories of secret data thus may remain a barrier, but the growing capacity of an increasing number of parties to a convention and its protocols to monitor recycling, trading, and destruction of offending substances is perhaps one offsetting and encouraging aspect of a maturing process of implementation.

Another theme in the implementation literature is illustrated here. When implementation activities are recognized as essentially political processes, purposeful ambiguity in marching orders rather than precision and clarity in mandates and commitments is to be expected. Such ambiguity may reveal itself in an inability or unwillingness to recognize or confront noncompliance directly. The need to deal with this feature of international environmental institutions shows clearly through efforts to manage the issue of ozone depletion.

32. National initiatives such as the chemical industry's Responsible Care program (Roczniak 1994) provided a body of experience on which to build, as did work of the International Organization for Standardization (ISO) in more technical standard setting and certification.

33. Whether this evolution of social institutions should be seen as a further development in the "civilizing process" (Elias 1982) or as the beginning of a more self-conscious "noosphere" (Vernadsky 1945) is a question that cannot be addressed here. We can hope that the evolution traced in this chapter can be seen as evidence that the research program is "progressive" (Lakatos 1986) and that indeed the record of implementation efforts traced here provides evidence of an increasingly informed sense of responsibility for the actions of humans as the most aware agents within the "noosphere."

References

Anderson, Robert H., T.K. Bikson, S.A. Law, B.M. Mitchell, C.R. Kedzie, B. Keltner, C.W.A. Panis, J. Pliskin, and P. Srinagesh. 1995. *Universal Access to Email: Feasibility and Societal Implications.* Santa Monica: RAND.

Axelrod, Robert. 1984. *The Evolution of Cooperation.* New York: Basic Books.

Bekhechi, Mohammed Abdelwahab. 1995. Comment on the paper by Alexandre Timoshenko. In Winfried Lang, ed., *Sustainable Development and International Law.* London: Graham & Trotman.

Brown Weiss, Edith. 1995. Environmental equity: The imperative for the twenty-first century. In Winfried Lang, ed., *Sustainable Development and International Law.* London: Graham & Trotman.

Calista, Donald J. 1994. Policy implementation. In Stuart S. Nagel, ed., *Encyclopedia of Policy Studies* (2nd ed.). New York: Marcel Dekker.

Campbell, Laura B. 1995. Comment on the paper by Robert Reinstein. In Winfried Lang, ed., *Sustainable Development and International Law.* London: Graham & Trotman.

Chayes, A., and A.H. Chayes. 1995. *The New Sovereignty: Compliance with International Agreements.* Cambridge: Harvard University Press.

Chayes, Antonia Handler, Abram Chayes, and Ronald B. Mitchell. 1995. Active compliance management in environmental treaties. In Winfried Lang, ed., *Sustainable Development and International Law.* London: Graham & Trotman.

Clark, William C., and Giandomenico Majone. 1985. The critical appraisal of scientific inquiries with policy implications. *Science, Technology, and Human Values* 10: 6–19.

Climate Action Network. 1995. *Independent Evaluations of National Plans for Climate Change Mitigation.* Washington: Climate Action Network.

Coase, Ronald H. 1960. The problem of social cost. *Journal of Law and Economics* 3: 1–44.

Collingridge, David. 1992. *The Management of Scale: Big Organizations, Big Decisions, Big Mistakes.* London: Routledge.

Daly, Herman E. 1991. *Steady-State Economics* (2nd ed.). Washington: Island Press.

Dawkins, Richard. 1989. *The Selfish Gene* (2nd ed.). Oxford: Oxford University Press.

de Beus, Jos. 1991. The ecological social contract. In D.J. Kraan and R.J. in 't Veld, eds., *Environmental Protection: Public or Private Choice.* Boston: Kluwer.

Dinkelman, Gerda. 1995. *Verzuring en Broeikaseffect, De wisselwerking tussen Problemen en Oplossingen in het Nederlandse Luchtverontreinigingsbeleid (1970–1994).* Utrecht: Jan van Arkel.

Dobell, Rodney. 1997. Complexity, connectedness, and civil purpose; public administration in the congested global village. *Canadian Public Administration* 40(2): 346–370.

Dobell, Rodney, and Edward Parson. 1988. A world atmosphere fund. *Policy Options* 9: 6–8.

Downing, Paul B., and Kenneth Hanf, eds. 1983. *International Comparisons in Implementing Pollution Laws.* Boston: Kluwer-Nijhoff.

Easterbrook, Gregg. 1995. *A Moment on the Earth.* New York: Viking Penguin.

Elias, Norbert. 1982. *The Civilizing Process: State Formation and Civilization.* Oxford: Basil Blackwell.

Environment Canada. 1976. Press Release, Minister of Environment, December 15, 1976. Ottawa: Environment Canada.

Frank, Robert H. 1988. *Passions within Reason: The Strategic Role of the Emotions.* New York: Norton.

GECR. 1990. *Global Environmental Change Report* (vol. 3, no. 18). Arlington, Mass: Cutter Information Corp.

Haas, Peter M. 1990. *Saving the Mediterranean: The Politics of International Environmental Cooperation.* New York: Columbia University Press.

———. 1992. Banning chlorofluorocarbons: Epistemic community efforts to protect stratospheric ozone. *International Organization* 46: 187–224.

Hardin, Garrett. 1968. The tragedy of the commons. *Science* 162: 1243–1248.

Hausker, Karl. 1992. The politics and economics of auction design in the market for sulfur dioxide pollution. *Journal of Policy Analysis and Management* 11: 4553–4572.

Hawken, Paul. 1993. *The Ecology of Commerce: A Declaration of Sustainability.* New York: HarperCollins.

Howlett, Michael. 1991. Policy instruments, policy styles, and policy implementation: National approaches to theories of instrument choice. *Policy Studies Journal* 19: 1–21.

Ingram, Helen. 1978. The political rationality of innovation: The Clean Air Act Amendments of 1970. In Ann F. Friedlaender, ed., *Approaches to Controlling Air Pollution.* Cambridge: MIT Press.

International Union of Air Pollution Prevention Associations (IUAPPA). 1988. *Clean Air around the World.* Brighton, U.K.: IUAPPA.

———. 1995. *Clean Air around the World.* Brighton, U.K.: IUAPPA.

Jakobeit, Cord. 1996. Nonstate actors leading the way: Debt for nature swaps. In Robert O. Keohane and Marc A. Levy, eds., *Institutions for Environmental Aid: Pit Falls and Promise.* Cambridge: MIT Press.

Johnston, Douglas M. 1997. *Consent and Commitment in the World Community: The Classification and Analysis of Economic Instruments.* Invington on Hudson, N.Y.: Transnational.

Keohane, Robert O., and Elinor Ostrom, eds. 1995. *Local Commons and Global Interdependence: Heterogeneity and Cooperation in Two Domains*. London: Sage.

Klarer, Jürg, ed. 1994. *Use of Economic Instruments in Environmental Policy in Central and Eastern Europe*. Budapest: Regional Environmental Centre for Central and Eastern Europe.

Kux, Stephan. 1994. *Subsidiarity and the Environment: Implementing International Agreements*. Basel: Europainstitut an der Universität Basel.

Lakatos, Imre. 1986. *Philosophical Papers*. Vol. 1. Cambridge: Cambridge University Press.

Lang, Winfried, ed. 1995. *Sustainable Development and International Law*. London: Graham & Trotman.

Lee, Kai N. 1993. *Compass and Gyroscope: Integrating Science and Politics for the Environment*. Washington: Island Press.

Levy, Marc A., Robert O. Keohane, and Peter M. Haas. 1994. Improving the effectiveness of international environmental institutions. In Peter M. Haas, Robert O. Keohane, and Marc A. Levy, eds., *Institutions for the Earth: Sources of Effective International Environmental Protection*. Cambridge: MIT Press.

Lichbach, Mark Irving. 1996. *The Co-operator's Dilemma*. Ann Arbor: University of Michigan Press.

Lipsky, M. 1978. Standing the study of implementation on its head. In W.D. Burnham and M.W. Weinberg, eds., *American Politics and Public Policy*. Cambridge: MIT Press.

Majone, G. 1983. *Regulatory Policies in Transition*. Laxenburg, Austria: International Institute for Applied Systems Analysis.

Majone, G., and A. Wildavsky. 1979. Implementation as evolution. In J.L. Pressman and A. Wildavsky, eds., *Implementation* (2nd ed.). Berkeley: University of California Press.

Mayntz, Renate. 1976. Environmental policy conflicts: The case of the German Federal Republic. *Policy Analysis* 2: 577–587.

Molina, Mario, and Sherwood Rowland. 1974. Stratospheric sink for chlorofluoromethanes: Chlorine atom catalyse destruction of ozone. *Nature* 249: 810–812.

Netherlands. 1995. *Environmental News from the Netherlands*, no. 1, p. 9. The Hague: Ministry of Housing, Physical Planning and the Environment, Government of the Netherlands.

Olson, Mancur. 1965. *The Logic of Collective Action*. Cambridge: Harvard University Press.

Organization for Economic Cooperation and Development (OECD). 1994. *Managing the Environment: The Role of Economic Instruments*. Paris: OECD.

O'Riordan, Tim, and Jill Jäger. 1995. Global environmental change and sustainable development. In Jill Jäger, Angela Liberatore, and Karin Grundlach, eds., *Global Environmental Change and Sustainable Development in Europe*. Luxembourg: Office for Official Publications of the European Communities.

Ostrom, Elinor. 1992. *Crafting Institutions for Self-Governing Irrigation Systems*. San Francisco: Institute for Contemporary Studies Press.

Palmer, Geoffrey. 1992a. The implications of climate change for international law and institutions. *Transnational Law and Contemporary Problems* 2: 205–257.

———. 1992b. New ways to make international environmental law. *American Journal of International Law* 86: 259–283.

Palumbo, Dennis J. 1987. Introduction: Symposium on implementation—What have we learned and still need to know. *Policy Studies Review* 7: 91–102.

Palumbo, Dennis J., and Donald J. Calista. 1990. Opening up the black box: Implementation and the policy process. In D.J. Palumbo and D.J. Calista, eds., *Implementation and the Policy Process*. Westport, Conn.: Greenwood.

Parson, Edward Anthony. 1992. Negotiating climate cooperation: Learning from theory, simulations, and history. Doctoral dissertation, Harvard University.

———. 1993. Protecting the ozone layer. In Peter M. Haas, Robert O. Keohane, and Marc A. Levy, eds., *Institutions for the Earth*. Cambridge: MIT Press.

Parson, Edward A., and Owen Greene. 1995. The complex chemistry of the international ozone agreements. *Environment* 37: 16.

Putnam, Robert D. 1988. Diplomacy and domestic politics: The logic of two-level games. *International Organization* 42: 427–460.

Rabe, B.G. 1995. Integrating environmental-regulation: Permitting innovation at the state-level. *Journal of Policy Analysis and Management* 14: 467–472.

Roan, Sharon. 1989. *Ozone Crisis*. New York: Wiley.

Roczniak, Dan. 1994. Responsible care initiative from the business community. In Rodney Dobell and Michael Neufeld, eds., *The North American Commission for Environmental Cooperation: Early Implementation*. Victoria, B.C.: North American Institute.

Ronfeldt, David, and Cathryn L. Thorup. 1994. NGOs, civil society networks, and the future of North America. In Rodney Dobell and Michael Neufeld, eds., *Trans-border Citizens: Networks and New Institutions in North America*. Lantzville, B.C.: Oolichan Books.

Sand, Peter H. 1988. *Marine Environmental Law in the United Nations Environment Programme*. London: Tycooly.

———. 1990. Innovations in international environmental governance. *Environment* 32: 16–45.

Sands, Philippe. 1995. *Principles of International Environmental Law*. Vol. 1, *Frameworks, Standards, and Implementation*. Manchester, U.K.: Manchester University Press.

Schmidheiny, Stephan. 1992. *Changing Course: A Global Business Perspective in Development and the Environment*. Cambridge: MIT Press.

Shrader-Frechette, K.S. 1991. *Risk and Rationality: Philosophical Foundations for Populist Reforms*. Berkeley: University of California Press.

Simon, Julian L. 1996. *The Ultimate Resource 2*. Princeton: Princeton University Press.

Skjaerseth, Jon Birger. 1992. The "successful" ozone-layer negotiations: Are there any lessons to be learned? *Global Environmental Change* 2: 292–300.

Soroos, Marvin S. 1991. The evolution of global regulation of atmospheric pollution. *Policy Studies Journal* 19: 115–125.

Spector, Bertram I., and Anna R. Korula. 1993. Problems of ratifying international environmental agreements: Overcoming initial obstacles in the post-agreement negotiation process. *Global Environmental Change* 3: 369–381.

Stone, Christopher D. 1993. *The Gnat Is Older Than Man: Global Environment and Human Agenda*. Princeton: Princeton University Press.

Susskind, Lawrence E. 1994. *Environmental Diplomacy: Negotiating More Effective Global Agreements*. New York: Oxford University Press.

Széll, Patrick. 1995. The development of multilateral mechanisms for monitoring compliance. In Winfried Lang, ed., *Sustainable Development and International Law*. London: Graham & Trotman.

United Kingdom. 1954. House of Commons Committee on Air Pollution (Chairman, Sir Hugh Beaver). *Final Report*. (Cmd. 9322. British Sessional Papers, 1953-54, November, 1954, Vol. 8, p. 663). London: House of Commons.

Vernadsky, V.I. 1945. The biosphere and the noosphere. *American Scientist* 33: 1–12.

Wood, B. Dan. 1992. Modeling federal implementation as a system: The clean air case. *American Journal of Political Science* 36: 40–67.

Wright, Martin. 1995. The EU's market instrument vision. *Tomorrow Magazine* 1: 78.

20
Evaluation in the Management of
Global Environmental Risks

Josee van Eijndhoven, Brian Wynne, and Rodney Dobell with
Ellis Cowling, Nancy M. Dickson, Gerda Dinkelman,
Peter M. Haas, Jill Jäger, Angela Liberatore, Diana Liverman,
Miranda A. Schreurs, Vassily Sokolov, and Ferenc L. Tóth[1]

20.1 Introduction

Evaluation asks the question "How are we doing at managing the risk?" Evaluations may be formal, such as the U.S. 1991 report on *The Experience and Legacy of the National Acid Precipitation and Assessment Program* (NAPAP), or informal, such as the evaluations by Mexican newspapers of the country's acid rain policy. They may focus on specific aspects of society's response to global environmental risks, such as the United Kingdom's 1992 review of its global atmosphere research program. Or they may range over environment policy as a whole, such as the RIVM evaluations in the frame of Dutch National Environmental Policy Plans or the National Report for the Federal Republic of Germany for the United Nations Conference on Environment and Development (UNCED). The goal of this chapter is to describe, understand, and appraise changes in evaluation practices and processes that were carried out for our global environmental risks during the period of this study.

The term *evaluation* derives from *value:* valuing things, experiences, situations. We can define evaluation as "assessing (some) one's achievement" in a certain area, or (more specifically in the context of this research) as an effort to evaluate the performance (as distinct from knowledge) or role of oneself and other actors. A sharp definition of evaluation is difficult to give, and part of the recent literature on evaluation no longer even thinks it apt to define the term (Guba and Lincoln 1989, 21). The editors of a recent handbook on evaluation state that the now classic answer to the question "What is evaluation?" is "Evaluation is about determining merit or worth." They add, "Although it is far from clear that we have consensus on that answer across the diversity of the evaluation profession" (Chelimsky and Shadish 1997, xii).

In this study, however, we need to be more specific about evaluation, not only because it is difficult to gauge the development of a nondefined entity but also because the practical application of the term *evaluation* usually extends over a number of other functions that we identified in this study and have been taken up in other chapters. This becomes clear from the description Chelimsky (1997) gives of evaluation. She mentions a number of roles for evaluation that can be seen as a descriptive definition of the objectives of evaluation:

• Determining past successes and failures;

• Identifying, empirically describing, and monitoring problems; and

• Developing and comparatively assessing proposed solutions.

In this study the items mentioned in the second list item have been taken up under risk assessment (chapter 15) and monitoring (chapter 16), and the items under the third list item are the subject of option assessment (chapter 17). Evaluation in the broad sense Chelimsky describes therefore covers all the knowledge functions (as opposed to the action functions "goal and strategy formulation" and "implementation") covered in this study. Of these, this chapter deals only with "determining past successes and failures."[2]

At the beginning of the study we focused mainly on evaluations (in terms of determining successes and failures) of *policies* to manage global environmental risk, but it quickly became apparent that in the relatively new and immature policy area of managing global environmental risks, evaluations tended to focus on the most well-developed parts of the activities: research initially and then evaluation of the quality of assessments of risks, responses, and strategies, with evaluations of policies and implementation processes only slowly emerging. The contents of the chapter reflect this. This holds even more true because the period we cover empirically ends in 1992. Also, our initial focus tended to be on formal evaluation of past performance, whereas in practice much evaluative activity is going on alongside ongoing knowledge development and action.

It is clear that evaluation, in terms of "determining successes and failures," is a normative activity. It has become increasingly evident that it is very difficult, if not impossible, to conduct an evaluation without bias. The solution advanced has been to admit that there is a potential for bias and to attempt to overcome it either by incorporating as many different biases as possible into the evaluation or by conducting the evaluation and admitting evident biases. Radin (1977), Weiss (1972), and Brandl (1978) all noted that political considerations intrude on evaluation. They argued that since policies are creatures of political decisions, evaluation of policies will feed into the political process and become subservient to political wills. This point led McLemore and Neumann (1987) to conclude that evaluations based on politics are normative and useful, since denial and avoidance of political considerations are not productive.

Another issue to be mentioned here is evaluation methods. Richard Sonnichsen (1989a, 1989b), argued that the trend in evaluations is to move away from strict devotion to a single methodology in favor of blending techniques that are "appropriate." He posited that there is no one best way to conduct evaluations and, consequently, that evaluations should be adapted to the type of program under examination and the needs of the stakeholders. Along the same lines Chelimsky (1995) argued that the policy question to be highlighted has become much more central in evaluation practices and therefore that methods have become complementary. The use of evaluation methods today involves a focus not just on the specific policy question being asked but also on the users' wider information needs.[3] According to Chelimsky the focus of evaluations has become simply to achieve a more complete picture of reality, with the prime objective to inform decision makers.

20.2 Researching the Evaluation Function

Our initial data-collection efforts sought instances of formal evaluation that were clearly distinct from assessment, monitoring, or implementation. We found so few of these that we revised our research protocol to look for information on more partial and less formal evaluation. In particular, we acknowledged that significant evaluation activities were likely to occur both in the context of other functions of risk management (such as part of risk assessment efforts) and embedded in larger societal evaluations of environmental policy and practices. Moreover, it became clear that evaluation was performed and interpreted even more differently among the various arenas we studied than were other management functions addressed in this study. Therefore, we expanded our efforts to include an overview of evaluation practices in general in our arenas. Starting from these overviews (landscapes of evaluation) we made a qualitative assessment of the characteristics of evaluative activities in each of the arenas. The qualitative assessment was generated by deriving the answers to a set of generic questions from the

arena studies:

• How extensive are the evaluative practices in the arena under consideration?

• Who is dominating evaluation activities in the arena?

• What kinds of impacts (such as institutional or policy changes) resulted from these evaluations?

The first set of overviews showed that the roles of and relationships among actors with respect to evaluation activities were widely different among the arenas studied. We decided that for each of the arenas considered we should look at the activities of a number of actors plus a number of specific activities to characterize evaluation practices.[4]

For each of the arenas we assessed whether there were evaluations conducted by each of the actors identified, whether the evaluation practices were relatively extensive, and whether they generally could be considered to have a strong influence on the policy process. We also tried to gauge whether the situation changed over time and over the three cases (acid rain, ozone depletion, and climate change). It was explicitly acknowledged that evaluations may not be public and may become visible only when they are being used. It was also important to consider the effects (positive and negative) of this limited availability on the evaluation process. As a result of this way of gathering data and the apparently different state of affairs with respect to evaluation in the arenas studied, this chapter has a somewhat more impressionistic style than most of the other chapters in the book.

20.3 Evaluation Practices

20.3.1 Introduction

It proved difficult to capture instances of relatively pure and formal evaluation systematically in our research. When we expanded the net drawn over our data to include less clear and formal evaluations and self-reflexive activities in the frame of other risk management functions, the situation became only partially clearer. In part this is a result of the blurring of the definition of evaluation to encompass other activities than those formally called evaluation of policy. In part it reflects the actual situation in which evaluation (as the normative assessing of successes and failures of activities) is embedded in the complete activity of generating and disseminating knowledge on the management of global environmental risks. This chapter discusses, for each of the issues we analyzed in this study (ozone depletion, acid rain, and climate change), one or two examples of activities that clearly entailed evaluation drawn from arenas where this type of activity was most visible. We have added an example of a more general evaluative activity in which these three issues were subsumed. The cases discussed are

• The Montreal Protocol and ensuing national activities,

• Acid rain evaluations especially concentrating on the NAPAP program in the United States, including a discussion on the relative (and unexpected) absence of evaluation in the frame of the LRTAP Convention,

• Evaluations in the frame of global climate change,

• The evaluation activities in the frame of the Dutch national environmental policy plan.

These examples suggest that evaluation as a function can only in some instances be seen as an activity apart from other functions. The examples also show how rare well-embedded evaluation of policies is. We start with such a rare example (evaluation in the frame of the Montreal Protocol) as a contrast to the situation in other areas and end with an example (the environmental planning cycle in the Netherlands) of the direction in which evaluation practice may evolve more generally.

20.3.2 The Montreal Protocol and Related Activities

The Montreal Protocol and the coordinating activities established by the Protocol can be viewed as a complex assessment and evaluation procedure. Under Article 6 of the Montreal Protocol, parties to the Protocol are required to regularly assess and review control measures and report data. Compliance with obligations under the Protocol is monitored through specific reporting requirements and through consultations among the parties and the Secretariat, through the Executive Committee of the Multilateral Ozone Fund (comprising seven representatives each from developing and industrialized countries), and through deliberations of the annual meetings of parties. Parties to the Protocol provide to the Ozone Secretariat of the United Nations Environment Programme (UNEP) annual statistical data on production, imports, and exports of controlled substances, including imports and exports to parties and nonparties.

The Conference of Parties to the Protocol established a number of scientific and technical bodies, including four panels of experts to assess the Montreal Protocol's control measures: the Scientific Assessment Panel, the Environmental Effects Panel, the Technology Assessment Panel, and the Economic Assessment Panel. The Technology and Economics Panels were merged in 1991. In 1989 and 1991 these panels analyzed the scientific knowledge of the causes, the impacts, and the technical

and economic aspects of the issue. Apart from being reassessments of risks and eventual responses, these studies can be seen as (indirect) evaluations of the adequacy of the control measures previously agreed on. And this holds more explicitly for the synthesis summaries of the studies—the 1989 UNEP Assessment Synthesis Report and the 1991 UNEP Assessment Executive Summary. These evaluations have been extremely important in the management of the risk of ozone depletion. It is unlikely that the London and Copenhagen Amendments to the Protocol would have been as far-reaching in the absence of the 1989 and 1991 analyses that were conducted in accordance with Article 6 of the Protocol. The 1989 and 1991 evaluations were targeted toward the parties to the Protocol with the specific aim of guiding future policy response. The conclusions of these evaluations regarding the increasing extent of ozone depletion, the control measures necessary to reduce ozone emissions, and the feasibility of further reductions contributed to the strengthening of the Protocol in 1990 and 1992.[5]

The Montreal Protocol and follow-up activities not only entailed assessments and evaluations in the international community but also evaluative activities at the national level and in nongovernmental organizations (NGOs). Friends of the Earth (FoE) International evaluated the extent to which (and thereby the way in which) the Montreal Protocol requirements were implemented by the parties (FoE International 1992).

The activities of two countries in relationship to the Protocol illustrate some of the dimensions of the possible roles of evaluation. In the Netherlands the Montreal Protocol was taken as the point of departure for a covenant with the chlorofluorocarbon (CFC)-using industry on the phasing out of CFCs and halons. Because the Netherlands was implementing the Protocol not by regulation but by voluntary agreement, it was important to have a reliable monitoring system to assess the implementation process and evaluate whether the phaseout goals are likely to be met.

The Netherlands The Montreal Protocol was ratified by the Netherlands on December 16, 1988. Two covenants on the reduction of CFC use had already been signed before that date between the Dutch government and the aerosol industry: the first in October 1980 and the second in January 1988. On June 21, 1990, the CFC action program was signed between the national government, provinces, the Union of Netherlands municipalities, and industry. Parties in the covenant agreed to reduce CFCs and halons by 99 and 100 percent, respectively, in 1995. In 1998 the production and import of "hard" CFCs and halons would be forbidden. The covenant was to implement the Montreal Protocol.

To evaluate the implementation of the covenant a CFC committee was installed, with representatives of the parties and members of environmental and consumer organizations involved. To assess the phasing out of ozone-depleting substances, an annual report was to be made every year before the first of May by an external accountant, who was granted leave to inspect the books of producers, importers, and industrial consumers (CFC action program, The Hague, June 21, 1990). Using this data report the Committee evaluated the situation in its annual report. By the end of 1992, two such annual reports had been published. In the first, the reduction program was said to be on schedule generally: domestic CFCs were phased out a little more quickly than scheduled, whereas halons lagged behind a little. The report pointed to new international evidence that the ozone layer was being affected more seriously than was previously thought.

In the report for the year 1991, the committee stated that the reduction numbers were disappointing. They gave two reasons. The first was the fact that the problem was considered to be so severe that there were demands for a quicker reduction. The second was that there were reasons to fear that the goals stated in the action plan would not be reached. The main problems mentioned were

• The Dutch exporters' alternatives for CFCs were not well enough accepted abroad;

• Too many refrigerators showed an excessive use of CFCs and leaked more than expected;

• The committee had no complete overview of some parts of the market of ozone-depleting compounds.

In the first report the committee member from environmental and consumer groups presented a more pessimistic evaluation of the situation after one year into the program than the majority of the committee.

The implementation of the CFC action program (the sales of CFCs and the other ozone-depleting compounds involved in the program) was supervised by an independent accountancy office, as already mentioned. The 1991 report showed the importance of such an independent procedure. The use of ozone-depleting compounds was much higher according to the numbers of the accountancy bureau than according to the numbers provided by the organizations cooperating in the covenant. The numbers of the accountancy office showed a much smaller decline in the use of CFCs as a cleansing agent than did the numbers provided by the branch organizations. In its report the committee stated that it would investigate the reasons for the differences.

At a more integrated level international and Dutch ozone policies were evaluated by the National Institute of Public Health and the Environment (RIVM) in its *Nationale Milieuverkenning* (RIVM 1991). In this report an assessment was made of the likely ozone reduction and its associated risks. Furthermore, the report used three scenarios (ongoing trends, a Montreal Protocol implementation scenario, and a London Amendments implementation scenario), developed within the organization, to assess the likely development of the chlorine concentration in the stratosphere. The likely development of CFC reduction (and threats to the reduction) was mainly discussed in the frame of European and worldwide trends and threats. In a separate paragraph reference was made to the national CFC action program, where the report stated that the national reduction target of 1990 had been easily reached but that there were several reasons for a more pessimistic outlook for future years.

Germany Germany established an Enquete Commission on Precautionary Measures for the Protection of the Earth's Atmosphere in December 1987, which covered the ozone issue in its activities. (Enquete Commission, 1990). The Commission assessed likely developments of ozone-depleting emissions including nonsignatory states, using locally developed models of ozone reduction, and in that light evaluated the Montreal Protocol. The Commission concluded that the Protocol, although a first step in the right direction, did not do justice to the aim of protecting the stratospheric ozone layer. Reasons for this verdict were that some potentially important causes had not been taken into account well enough and that there was still considerable latitude in the future development of the ozone emissions because of some of the provisions in the agreement (like the interpretations of "the needs of the parties"). The basis of the Commission's evaluation was a range of CFC emissions scenarios based on the provisions of the Montreal Protocol that the commission had asked experts to provide. The evaluation by the Commission played a clear role in determining the German response to the issue of ozone depletion. Internationally, Germany pushed strongly for the revision of the Montreal Protocol targets; nationally, Germany adopted targets that were stricter than those of the Montreal Protocol.

Less than six months after the London meeting of the parties to the Montreal Protocol the results were again evaluated by the Commission. The general conclusion then was that the strengthening of the Protocol at that meeting meant "major qualitative progress" in dealing with stratospheric ozone depletion.[6] But the Commission also regretted that the more drastic proposals of the European Community (EC) and Scandinavian countries had not found the support of the majority of the countries, that no binding and adequate provisions had been adopted for partially halogenated CFCs, and that provisions on financial and technical support for developing countries were not far-reaching enough. Another of its conclusions was that it would be possible to bring forward the phaseout deadlines by another three years at the next meeting of parties in 1992 because industry had stepped up its efforts to develop and use alternative substances and technologies, due to the considerable tightening of national reduction obligations in some countries and the tightening of the Montreal Protocol.

We can conclude that the pre- and post-London meeting evaluations by the Enquete Commission looked critically at advantages and disadvantages of the agreements reached and the ensuing progress. The evaluations were very sensitive to the different positions of developing and developed countries and to the role of industry, but the implementability of the proposed levy scheme was not addressed. A number of the conclusions of the evaluations were adopted at the 1992 Copenhagen meeting (especially accelerating the phaseout and including binding agreements for partially halogenated CFCs). However, this adoption was certainly more a result of the observed ozone depletion and of observed increases of chlorine oxide concentrations in the Northern Hemisphere than a direct consequence of the critical evaluation by the Commission.

In Germany the Montreal Protocol and its measures were evaluated not only by the Enquete Commission but also by the German Federal Environmental Agency (UBA) (Umweltbundesamt 1989). This evaluation was carried out between the signing of the Protocol and the London meeting. It looked at both the positive and the negative aspects of the Montreal Protocol and at all of the measures involved and evaluated their potential effectiveness. It provided a balanced assessment and addressed implementability in some detail. The conclusion was the same as that of the Enquete Commission—strengthen the Protocol as soon as possible—so it is difficult to decide whether the evaluation by the Enquete Commission or by UBA was more influential for policy development. But it should be acknowledged that the UBA report was comprehensive and became a standard reference for those looking at measures for responding to the issue of ozone depletion. Therefore, it may well be that the similar conclusions of the Enquete Commission and the UBA report served to strengthen the legitimacy of both.

The German activities surrounding the Montreal Protocol were also more informally assessed by NGOs and industry. One such assessment was the evaluation by the Working Group on Ozone by the NGO Robin Wood (RobinWood 1990). They assessed in particular the (German) Cabinet Proposal of May 30, 1990, which was to be the basis for new regulations. The Working Group concluded that the government ozone policies had been dominated by industry interests. The only solution they offered was a ban on the production of CFCs. This conclusion should be viewed in the light of a voluntary agreement with the Association of the Chemical Industry (VCI) on the *reduction* of CFC production in May 1990, while on the same day the German Cabinet agreed on a CFC ban that controlled the use and import of CFCs.

In the wake of large amounts of publicity about possible ozone depletion in the Northern Hemisphere in early 1992, Professor Nader of the VCI issued a statement that addressed the question "How well is the chemical industry responding to the issue of stratospheric ozone depletion?" His conclusion was that a fast global response is necessary, but he felt strongly that a national quick response in the form of an immediate CFC phaseout would not solve the problem. Thus he concluded that a fast phaseout with convincing solutions for substitutes was necessary. On the other hand, he pointed out the progress made by industry in the previous five years. The evaluation looked at all sectors of the CFC user and producer industries, with emphasis on the use of CFCs. It agreed with other evaluations that fast international responses were required but disagreed with some of the national responses. It therefore addressed important questions from the industry point of view. It provided proposals on next steps, including their implementability and the obstacles to further progress. The evaluation was presented at the opening of an international conference held in Berlin on substitutes and alternative technologies. The statement was made on behalf of all the companies belonging to four trade associations representing CFC producers and users.

20.3.3 Evaluations of Acid Rain Policy and Research

One might expect to find a parallel to the above-described highly integrated assessment and evaluation process in the case of acid rain, although not necessarily on a global scale because of the more regional character of the acidification process. However, searching for evaluations of acid rain policy one finds that formal evaluation activities in this area are dominated by evaluations of research (especially of the U.S. NAPAP program) more or less unconnected to evaluations of policy. Evaluations of policy developments have mainly been provided in presentations to conferences and professional journals by actors involved and by academic analysts.

At first sight this finding looks surprising in the light of the long history of international cooperation in the area of acid rain especially in Europe and the signing of the Convention on Long-Range Transboundary Air Pollution (LRTAP) by thirty-three states (mainly from Europe, but including Canada and the United States) in 1979. However, the Convention committed parties to broad principles and joint research activities but not to any concrete measures to reduce acid rain (Levy 1995). As Levy remarks, for the first few years of LRTAP's existence there was little indication that Europe was going to behave any differently just because a convention had been signed. For the 1985 Sulfur Protocol (signed in Helsinki) as well as the 1988 Nitrogen Oxide (NO_x) Protocol (signed in Sofia) there is, according to Levy, no evidence (with the possible exception of the Soviet Union) that any state signed without first determining that already planned policies would bring it into compliance more or less automatically.[7] This is not to say that activities related to the LRTAP Convention cannot be considered successful, but the successes centered around the way people saw the problem and not around the *actions* they took for abatement of acid rain (Levy 1995). Until 1992, the end of our study period, the LRTAP Convention was more successful in building a knowledge network than an implementation process. Partly this has to do with the fact that the 1979 Protocol did not contain targets and timetables nor commitments to further negotiations (Munton and Castle 1993). One might expect that the situation improved in the period after 1992 with the establishment of the Working Group on Strategies within the LRTAP Convention (Levy 1995). However, the history of acid rain in North America may be sobering in this regard. In August 1980 the United States and Canada signed a Memorandum of Intent (MOI) that committed each country to develop and put into effect new domestic air-pollution-control policies as necessary, and appropriate, to give effect to these, and to promote "vigorous enforcement of existing laws and regulations . . . in a way which is responsive to the problems of transboundary air pollution." (Munton and Castle 1993, 94). Although the MOI did not contain timetables for implementation, it did commit both countries to negotiate a formal bilateral transboundary air-pollution agreement and to commence these negotiations by June 1981. However, the coming of the Reagan administration with its lack of commitment to environmental matters led to an effective halt to the developments for some time.

National Acid Precipitation Assessment Program
Compared to the relative absence of evaluation of acid rain policy (the "action side"), there is an abundance of evaluations of acid rain research. The United States in

particular stands out here. NAPAP (see chapter 11 on the United States) was evaluated several times following initiatives that came up internally as well as externally. Most of these evaluations were initiated by the six agencies involved in conducting NAPAP. The internally initiated evaluations included public evaluation meetings in 1988, 1989, and 1990 (NAPAP 1989a, 1989b, 1989c) and national and international scientific peer-review meetings, culminating in the NAPAP international conference of 1990 in South Carolina (NAPAP 1990).

External assessment of NAPAP already started shortly after the beginning of the program. The 1983 Ad Hoc Committee to Review the National Acidification Policy Assessment Program (usually called the Deutch Committee after its chair) recommended changing the management of the program to give greater emphasis to integrated assessment. This became a recurring theme in evaluations of NAPAP, such as in the evaluation of the Oversight Review Board in 1991.

During the last two of the first ten years of the NAPAP, research and assessment program director James Mahoney recommended and the NAPAP Joint Chair Council approved the creation of an Oversight Review Board (ORB) for NAPAP. To install such a board was an innovative idea. Ten internationally distinguished scientists from outside the agencies of government were appointed by the agencies of government to provide scientific oversight for a major interagency program of research and policy-focused assessment. The principal evaluation document produced by the ORB was a summary report that was published by NAPAP in April 1991—a few months after the first ten years of the NAPAP program. The report was entitled *The Experience and Legacy of NAPAP*. The principal conclusions of the ORB report are quoted below:

The nine lessons derived by the ORB with regard to programs at the interface between science and policy:

1. Match institutional remedies to problems,
2. Obtain and maintain political commitment,
3. Take steps to assure continuity,
4. Configure organization and authority to match responsibility,
5. Give assessment primacy,
6. Provide for independent external scientific oversight,
7. Understand the role of science and how to use it,
8. Take special care with communication,
9. Prepare early for ending the program.

The conclusions and recommendations were formally published and widely distributed among federal and state government agencies, industrial research organizations, universities, and both for-profit and not-for-profit research organizations in the United States, Canada, and Europe. After publication of the final ORB report (NAPAP 1991), several meetings with selected leaders in the principal executive and legislative branches of the federal government were arranged by the ORB on its own initiative.

External assessments included not only evaluation reports that critically assessed what NAPAP had accomplished but also some reports directed at trying to find out what the implications for specific actors could be. One such report was the evaluation of NAPAP and EPRI activities by Ralph Perhac of the Electric Power Research Institution (EPRI) (Perhac 1991a, 1991b). In his evaluation he examined the roles played by the perceptions and influences of various individual actors. He also addressed the value and impact of scientific work and, specifically, how the research and assessments conducted by NAPAP and EPRI influenced the development of the acid-deposition-control provision (Title IV) of the Clean Air Act Amendments of 1990. Like other evaluators he concluded that the scientific quality of the research was excellent but of limited value in public decision making about acid deposition. He also recommended that industry be more proactive and work in a more collaborative manner. Another of the recommendations was to emphasize policy-related research. In this connection he considered integrated assessments and benefit-cost analyses especially important.

The above-described highly visible and very critical process of evaluation of NAPAP in the United States can be contrasted with the evaluation activities conducted in relationship to the parallel acidification research program in the Netherlands. In that case there was no predetermined view of how the program should be evaluated, but the program coordinator (T. Schneider) developed a procedure for evaluation consisting of internal assessment via symposia coupled to an external review by an invited team of experts. The internal procedure consisted of policy-oriented presentations of the research results by program coordinators. The external reviewers operated on the basis of broad questions directed at assessing the scientific quality of the research on the one hand and the policy relevance on the other. Schneider[8] commented, however, that the attention that policy makers paid to the evaluations was minimal. The relative amount of money spent for evaluation was much less than that spent in the United States—about 1 percent of the budget for the program as a whole (which was 50 million guilders in the acidification program). The evaluation process of the later climate change research program was an adaptation and refinement of the evaluation process in the case of the acidification program.

20.3.4 Evaluations and Climate Change

At the end of the period of our empirical research climate change policy was in a much more immature state than policies in response to ozone depletion and acid rain. Therefore, it is understandable that little policy evaluation was available, other than evaluation of research and evaluation of assessments. Some arenas also evaluated their policies as a preparation for UNCED.

As for the acid rain issue, formal and informal evaluations of research were most abundant. In the United Kingdom, for example, the earliest such reviews were those conducted on applied climatology and on world climatology by separate working groups of the Natural Environment Research Council (NERC). They were published together in 1976 (NERC 1976). The reports noted the very small element of NERC's budget that was devoted to climatology (the U.K. Meteorological Office was the dominant research body on climate and therefore was funded separately, whereas NERC's budget was the main source of funding for university-based research) and lamented the very narrow (meteorology) disciplinary base of the research funded.

In 1991 the U.K. government Advisory Committee on Science and Technology (ACOST) commissioned a review from an external consultant that evaluated the balance of U.K. public-sector environmental research and development (R&D) as a whole, reviewed existing definitions of environmental R&D (for example, global, regional, and local distinctions) used in different agencies, and made recommendations about future overall spending. One conclusion of this evaluation was that research in the field-biological as opposed to laboratory-biological aspects of climate R&D was underfunded (U.K. Cabinet Office 1991).

In 1980 a special U.K. Cabinet Office working group consisting of the chief scientists of government departments, and the director of the Meteorological Office reviewed existing research and policy relating to climate change, concluding that the "watching brief" strategy of existing research and policy was still appropriate and that the United Kingdom was well involved internationally, so that it would be informed early of any thoughts in the global scientific community about changes in the view of the risks. At a more detailed level the U.K. Department of Environment commissioned external consultants to evaluate its global atmosphere research program in 1992. Much of the focus was on the integration between this and other environmental research programs. It was based on not just documentary research but interviews with researchers and users.

An international activity that stimulated considerable evaluation effort was the Intergovernmental Panel on

Climate Change (IPCC) report of 1990. The IPCC was established by the World Meteorological Organization (WMO) and the United Nations Environment Programme (UNEP) in 1988 to assess the risks from climate change. The IPCC released its first major assessment report in August 1990 and a supplement in 1992.

There were several more formal evaluations of IPCC findings. The U.S. Electric Power Research Institution (EPRI) produced a review document of the IPCC documents. The intent of the review was to assist executives and environmental affairs managers at EPRI-member electric utilities in formulating a perspective on the principal findings of the three working groups. The report also concluded that assessments in the three reports did not attempt to integrate its findings within a common analytical framework or use the conclusions from one report as the basis for the analysis in another. Likewise, the conclusions in the policy-makers' summary were criticized because it was felt that they did not follow the body of the technical reports.

In the Netherlands the IPCC reports were evaluated to assess whether they could form the basis of policy in the Netherlands. To this end three working groups discussed the three parts of the IPCC report, and the conclusions of these groups were presented and discussed in a closing conference (Bakker, Dronkers, and Vellinga 1991). The report of IPCC I was discussed in a workshop in which the main skeptics in the Dutch climate debate were present.[9] The consensus of this meeting was that global warming was likely but that the estimated range of likely temperature change should be seen as greater than suggested by the IPCC (Turkenburg and Van Wijk 1991). In the conference there was some criticism of the IPCC report, but the general direction of the conclusions was accepted. The general conclusion of the conference was then adopted in the Climate Change Report of 1991 (Tweede Kamer 1991), which became the basis of Dutch climate change policy.[10]

The last type of evaluative activities to be mentioned in the climate case are those directly related to the UNCED conference in Rio in 1992.[11] We take Germany and Canada as examples here. The National Report of the Federal Republic of Germany for the UNCED conference provided a detailed analysis of the environmental situation at the beginning of the 1990s together with the policies implemented over the last twenty years (or more). The evaluation, published by the Environment Ministry in 1992, was made by the Ministry and the Federal Environment Agency.

In Canada, in the preparations for the Rio Earth Summit, an elaborate consultative structure was created. The effectiveness of this process has been extensively evaluated, at least at the level of conscious retrospective examination through a series of structured interviews. In a report to the Atmospheric Environment Service (AES) dated October 1992, the independent consulting firm Synergistics examined and commented favorably on the consultative process created by the federal government to advise on the development of Canadian positions in the negotiation of the climate change convention. Similar approval was voiced by the House of Commons Environment Committee in a report in 1993 in which the Committee also urged extension of the process to a continuing mechanism. The evaluations of the structure reveal generally positive reactions and strong claims that it must be maintained for future consultation both on international negotiations and on implementation of action to fulfill commitments undertaken. As a result, a continuing structure was developed, involving interdepartmental and interjurisdictional representation as well as industry, NGOs, and others. Thus what the House of Commons Committee and other Canadians have come to call the "Rio way" (openness, transparency, and inclusiveness) became-at least momentarily-the "Canadian way" nationally and internationally, with the mandates of the International Development Research Center and the International Institute for Sustainable Development both being extended to cover responsibilities in spreading roundtables round the world as mechanisms for integrating environmental and economic concerns.

20.3.5 Evaluation as Part of a Planning Cycle: The Netherlands Case

At the end of this descriptive section we discuss an example of evaluative practices at the national level much resembling the international procedures that developed around the Montreal Protocol, as a possible direction in which evaluation practices for global environmental risks could develop: evaluation as part of a planning cycle as developed in the Netherlands.

In the Netherlands evaluations of environmental laws started to become common practice around 1980. These evaluations were commissioned by policy managers and were strongly directed toward assessing goal attainment (implementation). Broader evaluations of environmental policy were conducted by environmental organizations[12] and, for instance, Interprovinciaal Overleg (IPO), a consultation group of the provinces. At the end of the 1980s the National Institute of Public Health and the Environment (RIVM) started to conduct national environmental foresight studies. After the first of these studies (RIVM 1988), RIVM was asked to conduct such a study every two years as part of national environmental policy planning. It was proposed that national environmental policy plans would in principle be published every four years. In its foresight studies RIVM was requested to assess the expected quality of the environment when the policy plans are prepared in the year after publication of the National Environmental Policy Plan, whereas in a year before a new policy plan possible additional policy options would be assessed.[13] Over the period of our study, therefore, evaluations of environmental policy first became common practice and then became part of a planning cycle.

20.4 Changes and Differences in Evaluation Practices

The discussion in the preceding section was very uneven in terms of the time periods discussed, the arenas selected, and coverage of types of evaluation. Apart from the fact that we put more emphasis on formal evaluations than on informal criticism of policy, the emphasis reflects the differences in evaluation practice over time, arenas, and issues that are discussed more systematically in this section.

20.4.1 Changes in Evaluation over Time
Formal, public evaluations of past policy performance with respect to global environmental risks remained rare throughout the study period but became less so during the second half of that period. Over the three decades surveyed in this study, there was no unambiguous linear flow through the successive functions of risk management, but there was a natural shift in the focus of concern and attention from the early stages of a struggle to identify the nature and magnitude of risk, toward the identification, selection, and ultimately putting into play of possible responses and inquiry as to the results. This flow is reflected by the shift of emphasis away from the appraisal of research efforts and toward commentary on the quality of option assessment and strategy formation, judgments about the adequacy of implementation efforts, and examination of the acceptability of decision processes themselves.

Evaluation of Research   Evaluation of research programs became a regular phenomenon over most of our arenas. Research programs set up in response to our issues were also evaluated, but the ozone case—being relatively limited in scope—did not lead to dedicated national programs, whereas climate change programs were still evolving in many arenas, and therefore not many evaluations were found over the period until the end of 1992. However, for instance, positive evaluation of the first climatology program in the European Community that addressed both climate change and ozone issues resulted in increasing funding and an expansion of the areas

to be covered (European Commission 1986). Although evaluation of science programs was accepted practice in principle, the extent to which such programs were evaluated and the type of evaluation process differed appreciably between arenas. The evaluations of acid rain research are a clear case in point. These differences are discussed in section 20.4.3.

Extent of Policy Evaluation   Evaluation activities of policies related to global environmental change, especially formal institutionalized evaluations, were relatively scarce. Examples of these institutionalized activities were all from the second half of the period we covered in this study. This is partly because evaluation has become more common in general, but certainly also because the issues matured, policies were implemented, and policy evaluation became feasible. Nevertheless, evaluation was relatively rare for global environmental issues compared with other issues on the public agenda, such as welfare programs. By the end of our study period, however, the relative rarity of policy evaluation related to global environmental risks appeared to be changing quickly, and possibly in a direction seen in the Netherlands, where evaluation of environmental policy became part of the planning cycle. For global environmental risks such a tendency was possibly strengthened by international evaluation cycles imposed by international treaties.

Style of Evaluation   There are some indications that the style of evaluation has changed over time, in accordance with more general tendencies in thinking about possible roles for evaluation, from a preoccupation with technical adequacy toward a greater emphasis on questions of process and legitimacy. Evaluation processes in general tended to become more comprehensive in the sense that more actors in society were involved and that, partly as a consequence, evaluation activities were becoming part of a diffuse process, of which the formal boundaries were less clear. This pattern does not seem to be a characteristic of the management of global environmental risks in particular but more related generally to the opening of societies to the influences of actors outside direct bureaucratic circles. Again, the differences between arenas are considerable, as is shown in section 20.4.2.

20.4.2 Cross-Arena Differences
All that has been said above must be seen against the background of huge cross-arena differences with respect to the way in which evaluation was embedded in society. In some arenas evaluation activities were relatively abundant, visible, and open, and in these arenas a relatively large number of formal evaluations were found. The United States and the Netherlands best exemplify

this situation. At the other extreme, represented by the United Kingdom, Hungary, the former Soviet Union, and—in part—the European Union, evaluation was typically quite exclusive with respect to the actors involved. Superimposed on this basic pattern, however, we find differences among the arenas in the amount of criticism involved in evaluations or (alternatively) in the constructiveness of evaluation activities.

It is instructive to contrast the highly visible and very critical process of evaluation of NAPAP in the United States to the evaluation activities conducted in relationship to the parallel acidification research program in the Netherlands. The latter was much less visible and costly, but it did feed back into the organizational structure for the climate research program.

In the United Kingdom evaluation was typically quite informal, as with other aspects of policy. However, formal and systematic reviews of research did take place, in each of the issue-domains as well as more comprehensively, of public-sector environmental research as a whole. Many of these reports were not published.

Not only with respect to research programs but also more generally, the extent to which management activities were evaluated differed among arenas. International institutions conducted few evaluations of global environmental risks, but this is germane to the whole evaluation atmosphere in which international institutions operate. UNEP conducted a few evaluations of its monitoring programs. Although the results were critical, the reports were tightly controlled by the organization and received no attention outside of the unit of the organization. There was seldom even an appreciation of the effectiveness of the activities sponsored by the institutions.[14] Although member governments may have challenged the institutions for appraisals and evaluations to justify annual budget requests, the annual reports were seldom specific enough to be able to provide any real insights into the quality of their activities. Evaluations of institutions' activities were occasionally conducted by external groups. These evaluations, particularly if conducted in the United States, tended to be highly informed by the political orientation of the evaluating organization.[15] Given the apparent difficulty of evaluation activities in the international context, the apparent success of evaluative activities in the frame of the Montreal Protocol is surprising.

Evaluations were much more open processes in the United States than in either the United Kingdom or international organizations, but in the latter cases evaluation activities could still be found, in many instances resulting in evaluation reports. This is different from the situation in some other arenas, like the former Soviet Union and the European Union, where (with the exception of research

evaluation in the Community, such as the European Commission in 1986) the evaluation process was completely internal to the system and did not lead to visible outside results. The exception, again in the case of the European Union, was evaluations implicitly or explicitly referred to in preambles of proposals for new or amended directives,[16] evaluations proposed by members of the European Parliament, or evaluations presented by European Union officials during conferences or interviews.

Hungary is an extreme example of a situation in which evaluation stayed internal to the policy process. In Hungary ministerial documents dealing with global environmental risks regularly contained a self-evaluation of earlier steps made by interministerial committees set up to coordinate action or by working groups that were sent out by these committees to investigate specific aspects of the problem (such as baseline emissions and technological options for emissions reductions and the associated costs). The most typical of these government evaluation documents were those that, from time to time, reviewed the progress made in complying with international environmental commitments.[17] In these cases evaluations were not open to outside review because there was no way in which outside views could be solicited other than by informal exchanges of opinion.

Lack of openness to outside review also existed in arenas where the process was more formal and evaluation reports were produced, but the final products of the evaluation in typical cases were embedded in confidential reports, as was often the case in the United Kingdom. In the United Kingdom outside views were solicited (for example, from industry people or from representatives of the scientific community). The comments from these actors could become part of the conclusions but were not normally visible in any externally available report. In a number of respects the situation in Canada reflects the situation in the United Kingdom, but apparently more so in the early period of our study than in later periods.

The clearest example showing that the style of evaluation in an arena is very dependent on the more general policy culture is Canada. This country underwent an integral shift in its policy culture in the twenty-five years from 1967 to 1992. Canada moved from a policy culture much like the British tradition of expert advice to one much more like the U.S. populist approach but with interesting differences. The early emphasis on administrative processes of policy formation and regulation founded on expert advice and an influential scientific elite gave way to a culture dominated by multisectoral, interest-based consultation, concern for participatory mechanisms, and the interaction of advocacy groups advancing scientific argument to further their particular interests. The creation of the Science Council of Canada in 1966 as a federally funded agency to provide independent scientific and analytical advice to governments, industry, and educational institutions and its demise (along with a wide range of other independent advisory bodies) at the hands of the Conservative government in the Mazankowski budget of February 1992 thus neatly brackets not only an extraordinary swing in political outlook but also a dramatic shift in policy culture, including the creation of a massive consultative apparatus of roundtables and advisory committees. The elaborate consultative process established in preparation for the Rio summit is an excellent example of this shift.

A completely different but equally interesting example is provided by the Japanese case, where evaluations highlighted the possible technological contributions of Japan. Consistent with the general observation in the Japanese case study related to involvement in global environmental issues, the Japanese were also relatively late in starting evaluations on these issues. However, the earliest evaluative statements we have from Japan related to global environmental risk issues immediately point in the direction that was later taken by Japan. The eighty-six pages devoted in the 1988 report on the *Quality of the Environment in Japan* (Environment Agency of Japan 1988) to global environmental problems represented a dramatic break with the strong domestic focus of earlier volumes. Here considerable discussion was given to the nature of various transboundary and global pollution problems (including acid rain, the extinction of wildlife species, atmospheric pollution problems, and marine pollution) and their sources, including expanding socioeconomic activities and poverty. This volume also reflected on how Japan could make contributions to solving global environmental problems focusing in particular on Japan's technological capabilities.

20.4.3 The Involvement of Actor Groups

An interesting difference between our arenas that warrants specific attention is the involvement of actor groups in evaluative activities. It is clear that in those situations where evaluation stays internal, involvement of other actors than those who are directly in charge will be limited. But even beyond this, there are differences to be found in the extent of actor involvement. In the United States in all cases many actors were involved, mainly via committees, hearings, or other ways of soliciting the views of outsiders. Also other actors were indirectly involved via their own critical appraisals of an activity. In the NAPAP case in the United States, for example, not only the dozen or so formal evaluations were conducted, but there also was a flurry of more informal evaluations. Such a climate of

critical appraisal was nowhere as strong as in the United States, but it was increasing in some other countries, like Canada, from which we take the following example.

In 1990 Friends of the Earth Canada launched a Federal Budget Review Project with a pilot project focused on climate change and particularly on carbon dioxide (CO_2) emissions. This 1990 pilot project illustrated well the immense difficulty of tracking implementation of any one policy undertaking (such as a commitment to limit CO_2 emissions) through the maze of government programs and budget elements. Nevertheless the review did suggest that the government's stated policy of limiting Greenhouse Gas (GHG) emissions had not resulted in significant budgetary or regulatory changes and characterized the resources allocated to preventing climate change as "virtually insignificant." This scrutiny of the credibility of government action to implement commitments with respect to reducing pressures on the environment was given further extension in a 1994 "report card" in which Friends of the Earth gave to each provincial government a grade for their actions to reduce emissions of ozone-depleting substances.

Some western European countries provided examples of a mechanism by which the government intentionally raised the level of critical appraisal. For instance, in the United Kingdon one way in which evaluation was institutionalized was by contracted critical review by scientists or through subsidizing of research institutes involved in critical scientific inquiry. Likewise, in the Netherlands and Germany NGOs were subsidized to facilitate critical evaluation. The interesting difference between the United Kingdom on the one hand and Germany and the Netherlands on the other, however, was that in the United Kingdom criticism was again incorporated in closed-shop activities, whereas in the other two arenas the process was via the explicit buildup of outside pressure on the government. Mexico was an interesting intermediary case, in which the second route was also used but with limitations from the side of the Mexican government on the issues that were opened to outside criticism.

In some arenas the role of the media was very central. In Mexico, for instance, some of the most explicit evaluations of climate change policy were presented in the newspapers (for example, *Excelsior* on June 5, 1991, and *La Jornada* on May 30, 1989). These were both evaluation articles on global warming written by environment reporters.

20.4.4 Issue-Specific Differences

The three issues we are analyzing are relatively special compared to other environmental issues in that their effects spread over borders, and therefore the policies related to the issues may be more open to scrutiny by other countries than in most other environmental issues. There may be a closer relationship to the patterns in international trade. But the three issues differ in that in the acid rain case the pollution is transborder rather than global, thereby invoking bilateral or trilateral interactions, whereas the risks that have effects on a global scale are more immediately addressed at the international level. We see this reflected in the differences in the evaluative activities.

In the acid rain case, we see critical appraisal of one country with respect to the environmental damage inflicted on another country. In the mid-1980s there were numerous reports and hearings at the federal and state levels in the United States to assess and address the problem of transboundary pollution, especially because of sulfur dioxide emissions from copper mines and copper smelters in Mexico.[18] Likewise U.K. policy was critically assessed by Scandinavian countries. Evaluations of acid rain damages were not restricted to bi- or trilateral activities. Especially in Europe the LRTAP activities invoked evaluations of performance over a larger number of countries, but still the spread of these was regional rather than global.

The ozone and climate issues have in common that the effects strike globally and that the regional relationship between the emissions and effects is weak. As has already become clear for the other risk management functions, this was reflected in a stronger relationship than for many other environmental issues between national and international activities and in attempts at international coordination with respect to assessment as well as policy. Explicit examples are the Montreal Protocol, ensuing international commitments, and the activities around the IPCC.[19]

20.5   Trends and Patterns

As discussed in section 20.3, evaluation efforts span a variety of activities, from criticism to formal evaluation procedures, from single reports to extended activities. By focusing on some evaluation activities around ozone depletion, acid rain, and climate change, we illustrated the ways in which such activities were conducted in our arenas. A disadvantage of this approach is that section 20.3 discusses only those arenas where explicit and open evaluative activities were conducted. This is redressed in section 20.4, where we discussed the situation in all arenas.

Over the three decades surveyed by this study there was (even if there was no unambiguous linear flow through the functions of risk management) a natural shift in the focus of concern and attention from the early stages of a struggle to identify the nature and magnitude of risk, toward the identification, selection, and ultimately putting into play of possible responses and inquiry as to the results. This flow is reflected as emphasis shifts away from the appraisal of research efforts and toward commentary on the quality of option assessment and strategy formation, judgments about the adequacy of implementation efforts, and examination of the acceptability of decision processes themselves.

In this chapter we see much of this flow exemplified:

• In most arenas, traditions of scientific research led to extensive appraisal of research efforts themselves, and much of this appraisal involved international participants in conventional peer-review process following well-established practices;

• Emergence of the international institutions as leaders and innovators in evaluation, including periodic review of science, and implementation success;

• In some countries, NGOs became prominent in attempting to assess critically the process of implementation (Netherlands) or to issue report cards and "letter grades" (Canada) to governments in comment on the credibility of their implementation efforts (much of this commentary is obviously highly judgmental and qualitative);

• In at least one case (Canada) major evaluation efforts were undertaken to assess, through surveys and interviews with participants, the degree of satisfaction with the operation of the participative process.

Monitoring emissions reductions was, of course, a direct form of evaluating the success of policies designed to reduce human impacts on the environment (although the fundamental objective is presumably some improvement in measures of environmental harm or damage to humans). In a number of settings where regional impacts were of concern, this appraisal took the form of countries that monitored and commented on the emission levels and policy successes (or shortcomings) of other nations. Such international scrutiny of domestic environmental policy was evidently increasing and increasingly raising questions about conformity not just to international codes but to interactions with trade policy and commercial rules as well as other internationally negotiated commitments.

One overall conclusion from this survey of evaluation efforts and their evolution over the time period of this study, then, is that there was a fair amount of serious and sustained evaluation of research and development efforts (primarily risk assessment activities but in some cases extending to option assessment as well), with the evaluation efforts themselves following traditional processes of (increasingly international) peer review. In the more goal-rational contexts of government agencies in countries like the Netherlands and Germany, there were some efforts to assess policy outcomes critically against targets specified in legislation or policy declarations and more often a growing effort by NGOs to comment critically on the credibility of implementation efforts. This latter effort seems to have led to one of the few genuine innovations that can be identified in the arena discussions of evaluation to date—namely, the highly visible practice of issuing "report cards" with summary letter grades in appraisal of government policy measures.

In any case, however, the major conclusion may be simply that evaluation beyond the peer review of science came very late in the risk management cycle. In the case of global risks, arenas are still too early in that risk management process to have attempted much systematic appraisal of how well they are doing—or how much they are learning—overall in the management of global atmospheric risks.

Against these general reflections on the development of evaluation practices, it is also clear that evaluation practices varied widely among arenas. In arenas where evaluation activities were visible and open to outside comments as in the United States, Germany, and the Netherlands evaluations were also relatively abundant and formal. In another group of arenas, evaluations tended to be less visible (for example, because reports were secret or not publicly available) and were closed-shop operations (with exclusive involvement) as in the United Kingdom and, even more, Hungary. It is difficult to decide whether this is related to less evaluation taking place because the difference may well be that the processes involved were less formal, resulting in fewer evaluation documents only.

Openness and visibility of evaluation processes do not seem to be directly related to the integration in decision processes. The United States stands out as having many instances of evaluation that were very open to external review and embedded in critical appraisal, but relative to other arenas these activities were less integrated with the decision-making process. In some of the other arenas the evaluation process was almost completely integrated with the decision process. An example is the European Union, where the results of evaluations became visible mostly in the preambles of new policy proposals.

Also there does not seem to be a visible relationship among openness, visibility and the actual involvement of actor groups. The important difference here seems to be the way in which criticism was invoked and how it played a role in evaluations. In the United Kingdom and initially Canada, in evaluation activities selected outsiders were being coopted as part of the usual process of how closure was attained. But it is unlikely that such a mechanism would lead to a legitimate outcome in more open political cultures, in which the balancing process operates via a different mechanism—namely, by invoking outside evaluations (if these do not develop on their own account, as in the United States). Mexico is an interesting example of an attempt to regulate this mechanism.

Overall, evaluation tended to become a more important process over time, but in some arenas, especially Canada, there was a remarkable shift in the character of the process over the years. Initially Canada operated in much the same way as the United Kingdom, but its evaluative style changed over time and became a more open process.

Where evaluation of the policy related to global environmental risks or of environmental policy including these risks became a more regular activity, it could become part of a planning cycle. This is most visible in the Netherlands, where environmental policy plans and evaluations by RIVM alternated, but it could also be the underlying structure in those less open situations where results of evaluative activities became visible via their conclusions in policy plans only.

Perhaps more than some other risk management functions, evaluation efforts strongly reflected the underlying political culture of a nation. The shift in political culture and the accompanying shift in evaluation practices in Canada exemplifies this. Thus, in drawing out central themes in the experience of different arenas with efforts to evaluate their progress in the management of global atmospheric risks, it is crucial to take adequately into account trends and changes in this underlying climate or landscape for evaluation.

In particular, almost all arena accounts take note of the interplay between tendencies toward a "scientific" or "rational" model of political process and decision making in risk management and a "democratic" or "pluralist" concept. With the one approach comes an emphasis on measures of outcome and effectiveness; with the other comes a concern for dimensions like legitimacy and fairness.

20.6 Conclusions: Enhancing Learning Capacity by Evaluation?

Is it possible to draw conclusions on the extent to which the practice of evaluation improved over the period addressed in this study? What expectations may we have for the future?

Evaluation in the sense used in this study generates a normative interpretation of data gathered in the frame of any of the other risk management functions or of data that are specifically generated for the evaluation. What evaluation adds to the other types of information is explicit normative reflection on knowledge and action; it generates answers about failures and successes of past performance, with the aim to direct action (possibly directed at knowledge generation) more effectively and to avoid pitfalls encountered in the past.

What counts as a failure or a success is of course dependent on the criteria against which one evaluates, and it is clear that those criteria are not neutral. In section 20.1 two routes were discussed that have been taken in reaction to the recognition of this problem. The first is to explicitly acknowledge the inherent bias and make it explicit. A way of doing so is recognizing who the client of an evaluation is and directing the evaluation toward the biases of the client. The other route has been to involve as many stakeholders as possible in the evaluation process. Each route has a different effect on the content of the evaluation, on the effect of it, and hence on the quality of evaluation.

20.6.1 The Quality of Evaluation Activities

As in other chapters in part III, here we try to draw conclusions on the quality of evaluation activities using the four metacriteria suggested by Clark and Majone (1985): effectiveness, adequacy, value, and legitimacy.

Effectiveness The criterion of effectiveness should give an answer to the question of whether evaluations do solve problems. Of course, the whole idea of evaluation is that it can help to detect problems as a prerequisite to solving problems. Many of the informal evaluations (the unsolicited criticisms) were doing precisely that: detecting problems in the form of pointing toward data that have not been taken into account or toward threatening or real implementation gaps. But especially if these critical analyses were not integrated into the policy-development process, it is unclear whether and how such analyses fed back into solutions for the problems identified.

Formal evaluations, however, may surprisingly lead to parallel untoward effects due to two mechanisms. The first is that the detection of problems does not necessarily lead to directions in which a solution can be found. The other is that the questions addressed in the evaluation may not be the ones that are relevant for solving the problems encountered. We discuss the latter under the criterion "value." With respect to the first we found two very different situations. The United States is the example for the first and Hungary, and the European Union for the second. In the United States evaluations tended to be very critical but directed relatively little at problem solving. In a number of other countries it was only the solution that was being presented, as in the preambles of environmental action plans in Hungary and the European Union. In the latter cases it is clear that evaluations were used as an argument for policy plans to be implemented, although it is not clear for outsiders which selection was being made from the evaluative information present. In the open and critical atmosphere in the United States the available information was known in principle, but it was much less clear how and whether this information fed back into the process.

It was only in situations where evaluation was a standard integral part of a policy process that evaluation activities were structured in a way in which they could suggest solutions for detected problems and feed them back into the process as well. The Montreal Protocol process and in a certain sense also the environmental evaluations used in the planning process in the Netherlands fulfilled these conditions.

Adequacy We interpret adequacy primarily as avoiding known pitfalls and channeling discussions into well-defined categories. An important pitfall in the case of evaluation is the use of data that are not accepted by parties in the field. In the management of global environmental risks, data seldom are completely undisputed, but evaluations based on disputed data will tend to become ineffective or possibly simply be proven wrong. The above-mentioned two routes to account for the normative nature of evaluation both provide ways to handle the issue. Evaluation in close connection to a certain client is, however, more likely to lead to the biased use of data with possible loss of legitimacy of the evaluation. The route involving multiple stakeholders makes it feasible to at least assess the range of data suggested and the possible interpretation space introduced by making a selection. In a sense the IPCC procedure, although strictly not an evaluation in our sense but an assessment process, did this. The way the results of IPCC 1990 were evaluated in view of the policy preparation for Rio may be a case in point.

Whether an evaluation does channel the discussion around an issue into well-defined categories may be seen as an important criterion for a good evaluation and certainly of a regular evaluation process. In the management of global environmental risks it opens up the possibility to evaluate whether progress has been made in well-defined areas, as is the case with respect to the Montreal Protocol. Even if categories are found to be missing in the process and the related evaluation procedure, the categorization enables debate about additions to be made or elements to be left out. Although it is tempting to see the Montreal Protocol as a model here for managing other areas of global environmental risk, we should realize that in many instances (of which acid rain and climate change are examples) the questions to be covered are much more complicated than in the ozone case and as a result categories defined will be much more likely to have to be redefined regularly.

Value The criterion of value addresses the question of whether an evaluation is relevant in the sense that it leads to an assessment of the value of the activity evaluated beyond the immediate interests of the activity itself and those involved in the activity; it also addresses the question of whether it leads to suggestions for change that are feasible.

Many evaluations of research and especially many NAPAP evaluations exemplify this point. Although several evaluations of research have been conducted, many of these were primarily directed at evaluating research as research and not as an activity designed to help support the solution of environmental problems. This is not to say that evaluations directed at research programs should not evaluate research according to internal quality criteria, but if research programs are set up to help solve risk management problems, they should be gauged against those criteria external to research too.

A parallel problem can again be exemplified with NAPAP as an example. Many of the recommendations made in this context do not score high if we gauge them against the feasibility of their implementation. Although the advantage of a distanced critical evaluation clearly rests in the independence of the judgment, the level of criticism itself may reduce the chances of the suggested changes to become reality.

Legitimacy Clark and Majone relate legitimacy to two subcriteria that are relevant for evaluation—well analyzed and democratic. *Well analyzed* relates to the knowledge base of the activity, whereas *democratic* relates to the process. In relation to the practice of evaluation as we found it in the arenas we studied, it is relevant to discern two categories of actors for whom this question can be addressed: the direct client and the stakeholders. In many of the less open systems that we studied (including Hungary, the former Soviet Union, the United Kingdom, and the European Union) legitimacy of an evaluation primarily rested on the analytical quality. Little direct effect could be expected from a process in which democratic means were not being employed. The culture in these arenas led to evaluations that were based on nondemocratic procedures. In other arenas legitimacy was lost if no attention was paid to the democratic character of the process. We did not encounter many examples in the areas studied (in contrast to what has happened in many countries around nuclear energy or biotechnology), but in the Canadian case we saw how important a democratic procedure may be for enhancing legitimacy of policy.

The Canadian case also exemplified that the cultural style of a state may change rather drastically with ensuing effects for the legitimacy of certain evaluation procedures.

20.6.2 Prospects for Improvement

The first conclusion we can draw is that evaluation, although scarce, has become more abundant. Our initial view of evaluation was that it is a prerequisite to effective learning, since it provides a mechanism for self-reflection. From our results, however, we do have to make two qualifications. The first is that such reflection under some circumstances is more likely to generate *awareness* of problems rather than possible *routes* to handle the problems. Therefore, evaluation should at least have some link back to the policy process it evaluates. The second is that evaluation may become inefficient if it becomes standardized in ways that do not reflect the problems encountered. In those cases evaluation is unhelpful for policy learning, since it leads away from the questions to be addressed.

From the trends detected, we can conclude that there are now some interesting mechanisms in place in which evaluation is well embedded in a policy process and that it is likely that such evaluation procedures will become more abundant. With evaluations procedures becoming more regular, however, it is also likely that inefficient evaluation procedures become more common.

The second conclusion is that it is likely that evaluation in the sense defined in this chapter will become part of complex assessment and evaluation processes at the international as well as at national levels. Although evaluation is distinct from assessment, it would be unhelpful to strive for a complete separation of both. This is not to say that monitoring and risk and option assessment activities should not be undertaken independent of evaluation. They certainly should. It is only important that, related to evaluation, assessment (in most cases secondary assessment) of risk and option data can take place to generate an integrated process.

To make evaluation effective it is important that it does not relate only to the other knowledge functions, but also to action, as in the example of the Montreal Protocol. It is, however, not likely that the favorable circumstances that made the Montreal Protocol a success will be repeated in many instances around global environmental risks. The activities following from the Montreal Protocol may, however, serve as an example of what can be done in the global environmental area. The fact that such an example exists may be even more important than the exact details of that situation because it teaches that in the environmental area as in other areas (like weapons treaties) international cooperation can be effective and that evaluation procedures can support the effectiveness.

Appendix 20A. Acronyms

ACOST   Advisory Committee on Science and Technology (U.K.)

AES   Atmospheric Environment Service

CFC   chlorofluorocarbon

CO$_2$ carbon dioxide

EC   European Community

EPRI   Electric Power Research Institute (U.S.)

FoE   Friends of the Earth

IPCC   Intergovernmental Panel on Climate Change

IPO   Interprovinciaal Overleg (Netherlands consultation group of the provinces)

LRTAP   (Convention on) Long-Range Transboundary Air Pollution

MoI   Memorandum of Intent

NAPAP   National Acid Precipitation Assessment Program

NERC   Natural Environment Research Council (U.K.)

NGO   nongovernmental organization

NO$_x$ nitrogen oxides

OECD   Organization for Economic Cooperation and Development

ORB   Oversight Review Board (for NAPAP)

R&D   research and development

RIVM   Rijksinstituut voor Volksgezondheid en Milieu (Netherlands National Institute of Public Health and the Environment)

UBA   Umweltbundesamt (German Federal Environment Agency)

U.K.   United Kingdom

UNCED   United Nations Conference on Environment and Development (held in Rio de Janiero, June 1972)

UNEP   United Nations Environment Programme

UNGA   United Nations General Assembly

VCI   Verband der Chemischen Industrie (German Association of the Chemical Industries)

WMO   World Meteorological Organization

Notes

1. The authors are grateful for the contributions of all of the research teams that provided information on evaluation. For reviews of earlier drafts we also thank five anonymous reviewers.

2. Chelimsky mentions a fourth role: increasing public awareness of the current and likely future impacts of these problems. Increasing public awareness is a function evaluation can fulfill, especially if evaluation is conducted in ways involving various actors. But in contrast to the other items listed, it does not determine an object for evaluation. For that reason we do not include it under the descriptive definition of evaluation.

When referring to the other knowledge functions in this chapter, we denote them as *assessment* as opposed to *evaluation*.

3. This is, of course also related to the function that Chelimsky sees for evaluation in increasing public awareness of issues.

4. These were

• Research evaluation (peer reviewing, budget spending, policy-related tasks),

• Government and committees (ministerial committees, ad-hoc or standing advisory committees),

• Parliament and parliament-related committees, ad hoc or longer term (minority parties, backbenchers, noninfluential parliaments),

• Industry (industrial federations, individual industries),

• Institutionalized criticism and government invoked criticism (counterscience organizations, government funded NGOs, critical scientists),

• Social criticism (NGOs, media),

• Outside actors (OECD, other countries, NGOs from elsewhere), and

• Evaluation activities in relation to international treaties.

5. See Parson (1993) for a discussion of the influence of the assessment panels on negotiations. Benedick argues that the 1989 synthesis report of the assessment panels was the basis for the London revisions and a key to the bargaining process (see Benedick 1991, 129, 173).

6. The decision on the global phaseout of CFC production and consumption of fully halogenated CFCs by the year 2000 and a corresponding phaseout for other CFCs, halons, and carbon tetrachloride were greeted favourably by the Commission. They also found the solution reached for methyl chloroform to be acceptable. Further, they felt encouraged by the establishment of a fund to provide technical and financial aid to the developing countries and by India's and China's announcements that they were now prepared to accede to the Vienna Convention and Montreal Protocol.

7. To be able to do this means at least some kind of informal evaluation at the national level of the likely impact of policy measures or other changes likely to happen in society. Examples of the last occurred when Eastern European countries faced economic decline and the United Kingdom changed to fuels other than coal.

8. Interview, September 23, 1995.

9. Including also Böttcher, retired professor of physical chemistry and honorary member of the Club of Rome, who is one of the most vocal people in the Netherlands in calling the climate change theory into question.

10. This history entails an interesting example of the difference in the way criticism is dealt with in the Netherlands as opposed to the United States. In the Netherlands there are a number of people who actively discuss the likelihood of global warming, and these views are reported in newspapers and the like. The most important of those people are then made members of working groups that try to reach consensus. In many cases such a consensus is reached by subtle formulations, which make disagreement very hard to maintain. Although the substantive arguments are considered more important than consensus for its own sake, it is nevertheless considered improper not to try to reach consensus. In many cases such a process does not really end the controversy, but the fragile consensus may enable policy that has legitimacy to be developed.

11. In the earlier examples the relationship was indirect, although the criticism as informal evaluation became more relevant because of the approaching UNCED process.

12. The fact that some of these environmental organizations are subsidized by government leads to the installation of a critical function in society, in which government action is permanently assessed and evaluated.

13. Van Noort, Director-General of RIVM, in his introductory words in *Nationale Milieuverkenning 2* (see RIVM 1991, 7).

14. For instance, Secretariat members are happy to be able to conduct training seminars but have little idea about what happens to attendees when they leave.

15. For example, the (conservative) Heritage Foundation is highly critical of almost all international institutions, while liberal internationalist organizations (such as the United Nations Association) typically produce evaluations that are fairly laudatory.

16. One example can be found in the preamble of Regulation 1613/89 amending Regulation 3528/86 on forest protection, where the proposal for the amendment is based on the following evaluative statements:

Whereas forest decline is persisting in many parts of the Community, and whereas the Community scheme provided for in the regulation 3528/86 should therefore be reinforced, . . . Whereas in order to help stem the dieback of forests, Member States should be helped to take measures to maintain and restore them, aimed at reestablishing favorable soil conditions in forest stands where they have deteriorated, in particular through acid deposition.

17. Most of these documents were classified "for internal use only"; therefore, we were not able to obtain copies. But we have reviewed some of them in the archives of various ministries, and they were also mentioned by ministerial officials interviewed for the project. The key evaluation criterion was, almost without exception, how the country was doing in terms of achieving the emission targets as required by the country's international commitments (Ferenc Tóth, personal communication).

18. See, for example, Grimes (1985), Hutchinson and Varady (1986), Kamp (1986), U.S. Congress (1983, 1985).

19. The UNCED conference in 1992 is an outstanding example. We did not discuss the latter because the evaluative efforts involved largely stretch beyond the conference into a period we did not study in this research.

References

Bakker, H., J. Dronkers, and P. Vellinga. 1991. *Veranderend Wereldklimaat, Nederlandse Visies, reacties en commentaren (Changing World Climate, Changing Views: Reactions and Commentaries)*. The Hague: Delwel.

Benedick, R. 1991. *Ozone Diplomacy: New Directions in Safeguarding the Planet*. Washington: Georgetown University.

Brandl, J. 1978. Evaluation and politics. *Evaluation and Program Planning*. Special issue, 6–7.

Chelimsky, E. 1995. Where we stand today in the practice of evaluation: Some reflections. *Knowledge and Policy: The International Journal of Knowledge Transfer and Utilization* 8 (3): 8–19.

———. 1997. The coming transformations in evaluation. In E. Chelimsky and W.R. Shadish, eds., *Evaluation for the Twenty-First Century: A Handbook*. London: Thousand Oaks.

Chelimsky, E. and Shadish, W., eds. 1997. *Evaluation for the Twenty-First Century: A Handbook*. London: Thousand Oaks.

Clark, W., and G. Majone. 1985. The critical appraisal of scientific inquiries with policy implications. *Science, Technology, and Human Values* 10 (3): 6–19.

Electric Power Research Institute (EPRI). 1991. Review of the Intergovernmental Panel on Climate Change Working Group Reports. EN-7374 0061P, Electric Power Research Institute, California.

Enquete Commission. 1990. *Protecting the Earth: A Status Report with Recommendations for a New Energy Policy*. Bonn: Bonner Universitaets Buchdruckerei.

Environment Agency of Japan. 1988. *Quality of the Environment in Japan*. Japan: Environment Agency.

European Commission. 1986. *Evaluation Report on Community's Environmental Research Program*. Brussels: European Commission.

Friends of the Earth (FoE) International. 1992. *National Action Plans to Save the Ozone Layer: Countries' Restrictions on Ozone-Depleting Chemicals*. London: Friends of the Earth Limited.

Grimes, Maria. 1985. Nacozari, Mexico, Copper Smelter: Air pollution impacts on the US Southwest: A case study of transboundary air pollution. Congressional Research Service, The Library of Commerce, April 17.

Guba, E., and Y. Lincoln. 1989. *Fourth Generation Evaluation*. Newbury Park, Calif.: Sage.

Hutchinson, B., and B. Varady, eds. 1986. *U.S.-Mexico Conference on Border Smelter Emissions*. Tucson: University of Arizona Press.

Kamp, Dick. 1986. Health damage from SO_2 in southern Arizona. Comments of Dick Kamp, Director Border Ecology Project, Concerning S.2203 (to amend the Clean Air Act and the Clean Water Act to establish a national program to control acid rain) before the Senate Committee on Environment and Public Works, October 2.

Levy, M. 1995. International cooperation to combat acid rain. In H.O. Bergesen and G. Parmann, eds., *Green Globe Yearbook 1995*. Oxford: Oxford University Press.

McLemore, Jacqueline R., and Jean E. Neumann. 1987. The inherently political nature of program evaluators and evaluation research. *Evaluation and Program Planning* 10: 83–93.

Munton, Don, and Geoffrey Castle. 1993. Acid Rain, Basic Politics and Social Learning in Canada, working draft prepared for the Social Learning Group (appendix I), July 19, 1993.

NAPAP. 1989a. *Assessment Plan Update*. Washington: NAPAP.

———. 1989b. Plan and schedule for NAPAP assessment reports, 1989–1990. Washington: NAPAP.

———. 1989c. *Review of public comments on the October 1988 draft of the plan and schedule for NAPAP assessment reports: 1989–1990*. Washington: NAPAP.

————. 1990. *Proceedings of the NAPAP International Conference.* Washington: NAPAP.

————. 1991. *The Experience and Legacy of NAPAP: Report of the NAPAP Oversight Review Board to the Joint Chairs Council of the Interagency Task Force on Acidic Deposition.* Washington: NAPAP.

NERC. 1976. *Research in Applied and World Climatology.* London: HMSO.

Parson, E. 1993. Protecting the ozone layer. In P. Haas, R. Keohane, and M. Levy; eds., *Institutions for the Earth.* Cambridge: MIT Press.

Perhac, R. 1991a. Making credible science usable: Lessons from CAA, NAPAP. *Power Engineering* 95(9): 38–40.

————. 1991b. Usable science: Lessons from acid rain legislation, NAPAP. *Power Engineering* 95(10): 26–29.

Radin, B. 1977. Political relationships in evaluation: The case of the experimental schools. *Evaluation and Program Planning* 4: 201–204.

RIVM. 1988. *Zorgen voor Morgen (Concern for Tomorrow), Nationale Milieuverkenning 1985–2010.* Alphen aan den Rijn, Netherlands: Samson H.D. Tjeenk Willink.

————. 1991. *Nationale Milieuverkenning 2.* Alphen aan den Rijn, Netherlands: Samson H.D. Tjeenk Willink.

RobinWood. 1990. *Klima Aktionsbuch, Was tun gegen Ozonloch und Treibhauseffekt?* Bremen: Die Werkstatt.

Roe, E. 1992. Intertextual evaluation, conflicting evaluative criteria, and the controversy over Native American burial remains. *Evaluation and Program Planning* 15(4): 369–381.

Sonnichsen, R. 1989a. Methodology: A bridge or barrier between evaluators and managers? *Evaluation and Program Planning* 12(3): 287–290.

————. 1989b. *Program Managers: Victims or Victors in the Evaluation Process? New Directions for Program Evaluation.* San Francisco: Jossey-Bass.

Turkenburg, W., and A. Van Wijk. 1991. *Onze kennis over klimaatverandering (Our knowledge of Climate Change).* Utrecht, Netherlands: University of Utrecht.

Tweede Kamer. 1991. 1990–1991, *Nota Klimaatverandering (Climate Change Report)*, 22, 232, nos. 1–2. Netherlands Parliament.

U.K. Cabinet Office. 1991. *ACOST Report on Environment R&D in UK.* London: HMSO.

Umweltbundesamt. 1989. *Verzicht aus Verantwortung: Massnahmen zur Rettung der Ozonschicht.* Berlin: Erich Schmid Verlag.

U.S. Congress. 1983. House Committee on Interior and Insular Affairs. Hearing before the Subcommittee on Mining, Forest Management, and Bonneville Power Administration on U.S. Assistance to Foreign Copper Producers and the Effects on Domestic Industries and Environmental Standards. Tucson, Ariz., May 20.

————. 1985. House Committee on Energy and Commerce. Hearing before the Subcommittee on Health and the Environment on Acid Rain in the West. Albuquerque, N. Mex., June 28.

Weiss, Carol. 1972. Evaluation Research: *Methods of Assessing Program Effectiveness.* Englewood Cliffs, N.J.: Prentice-Hall.

21

Knowledge and Action: An Analysis of Linkages among Management Functions for Global Environmental Risks

Jill Jäger, Josee van Eijndhoven, and William C. Clark

This chapter provides a cross-cutting analysis of data on risk management functions that were developed in part III of this book: risk assessment, monitoring, option assessment, goal and strategy formulation, implementation, and evaluation. It is designed to complement Chapter 14's cross-cutting analysis of part II's data on agenda setting, issue framing, and actors. Taken together, these two cross-cutting analytic chapters provide the foundations for our synthesis, in chapter 22, of this study's overall conclusions and their implications for the future of global environmental risk management.

It should be noted again at this point that the management "functions" reported on here were selected as units of analysis on the basis of previous studies in the risk management and international institutions literature.[1] This literature suggested it would be worthwhile to complement the *actor* perspective summarized in chapter 14 (*Who* did what?) with the *action* perspective summarized here (*What* was done?). We found further value in the functional or action perspective because the long time span and multiple-arena nature of our study made problematical an analytic framework focused solely on institutions (some of which came into existence during the study period) or actors (some of which were not present in all arenas.) In our analysis of these management functions we made no prior assumption about how they were related to one another, rejecting in particular any presumption of a linear relationship in which efforts to gather systematic knowledge about a risk necessarily preceded action or goal setting bearing on that risk. We were also careful not to assume that each management function would be found, much less have the same characteristics, in each of our three case studies. In retrospect, we have found both the *actor* and *action* perspectives to be incomplete but helpful within their limitations. We attempt to synthesize their respective insights in the final chapter of this book.

The central question for this chapter is how the activities of the six management functions affected one another and how those relationships changed as the global environmental issues moved from their early condition of scientific curiosities onto the high political agenda.

We approach this question in stages. Section 21.1 summarizes relevant empirical findings from the individual management function chapters of part III. Section 21.2 reports the results of our analysis across those functions to identify common characteristics. Section 21.3 explores causal relationships among the functions: which risk management activities influenced others, when, and how. Finally, we conclude in chapter 22 by discussing how our empirical findings on linkages among management functions relate to the existing conceptual models of risk management and policy development outlined in chapter 1.

21.1 What Are the Most Significant Features of the Individual Risk Management Functions?

The preceding chapters in part III describe the major influences that individual risk management functions have had on the evolution of our global environmental issues. Major conclusions of each chapter are summarized here to provide a foundation for the linkage analysis that follows. Recall that the activities embodied in these risk management functions can be directed at one or more elements of the "hazard taxonomy" introduced in chapter 1 and reproduced as figure IIIA in the introduction to part III. As was shown in the previous chapters and summarized below, most functions have not addressed all elements but have concentrated on parts of the overall taxonomy.

21.1.1 Risk Assessment

Risk assessment, described in chapter 15, is concerned with understanding the nature, causes, consequences, likelihood, and timing of the risk in question. The empirical evidence shows that throughout the period of our study there was a general recognition of the potential benefits of "end-to-end" assessments addressing all of the elements of the hazard taxonomy. At the same time, however, persistent difficulties and tensions arose in efforts to actually implement such comprehensive analyses. In practice, most risk assessments concentrated on emissions and concentrations and paid some attention to environmental impacts. Few explored the underlying driving factors—the human needs, wants, and choices—that cause emissions. Few addressed the socioeconomic and ecological impacts that result from changes in concentrations and the environment.

Risk assessments of global environmental risks also demonstrated "stickiness": a tendency for findings, once stated, to resist modification even in the face of new evidence. Several examples are discussed in chapter 15. For example, in the climate change debate, the estimate was made in 1979 by the U.S. National Research Council that if the concentration of carbon dioxide in the atmosphere doubled, the global mean temperature would increase by 3 degrees Centigrade plus or minus 1.5°C—that is, in a range of 1.5° to 4.5°C. In most of the major risk assessments on climate change made over the next decade the range was retained, even though climate models changed and other aspects of climate science were evolving. Various reasons for this "stickiness" are discussed in chapter 15 and—since the phenomenon turns out to occur in other functions as well—are addressed in section 21.2 of this chapter.

Finally, our studies revealed a common trend over time toward multiagency design of risk assessments, both within countries and internationally.

21.1.2 Monitoring

Monitoring, addressed in chapter 16, played an important, if often underrated, role in shaping social responses to global environmental risks. Much of the monitoring that turned out to be relevant to the management of global environmental risks was originally motivated by concerns other than those risks. This was true for monitoring of emissions, concentrations, and other environmental variables. For example, monitoring of the ozone concentration in the stratosphere began long before there was a concern about anthropogenic stratospheric ozone depletion and was motivated by scientific curiosity about the physics and chemistry of the upper atmosphere. In other cases, monitoring began or was intensified as a result of assessments of global environmental risks. For instance, the measurements of the European Air Chemistry Network were expanded in Europe after the Swedish reports of acidification at the conference in Stockholm in 1972.

The majority of monitoring activities focused on natural science components of the relevant risks (emissions, concentrations, and other environmental variables). There was much less monitoring of human needs or social responses. This focus was, of course, not unrelated to the fact that risk assessments were also concentrating on central elements of our hazard taxonomy rather than on the outside elements.

For each of the global environmental risks studied here, monitoring played a major role in reframing the debate and in stimulating risk assessments, goal and strategy formulation, and implementation. The "discovery" of forest dieback and of the Antarctic ozone hole as a result of monitoring programs pushed their respective issues significantly higher on the public and policy agendas of many of the countries studied here. The observation that the early years of the 1990s were warm in comparison with the previous century also played an important role in stimulating debate about the risk of climate change.

21.1.3 Option Assessment

Option assessment, addressed in chapter 17, is concerned with exploring alternative actions for responding to global environmental risks. (We use the broader term *option assessment* rather than the narrower if more common one of *policy assessment* to avoid the implication of governmental action only that the latter term carries in many of our arenas.)

In the vast majority of the issues and arenas we studied, option assessments were systematically separated from risk assessments. In many cases, the risk assessments simply became stand-alone documents, with no institutional connection to any options studies whatsoever. Even ostensibly end-to-end studies such as the report of the Intergovernmental Panel on Climate Change (IPCC) (IPCC 1990) placed its "science" and "response" assessments in separate volumes and did everything possible to ensure that criticism of the latter did not undermine the consensus on the former.

Throughout most of our study period, the proportion of option assessments performing appraisals of actions to manage impacts declined, and the proportion restricting their attention to changes in emissions increased. Overall, option assessments exhibited a bias toward emission-reduction technologies, with a neglect—sometimes amounting to suppression—of options to change the environment or adapt to environmental changes.

21.1.4 Goal and Strategy Formulation

Goal and strategy formulation, addressed in chapter 18, is an important part of the process of managing global environmental risks. In our usage, goals are statements of objectives or of conditions that an actor wishes to bring about. Strategies are plans for how goals will be achieved.

For each of the global environmental risks, our empirical data show that goals to build capacity to understand or manage the risk were dominant during the early phases of management of the risk. Later on, goals to reduce the emissions of pollutants dominated.

Goals for global environmental risk management converged through time across arenas. In the early stages of debate about an issue, a wide range of goals and strategies was generally discussed. At some point, however, the range of goals in active discussion suddenly became much narrower, and the convergence of goals across arenas began. Thereafter, in a relatively short period of time the same goal was found in most arenas, whether it was a certain percentage reduction of emissions or goals of capacity building. For example, the goal of targeting the aerosol sector for immediate chlorofluorocarbon (CFC) reductions emerged in 1974 in the United States. After 1976, when the goal had been adopted by the U.S. federal government, the goal spread to other countries. The story told in chapter 18 of the 1988 Toronto Conference goal to reduce carbon dioxide (CO_2) emissions illustrates how varied and far-reaching the repercussions of certain goals can be. The goal was treated sufficiently seriously to spark national option assessments in several countries and implementation at both the national and municipal level.

21.1.5 Implementation

The empirical data on implementation developed in chapter 19 show, perhaps not surprisingly given the data on goal and strategy formulation, that in the regulatory action there was a clustering on actions that reduce emissions and build capacity. Our analysis of implementation actions shows a growing emphasis over time on the creation of institutional capacity at the international level and a growing significance of international agreements as a spur for national action. Furthermore, we observed that there was a creation of capacity at the international level for debate about frameworks and standards and devolution of responsibility for action. We also observed evolution toward an understanding of implementation as an organic, uncertain process of iterative negotiation and feedback.

Over the time period studied here, there was an increasing number of flexible regulatory measures introduced in response to the global environmental risks.

The implementation process opened to wider participation over time from the beginning of the 1970s to 1992 with increased nongovernmental organization (NGO) involvement in particular. NGOs accumulated increasing technical capacity to focus on scrutiny, audit, and disclosure of activities throughout the process of negotiating regulations and enforcement provisions. At the same time, the debate about implementation evolved from being very technical to being more political in terms of what was going to be implemented and by whom. NGOs learned that it is necessary to stay involved for a longer period of time, in the domestic process of policy formulation and implementation, in particular after implementation has

begun. Implementation experience in the arenas shows the growing interest in voluntary agreements, which require careful monitoring of compliance.

21.1.6 Evaluation

The function of evaluation is to ask, "How are we doing at managing the risk?" Evaluations may be formal reports on progress in a particular policy area or informal, such as newspaper reviews of government policy.

The evidence presented in chapter 20 suggests that there are very few explicit examples of evaluation in the history of global environmental risk management covered in this study.

Such evaluation practices as did exist were not universal or homogeneous during the time period examined in this study. Political culture in the different arenas studied here played a very important role in shaping how evaluation was done and indeed what evaluation was done.

The types of evaluation that were sought included evaluation of research programs and of integrated efforts at assessments of policy. Indeed, our survey of evaluation efforts and their evolution shows that there was a fair amount of serious and sustained evaluation of research and development efforts. This occurred after the first phase of scientific interest in the issue has developed. On the other hand, evaluation beyond the peer review of science appeared to come later in the risk management process, after the issue had moved to a high position on the policy agenda.

21.2 What Features Do the Risk Management Functions Share in Common?

The empirical data presented in the individual chapters on risk management functions and summarized briefly in the previous section indicate that certain features are common to most if not all of the risk management functions. Common features reflect overall characteristics in the management of global environmental risks, relatively independent of risk management functions, arena, or even issue.

21.2.1 Focus on Emissions and Concentrations

The first common feature was a focus on emissions and concentrations of pollutants. This matters because, as noted in chapter 1, both understanding and management response can in principle be targeted at any location in the "taxonomy of hazard," from human needs to emissions to impacts. It becomes particularly important in cases where, despite concern and regulatory efforts, some amount of global environmental change and its associated impacts are likely to occur. In such cases, which may very well characterize the most likely situation concerning the issues addressed in this study, a widespread failure to address impacts and adaptations can leave society poorly served by risk management efforts.

The focus on emissions and concentrations was noted, for example, in both risk assessment and option assessment for each of the global environmental risks, as well as goal and strategy formulation and implementation. Actors tended to concentrate on the emissions and concentrations rather than on the human driving factors leading to these emissions or the socioeconomic consequences of environmental changes.

As far as the assessment functions are concerned, this focus on natural-science aspects of the risk was to some extent a reflection of the fact that the human dimensions of global environmental change came onto the research agenda to a significant extent only toward the end of the 1980s. Social science research on topics of global environmental change is a relatively new phenomenon, though efforts had been mounted as early as the late 1970s, only to languish for lack of funding or interest.

As pointed out in chapter 17, the dominance of emission reduction measures was consistent with both the general value preference of many of the actor groups for prevention over remediation and the likelihood that the options bestowing concentrated benefits (for example, on the manufacturers of scrubber technologies or on the regulators of emissions) attracted relatively many and forceful advocates.

Once the initial dominance of emission reduction measures in the assessment of response options was established, several factors may have reinforced it. Possible reinforcements include the tendency for researchers to identify uncritically a small subset of the possible problems as worthy of attention and to disregard or even denigrate work on alternatives; the tendency of one or a few initially popular options to crowd out others in early stages of the policy debate leaving a narrow selection of alternatives as the debate evolves; as well as the possibility that users of assessments do not want to consider a wide range of alternatives.

Goals to reduce emissions dominated other types of goals for all three risks studied, in each arena and for the entire time period investigated. For example, in the ozone-depletion case goals of reducing the impact of stratospheric ozone depletion were quite rare, with only a few examples relatively late in the issue-development cycle. In the case of climate change, impacts and adaptation goals received somewhat more representation, with discussion of measures in response to sea-level rise and agricultural responses, but these goal statements were overwhelmed in the broader discussion by emission goals. Similarly in the acid rain debate, goals that target impacts, such as liming or breeding acid-resistant fish,

were overwhelmed by goals to reduce emission of acidifying compounds.

Chapter 18 looks for explanations for the dominance of emission goals over our study period. It points out that one possible explanation could be that since emission reductions dominated in option assessment, the dominance of emission-reduction goals was simply replicating the patterns in option assessment. This is not a satisfactory explanation, however, since many actors involved in goal setting did not routinely rely on formal option assessments. Furthermore, the evidence suggests that goals frequently were discussed and promoted before any formal option assessment had taken place. Other explanations include ideological bias (some goals were not socially legitimate), psychological bias (a preference to think that problems will be avoided), and a political economy bias (political and economic factors favor emission reduction).

In summary, the empirical evidence shows that for most of the risk management functions, over the time period studied here and for each of the global environmental risks there was an emphasis on emissions and emission-reduction elements of the overall hazard taxonomy. A number of possible explanations for this observation have been advanced. It should be noted, however, that by the end of our study period, the shortcomings of emission-oriented analysis and goal setting were widely recognized. Efforts to expand risk management to encompass a wider view of effects and impacts had begun to take shape. In the case of acid rain, there was a shift in the 1990s in many arenas toward an emphasis on achievement of environmental protection. In the case of ozone depletion, as the issue matured in the late 1980s there was a shift toward consideration of total chlorine levels in the stratosphere instead of emissions and concentrations of individual CFCs. Even in the case of climate change, while the focus of debate has largely remained on the reduction of carbon dioxide emissions, there has been discussion of the "global warming potential" as a total environmental impact to be dealt with. Whether these efforts to expand the scope of risk management strategies take hold against the practical and ideological pressures for focus on emissions remains to be seen.

21.2.2 "Stickiness"

The second common feature in our risk management functions is what we have referred to as "stickiness." Examples were given above of this feature in risk assessment. It is also evident in other risk management functions, especially in goal and strategy formulation.

Our evidence shows that goals can get locked in early, either on a particular target or a particular mix of options. Such lock-in blocked progress in risk management by making it harder to incorporate new scientific knowledge, to make use of experience with implementation, or to engage new actors. In the acid rain case in the United States, for example, the goal "to reduce emissions by 10 million tons of sulfur" was simple and effective in generating attention and serving as a rallying cry but had little justification and never was replaced by anything better. Similarly, in the case of stratospheric ozone depletion the goal of banning CFCs in spray cans was still being voiced after CFC use in spray cans was insignificant compared to use for solvents or refrigerants.

The potentially positive aspects of stickiness in risk assessment have been discussed by van der Sluijs, van Eijndhoven, Shackley, and Wynne (1997) using the term *anchoring device* from the behavioral science literature. They suggest that because the estimate of climate sensitivity remained stable over more than fifteen years, the fragile process of building a global policy community was held together. Moreover, the empirical evidence suggests that over time there were tacitly different local meanings attached to the range, so that while the stated range remained the same over time, the interpretation of what it signified changed.

Just as the stickiness in risk assessment could be beneficial to the risk management process, chapter 18 suggests that the stickiness in setting goals can have advantages. The empirical evidence suggests that simple, rigid, unscientific goals are almost always more effective in the early stages of development of an issue than more sophisticated goals. Sticking with these simple goals in the early stages of debate saves time and allows an early identification of supporters and opponents. The problems with stickiness arose when the issue matured and the need developed for more scientifically sophisticated, flexible, and consensual goals.

21.2.3 Broadening Participation

The broadening of participation in performance of risk management functions is the third common feature revealed through our cross-function analysis. For both risk and option assessment, most early work concerning global environmental risks was carried out on an ad-hoc basis by individual scientists or small expert groups. Through time, more individuals representing more disciplines, actor groups, and arenas joined in the assessment process.

In option assessment, the role of engineers, economists, and professional policy analysts grew throughout the period. By the time of the Rio Conference many of the most visible and influential option studies were being performed by experienced policy analysts within the context of formally institutionalized international assess-

ments. While option assessments from government have been most plentiful, other social actors have been increasingly present and influential.

In goal formulation there was also a broadening of participation. This was especially evident in the later stages of the issue development, when there was a move away from simple, rigid goals to more sophisticated, flexible goals. This move required a process of socialization with the building up of a set of working relationships across arenas and social actors. Participation in the second phase of goal formulation also broadened to include actors who were "opponents" in the first stage. One reason for this was that during the early stages of the debate the proponents of swift action concentrated their energy on mobilizing support for early, simple goals, while the opponents to swift action were promoting more sophisticated goals. For example, in the acid rain case in the early 1980s Scandinavia and Germany pushed for quick adoption of the 30 percent goal, while the United Kingdom argued that international targets ought to take in the differential impacts of alternative reduction scenarios. The latter became central to the more sophisticated goal of the second Sulfur Protocol.

Expanding participation was also noted in the case of implementation, particularly with the observed move of NGOs into the center of the implementation process. The pattern observed in the case of implementation was that initial commitments of resources were uncoordinated, reflected in the activities of individual scientists and isolated commentators. This was followed by a phase of contested science in which implementation efforts were concentrated generally in attempts by NGOs, industry, and government scientific establishments to influence the public debate of the issue. Over the time period studied here, some NGOs created significant capacity to review scientific research and began to direct their own implementation actions toward participation in regulation-writing processes, as well as monitoring of compliance. This had the effect of moving the implementation activity from an elite negotiation around technical concerns to a more open accommodation involving value-based differences. Participation in implementation also expanded over time to include more actors from business and industry.

21.2.4 Conclusion

This section has identified three features that are common to most of the risk management functions—a focus on emissions and their reduction, "stickiness" in assessment conclusions or goal formulation, and broadening of participation over time.

It is important to note how the issue-development cycle discussed in chapter 14 relates to these features. As the issues first began to receive sustained attention in the technical community, the management focus for each of the environmental risks considered here was on emission reduction: reduce sulfur dioxide (SO_2) emissions, ban CFCs in aerosols, reduce carbon dioxide emissions. At the same time, participation in the debate was limited, and there was a tendency to stick to simple, rigid goals.

As the issue matured and received increasing attention in the policy community, the focus broadened to include environmental impacts: critical loads for acidification, chlorine loading in the stratosphere, global warming potential. Participation in performance of the risk management functions also broadened.

The importance of the issue dynamics for the risk management, as illustrated for particular features in this section, is further developed in the following section.

21.3 What Are the Linkages between Knowledge and Action in the Management of Global Environmental Risks?

This section uses empirical evidence gathered in other chapters of this book to suggest plausible patterns in the linkage between knowledge and action in the management of global environmental risks. *Linkage* is taken here to mean causal relationships—that is, changes related to one risk management function influencing changes related to another.

21.3.1 Motivation

As noted in chapter 1, in conducting this study we have been well aware of the body of risk and science studies literature that argues that these two faces of management cannot be separated: knowledge always informs action; action, together with its associated interests and agency, always structures knowledge. We have no quarrel with this perspective, as should be evident in the substantial attention we have devoted to questions of issue framing, agenda setting, and agency in the research protocol for this study and the analysis presented in part II. We also believe, however, that our extensive empirical evidence can be usefully illuminated by a perspective that explores the conditions under which, and ways in which, management activities such as risk assessment (which emphasizes scientific data and modeling) may exert a heavy influence on activities such as goal setting (which emphasizes politics and values). The same approach is interested in the conditions under which the reverse is true—when action-intensive activities such as implementation exert a particularly strong influence in the subsequent framing and conduct of scientific assessments. With most scholars, we thus assume the intercalation of

knowledge and action in risk management but have found it helpful to ask why the causal currents within this system sometimes run more strongly in one direction, sometimes in the other. We believe that answers to these questions are of some relevance to those actually engaged in the practice of risk management. And we hope that even those scholars with other theoretical predilections will find our empirical evidence on these questions to be of interest for pursuing their own agendas.

For all of the cases studied here, we found evidence of a common pattern of linkages. This pattern can be most economically described by dividing our risk management functions into two groups based on the empirical evidence summarized in part III: the more *knowledge-intensive* functions of risk assessment, option assessment, and monitoring and the more *action-intensive* functions of goal and strategy formulation, implementation, and evaluation. Within these classifications, the general pattern we observed has the following basic characteristics. During the long periods—sometimes several decades—before the issues were taken up on the policy agenda, knowledge-intensive and action-intensive functions of risk management were largely disconnected. The risks were largely treated as scientific issues, with any goals for management action posed in general terms more likely to be shaped by debates in other issue areas and reflecting other agendas than by any close reading of the state of the science. As the issues burst onto the policy agenda, however, the knowledge-intensive management functions began to influence performance of the more action-intensive functions. In particular, risk assessments stimulated goal statements (in the cases mentioned here, most often with respect to emission reductions). Finally, as political attention peaked and fell, there was an intensification of two-way linkages between knowledge and action functions. Assessment and monitoring activities were driven by goal statements and implementation measures. The beginnings of self-conscious evaluation of progress in responding to the issue at hand also emerged at this time.

As with many efforts to comprehend management processes, this pattern or model, once explicated, may seem a complicated way to say the obvious. Why, then, is it worth exploring? It is worthwhile because most of the efforts we have encountered to understand or conduct activities supporting the management of global environmental risks have focused on one part of the pattern we describe here while ignoring the others. Thus, scholars interested in "crises" and "focusing events" tend to focus almost largely on the middle stage of the pattern we have observed and to see disproportionately a one-directional influence of knowledge on action. Those looking at mature science advisory systems, in contrast, emphasize the ongoing negotiations between the knowledge-intensive and action-intensive aspects of risk management, while missing that for most of the issues' history such negotiations are unlikely to have occurred. And scientists participating in or running assessments often continue with the same exclusively knowledge-focused approaches and designs even when the issue they are addressing has developed to a stage at which interactions with the action-intensive aspects of risk management have become intense. All of these partial perspectives can

thus be misleading in ways that we hope our explication of the larger pattern of interactions can help remedy. The following sections explicate and illuminate this general pattern of linkages between knowledge and action for each of the three cases we studied.

21.3.2 Acid Rain

Figure 21.1 summarizes key events in the six risk management functions for the case of acid rain in Europe from the beginning of the 1970s until 1992. In 1948 the establishment of the European Air Chemistry Network

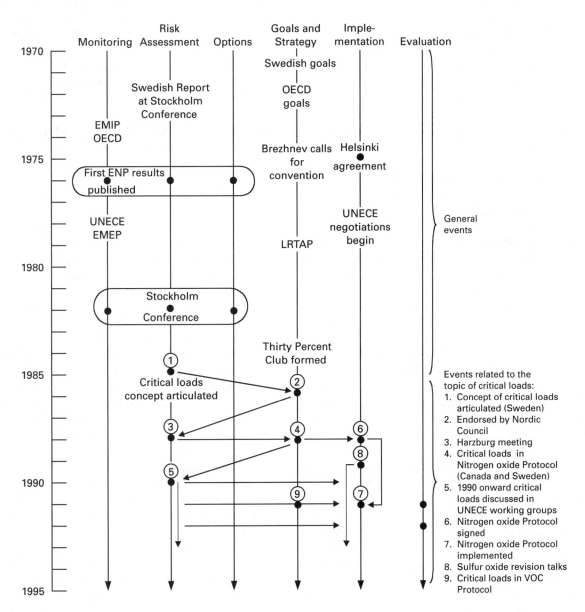

Figure 21.1
The issue of acid rain in Europe: Risk management events and linkages

marked the beginning of monitoring activities. In 1967 Oden was able to use the results of this network to demonstrate that acid deposition in Scandinavia was the result of long-range, transboundary transport of pollutants. The time between 1948 and 1967 was thus one in which scientific information was being collected and analyzed with virtually no connection to goals or measures for management. In the 1970s the establishment of the European monitoring program under the auspices of the Organization for Economic Cooperation and Development (OECD) and its transfer to the United Nations Economic Commission for Europe (UNECE) in 1978 were parts of the ongoing effort that built up scientific capacity to address the issue of acid rain in Europe.

In the 1970s there also emerged ongoing streams of goal and strategy formulation and implementation activities. These, however, were only weakly and generically tied to the state of the science and monitoring activities. A significant goal was the Swedish decision to bring up the issue of long-range, transboundary air pollution at the 1972 Stockholm Conference. This was followed by the goal established in 1972 to determine the real deposition and transport patterns. Another important goal in the mid-1970s was set by the Soviet leader Brezhnev, who wanted a multilateral agreement as a follow-up to the Helsinki Convention in 1975. This matched well with the goals of Scandinavian countries, which were interested in reaching agreements dealing with acid deposition.

Parallel to the monitoring and assessment activities and to the formulation of goals, there was an ongoing stream of implementation activities in the 1970s. The signing of the Helsinki Convention in 1975, although unrelated to the issue of acid deposition in Europe, proved important in the management of the risk by providing a framework for further discussions. This was followed by the beginning of the UNECE negotiations for a convention on long-range, transboundary air pollution and the signing of the Long-Range Transboundary Air Pollution (LRTAP) Convention in 1979.

The period until 1979 in Europe was thus characterized by ongoing streams of activities both in the building up scientific capacity to deal with the risk of acid deposition and in goal setting and implementation. At the national level, particularly in Scandinavia, there were tight linkages between knowledge and action—with scientific knowledge based on monitoring and assessment in Scandinavia stimulating calls for action. At the international level until 1979 the linkages between knowledge and action remained weak, however, as evidenced by the LRTAP Convention signed in 1979, which did not contain any commitments to actions such as emission reductions.

The issue of acid rain moved onto many more national policy agendas and was pushed higher on the international policy agenda at the beginning of the 1980s with the "discovery" of forest dieback in Germany and lake acidification in Canada. The results of monitoring of forest and ecosystem damage combined with assessments of the risks of acid deposition led to an increase in the number and ambition of goal statements. Most important, the monitoring results and the political outcry they created led to a reversal of the German position with regard to the LRTAP and to implementation of the Federal Large Combustion Plant Directive. The scientific knowledge was enough to stimulate agreement first of all on the Thirty Percent Club (countries willing to reduce sulfur dioxide emissions by 30 percent) and by 1985 on the Protocol to the LRTAP Convention formalizing this 30 percent goal.

After a short phase at the beginning of the 1980s, when knowledge (in particular, risk assessment activities) was stimulating the formulation of goals and a certain amount of implementation, the interactions between knowledge and action intensified and became bidirectional. The streams of assessment, monitoring, goal and strategy formulation, and implementation continued but were linked more frequently.

This is illustrated schematically in figure 21.1, which picks out the story of "critical loads" in Europe to illustrate the cross-functional linkages. This figure is not intended to be comprehensive but includes important events and interactions. Thinking about a critical-loads concept was stimulated in 1982 by Swedish scientists who introduced the idea of a deposition threshold (point 1). This was a departure from thinking about emission reductions to considering thresholds for deposition, above which damage to ecosystems would occur. In 1985 the concept was articulated by Sweden in terms of critical loads, and this is the starting point of the linkages shown in figure 21.1. The concept of critical loads was endorsed by the Nordic Council, a policy-making body, in 1986 (point 2), which agreed that the concept would be useful as the basis for further negotiations. As a result of the growing interest on the policy side in the concept of critical loads, a scientific assessment of the critical-load concept was made at the international meeting in Harzburg in 1988 (point 3).

By 1988 there was a consensus within the negotiating body that future protocols to the LRTAP Convention should use critical loads, and Canada and Sweden proposed the inclusion of critical loads in the Nitrogen Oxide Protocol signed in Sofia in 1988 (points 4 and 6). The Nitrogen Oxide Protocol itself called for uniform emission reductions but called for subsequent negotiations on nitrogen oxides (NO_x) to be based on critical loads. Similarly in 1991, the Volatile Organic Compound (VOC) Protocol included the call for future VOC negotiations to be based on critical loads (point 9).

From 1990 onward the concept of critical loads was discussed in the UNECE Working Groups (point 5), and the concept was included in subsequent assessments. The Regional Acidification Information and Simulation (RAINS) model, used in the negotiations of the second Sulfur Protocol, included critical loads by 1992. The talks on the second Sulfur Protocol began in 1989 (point 8) and picked up on the concept of critical loads, since it had already been accepted as a goal for future negotiations on nitrogen oxides. During the negotiations of the second Sulfur Protocol there was regular input from the Working Groups, including calculations using integrated assessment models to determine whether particular strategies could achieve critical loads and what the costs would be (indicated by the arrow from assessment to goal and strategy formulation between 1990 and 1992).

In 1992 the negotiators of the second Sulfur Protocol decided, on the basis of results from the RAINS model, that achievement of protection of 100 percent of the ecosystems in Europe would be too expensive. They agreed instead to reduce emissions to the extent necessary to close the gap between present deposition and critical loads by 60 percent. This became the basis of the Protocol signed in 1994 in Oslo.

Figure 21.1 thus illustrates for the case of acid rain in Europe that knowledge (about critical loads for acidification) influenced action (inclusion of the concept in negotiations, formulation of a protocol on the basis of integrated assessment model results using the concept). On the other hand, once the concept had been endorsed politically, the goal statements influenced the assessments that were performed (at international meetings and in the UNECE working groups), since the assessments had to be based on the concept and answer related questions that were arising in negotiations.[2]

21.3.3 Ozone Depletion

As in the case of the other two global environmental risks studied here, the pattern of linkages between knowledge and action in the ozone case began with a phase in which there was a build-up of scientific capacity. During this phase the issue did reach the attention of some people involved in policy development, but the issue did not move onto the broad political agenda.[3]

In the case of stratospheric ozone depletion, this build up of scientific capacity (monitoring of the ozone concentration and assessment activities with respect to the knowledge about the physics and chemistry of the upper atmosphere) began well before the period studied here. The first European network for stratospheric ozone monitoring began in 1926, motivated by scientific curiosity and not by any suspicion about depletion due to human activities. The interest of atmospheric scientists in the

physics and chemistry of the upper atmosphere increased, however, in the late 1960s. The U.S. Climate Impact Assessment Program in the early 1970s took place in response to a fear that supersonic transport (SST) could lead to stratospheric ozone depletion. The program provided an impetus for increasing scientific capacity in this area.

Stratospheric ozone depletion began to receive increasing attention by policy makers after the paper by Molina and Rowland was published by *Nature* in 1974, showing that chlorofluorocarbons can reach the stratosphere and that the chlorine that is released as these chemicals are broken down in the upper atmosphere catalytically destroys ozone. As a result of the wide publicity given to these results, public attention to the issue increased in some countries, and scientists, NGOs, and some policy makers made goal statements calling for banning nonessential aerosols, boycotting spray cans, introducing production ceilings, and so forth. These early actions did not, however, lead to significant changes in the timing or terms of reference of assessments and monitoring activities. Assessments were made regularly and perhaps more frequently as a result of increasing concern (e.g., NRC 1979; NASA 1981; NRC 1984). Monitoring also continued, including the introduction in the late 1970s of satellite monitoring.

The year 1985 marked the beginning of a second phase in which monitoring and assessment stimulated actors to formulate goal and strategy statements. A major step in the assessment process was the publication in 1986 of the three-volume assessment *Atmospheric Ozone 1985: Assessment of Our Understanding of the Processes Controlling Its Present Distribution and Change* by the World Meteorological Organization (WMO), the U.S. National Aeronautics and Space Administration (NASA), and a number of other national and international organizations (World Meteorological Organization 1986). The scientific evidence published in these reports stimulated a number of goal statements, including an endorsement of CFC controls by industry in 1986 and the formulation of the Montreal Protocol, signed in 1987. The 1985 assessments also laid the basis for a continuing series of scientific assessments.

Beginning around 1987 the interactions between knowledge- and action-intensive management functions intensified significantly. Moreover, these interactions were in both directions. After 1987, in particular with the signing of the Montreal Protocol, the issue of stratospheric ozone depletion was firmly on the policy agenda. The discovery of the Antarctic ozone hole published in *Nature* in 1985 played an important role in this intensification. The signing of the Vienna Convention on the protection of the stratospheric ozone layer, while an important link in the chain of implementation activities, appears to have been less important in stimulating an intensification of the interactions between science and policy on the ozone issue at this time. In contrast, the Montreal Protocol, signed in 1987, was significant in forging linkages between knowledge and action as well as in introducing subsequent regular evaluation of the management of the risk of stratospheric ozone depletion.

The negotiation of the Montreal Protocol led to a need for further assessments, and in this respect the conference in Würzburg in 1986 played an important role in establishing a scientific consensus.

The Montreal Protocol's goal of reducing chlorofluorocarbon emissions by 50 percent led to

• Response assessments by industry into CFC alternatives and the work of the Technology Assessment Panels (TAP),

• Implementation of industry phaseouts, and

• National evaluations of whether the goal and its implementation would be sufficient to eventually stop and reverse stratospheric ozone depletion.

Figure 21.2 illustrates, schematically, important assessments, goals, and implementation after the Montreal

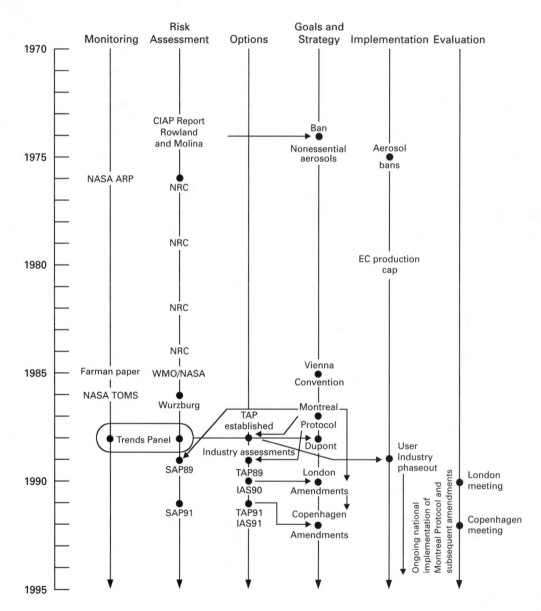

Figure 21.2
The issue of stratospheric ozone depletion: Risk management events and linkages

Protocol was signed in 1987. Clearly, the Protocol stimulated linkages to all risk management functions; it established a Scientific Assessment Panel (SAP) and a TAP and stimulated industry to implement phaseouts and national plans for the implementation of CFC phaseout. The reports of the Assessment Panels were important in stimulating further goals at the London Conference in 1990 and the Copenhagen Conference in 1992.

The Montreal Protocol was set up in a way that ensured regular formal evaluation of the ongoing management of the risk of stratospheric ozone depletion. At the London Conference in 1990 it was agreed that stricter reduction timetables were required, and the Agreement drawn up at the meeting set new goals. In particular, the London meeting set new goals by discussing the stabilization of the chlorine content of the stratosphere and not the elimination of certain CFCs. Also, the London meeting introduced the goal of financing for the developing countries, which led to the implementation of the Montreal Fund. The Copenhagen meeting in 1992 similarly added new goals, including adding hydrochlorofluorocarbons (HCFCs) and shortening the phaseout for other CFCs). The London and Copenhagen meetings evaluated progress in dealing with stratospheric ozone, using new assessments, and set new goals that were implemented.

21.3.4 Climate Change

Figure 21.3 shows for the climate change case the major events related to each of our management functions. This is a schematic representation of contributions to the management of the risk. It is not comprehensive and does not include all assessments and actions that took place during the 1970 to 1992 period. Moreover, it reflects only the broad international level rather than national activities.

Three features of figure 21.3 are notable. First, for each of the risk management functions there was a continuous stream of events. For example, a series of risk assessments started with the Study of Man's Impact on Climate (SMIC 1971) and ended with the IPCC reports of the 1990s. Similarly, a series of goal statements started with those at the end of the 1970s and ended with those incorporated in the U.N. Framework Convention on Climate Change (FCCC) signed in Rio in 1992.

The second notable feature is the long time period during which activity was largely science based (that is, where assessment and monitoring dominated). This phase essentially ended in 1985, when goal statements and implementation activities began to occur regularly.

The third feature is the linkages among the risk management functions when events in one function influenced events in another. Looking in more detail at the time period between 1985 and 1992, the ongoing series of

activities and the linkages illustrate patterns found also in the cases of acid rain and stratospheric ozone depletion.

Already in 1985, continuing monitoring activities were showing that the average temperature of the Northern Hemisphere had risen by 0.3 to 0.6 degree Centigrade over the past 100 years. Risk assessments were also carried out regularly, including, as noted, the assessments in Villach 1985 (Bolin, Döös, Jäger, and Warrick 1986), Villach and Bellagio 1987 (Jäger 1988), and the IPCC (1990, 1992). At the same time, assessments were made of the options available for responding to the risk of climatic change. (The circles in figure 21.3 around events like the Villach 1985 report are included to indicate that monitoring, risk assessment, and option assessment were all reported on.)

During the same period of time, there was an ongoing stream of goal and strategy formulation activities. The first goal, included in the conference statement from the Villach Conference in 1985, was to set up an Advisory Group on Greenhouse Gases (AGGG) to regularly advise the heads of WMO, the United Nations Environment Programme (UNEP), and the International Council of Scientific Unions (ICSU) on the climate change issue. Further goals as time progressed included setting up the IPCC, attaining the "Toronto goal" of reducing carbon dioxide emissions by 20 percent, setting up the Intergovernmental Negotiating Committee (INC) for the United Nations Framework Convention on Climate Change (FCCC), and preparing the goal statements incorporated in Articles 2 and 4 of the FCCC.

Likewise, there was a series of implementation activities between 1985 and 1992. The AGGG was set up in 1986, the IPCC was established in November 1988, the INC met for the first time in February 1991 following the call for its establishment in the U.N. General Assembly in December 1990 and the FCCC was signed at the Rio Conference in June 1992.

For the climate change issue, the first evidence of an evaluation of other than research programs at this international level comes with the reorganization of the IPCC after the first report in 1990. The IPCC discussed progress in the management of the risk of climate change and reorganized its working groups and questions of participation in the working groups as a result of this self-evaluation. It should be noted here that after the FCCC became legally binding, evaluation of progress in dealing with the risk of climate change became institutionalized.[4] The country reports required by the FCCC are a formal mechanism for evaluating progress.

Figure 21.3 thus shows that at the international level during the time period 1985 to 1992 there were ongoing streams of activities involving monitoring, risk assessment, option assessment, goal and strategy formulation,

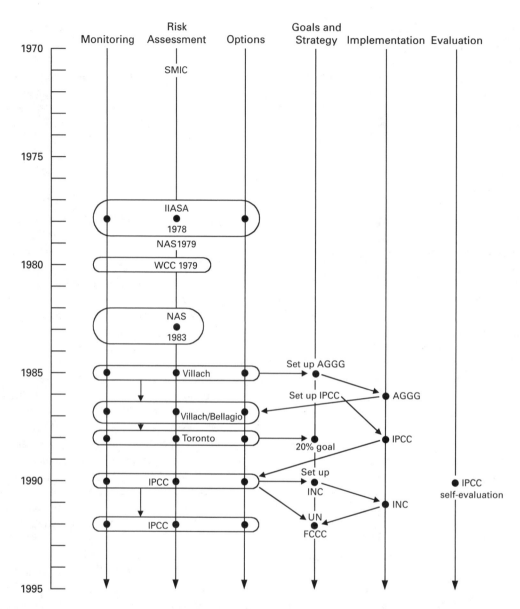

Figure 21.3
The issue of climate change: Risk management events and linkages

and implementation. Evaluation began only toward the end of the period but became more common at national and international levels after the FCCC became legally binding.

The streams of activities shown in figure 21.3 were also linked. The direction of these linkages is of importance for this chapter and for the thesis that recognizable patterns exist in the linkages between knowledge and action in the management of global environmental risks. The consensus reached at the Villach Conference in 1985 (an assessment of the state of knowledge about the risk of climate change) stimulated the participants to include a goal

statement in the conference statement (to establish the AGGG and to start a dialogue between the scientists and the policy makers). The risk and option assessment discussions at the Toronto Conference in 1988 likewise led the participants to propose a goal formulation (to reduce carbon dioxide emissions by 20 percent by 2005). In general, during the time period between 1985 (when the climate change issue began to appear on policy agenda) and 1988 (when the IPCC was established), science was providing much of the impetus for goal and strategy formulation and some implementation. During this phase when the issue was moving up on the policy agenda, knowledge-intensive management functions were driving the development of action-intensive functions.

After the establishment of the IPCC in 1988 the linkages between knowledge and action intensified and were in both directions. Not only did knowledge-intensive functions stimulate goal and strategy formulation and some implementation, but action-intensive functions such as goal setting began to have an influence on the kind of assessments that were carried out, in particular, by influencing the terms of reference of major assessments. This occurred, for example, at the international level with the formation of an intergovernmental panel to provide an assessment of the climate change issue. At the intergovernmental level, policy makers were directly involved in the setting up of the terms of reference of the assessment activity, as well as in the negotiation of the final wording of the Summary for Policy Makers of the assessment. However, even at the national level this intensification of activity and bidirectional linkages between science and policy was increasingly evident after 1988. Such linkages show up clearly in several of the individual arenas studied here (including Germany, the Netherlands, and the European Community).

21.3.5   External and Cross-Case Influences

Although the empirical results of the study show that linkages between knowledge- and action-intensive risk management functions were important in shaping issue development, external influences can also play a role.

For example, in the acid rain case the push for a convention on acid rain was stimulated by the Soviet Union, which was not concerned about acid rain as a global environmental risk but saw it as a potentially useful vehicle for furthering the détente process. In the climate change case, the "Toronto goal" was not a result of a major new assessment of the risk of climate change but was strongly influenced by the occurrence of a serious drought in the U.S. Midwest at the time that the conference was held in Toronto, by the Opening Speech at the Conference by Gro Harlem Brundtland, and by strong presence of both NGOs and the media. For the case of stratospheric ozone depletion, a good example of an external influence is found in the preparations for the Montreal Protocol: on the U.S. side, the idea of a ban on CFCs was introduced by Lee Thomas of the U.S. Environmental Protection Agency because he had been involved in developing legislation to ban asbestos, so the idea of a ban as a useful response option was transferred from experience in another problem area.

A special kind of external influence particularly relevant to our study of social learning is due to developments in the management of other global environmental issues.

We found such cross-issue influences to occur largely between the climate change and stratospheric ozone-depletion cases. Developments in the acid rain case have not been heavily influenced by the developments in the two other cases studied here, not only because acid rain is a "different" kind of global issue (transboundary and universal rather than global) but also because it moved onto the policy agenda and the phase of intense linkages between knowledge and action began earlier than in the other two cases. Perhaps more surprisingly, the extensive functional experience in many arenas with acid rain seems to have had little influence on how those arenas dealt with management functions for climate change or ozone depletion.

An example of the climate change issue influencing developments in the stratospheric ozone–depletion issue is found in the mid-1980s, where Hoffman of the U.S. Environmental Protection Agency (EPA) found that consumption of CFCs was going up despite the ban on the use in aerosols. This finding led to a goal to cut consumption by 85 percent to stabilize CFC concentrations. The concept of cutting consumption to stabilize concentrations was taken by Hoffman from work done in the EPA by Seidel and Keyes (1983) on the climate change issue.

In several instances developments in the climate change case were influenced by the ozone case. In general, the discovery of the Antarctic ozone hole, which had not been predicted, enhanced concern that "meddling with the earth's atmosphere" could have disastrous consequences. This certainly influenced some of the national goal statements on climate change and the adoption of precautionary strategies at the national and international levels. Furthermore, the approach of adopting a framework convention with protocols in both cases suggests some carryover from the ozone issue to the climate issue. In the assessment area, the success in the use of the "ozone-depletion potential" in the ozone case, allowing flexibility in emission reductions, was picked up in the IPCC assessments through the development of "global warming potential."

Appendix 21A.   Acronyms

AGGG	Advisory Group on Greenhouse Gases
CFC	chlorofluorocarbon
CO_2	carbon dioxide
EC	European Community
EMEP	European Monitoring and Evaluation Programme
EPA	Environmental Protection Agency (U.S.)

FCCC	Framework Convention on Climate Change (U.N.)
HCFC	hydrochlorofluorocarbon
ICSU	International Council of Scientific Unions
INC	Intergovernmental Negotiating Committee
IPCC	Intergovernmental Panel on Climate Change
LRTAP	(Convention on) Long-Range Transboundary Air Pollution
NASA	National Aeronautics and Space Administration (U.S.)
NGO	nongovernmental organization
NO$_x$	nitrogen oxides
OECD	Organization for Economic Cooperation and Development
RAINS	Regional Acidification Information and Simulation (model)
SAP	Scientific Assessment Panel
SO$_2$	sulfur dioxide
SST	supersonic transport
TAP	Technology Assessment Panel
UNECE	United Nations Economic Commission for Europe
UNEP	United Nations Environment Programme
VOC	volatile organic compound
WMO	World Meteorological Organization

Notes

1. See chapter 1 and the introduction to part III.

2. The examples shown here of linkages between knowledge and action for the acid rain case are based on the international activities that culminated in the LRTAP Convention and Protocols, with the European Monitoring and Evaluation Program (EMEP) playing a central role. A similar set of linkage patterns could presumably be derived by looking at activities within the auspices of the European Community (EC), where assessments were funded by the Commission of the European Community and policy action took the form of formulation and implementation of Community regulations, such as the European Community Large Combustion Plant Directive (agreed to in 1988 and put into force

in 1991). At the same time, there were both activities at the national level within Europe and linkages between the national and international levels. These are discussed further in a later section of this chapter.

3. In fact, during this early phase where scientific capacity on stratospheric ozone was building up, the issue of ozone depletion did enter into the policy debate on SSTs in some arenas studied here, especially the United States and the United Kingdom. However, this appearance was short-lived and had the effect of further enhancing scientific capacity.

4. The Convention became legally binding in December 1992 after fifty states had submitted their ratifications, so further actions took place after the data gathering for this study stopped.

References

Bolin, B., B.R. Döös, J. Jäger, and R.A. Warrick, eds. 1986. *The Greenhouse Effect, Climatic Change, and Ecosystems*. Scope 29. Chichester: Wiley.

Intergovernmental Panel on Climate Change (IPCC). 1990. *Climate Change: The IPCC Scientific Assessment*. Cambridge: Cambridge University Press.

————. 1992. *Climate Change: The IPCC 1990 and 1992 Assessments*. Geneva: World Meteorological Organization.

Jäger, J. 1988. *Developing Policies for Responding to Climate Change*. WMO/TD No. 225. Geneva: World Meteorological Organization.

National Aeronautics and Space Administration (NASA). 1981. *Ozone Trend Detectability*, edited by Janet W. Campbell. Washington: NASA.

National Research Council (NRC). 1979. *Stratospheric Ozone Depletion by Halocarbons: Chemistry and Transport*. Washington: National Academy of Sciences.

————. 1984. *Causes and Effects of Changes in Stratospheric Ozone: Update 1983*. Washington: National Academy Press.

Seidel, S., and D. Keyes. 1983. *Can We Delay a Greenhouse Warming? The Effectiveness and Feasibility of Options to Slow a Buildup of Carbon Dioxide in the Atmosphere*. Washington: U.S. Environmental Protection Agency.

Study of Man's Impact on Climate (SMIC). 1971. In W.H. Mathews, W.W. Kellogg, and G.D. Robinson, eds., *Inadvertent Climate Modification: Study of Man's Impact on Climate*. Cambridge: MIT Press.

van der Sluijs, J.P., J.C.M. van Eijndhoven, S. Shackley, and B. Wynne. 1998. Anchoring devices in science for policy: The case of consensus around climate sensitivity. *Social Studies of Science* 28(2): 291–323.

World Meteorological Organization. 1986. *Atmospheric Ozone 1985: Assessment of our Understanding of the Processes Controlling Its Present Distribution and Change*. WMO Global Ozone Research and Monitoring Project, Report No. 16. Geneva: WMO.

IV
CONCLUSION

22

The Long-Term Development of Global Environmental Risk Management: Conclusions and Implications for the Future

Josee van Eijndhoven, William C. Clark, and Jill Jäger

In this final synthesis chapter we return to the questions that motivated the study: How has the management of global environmental issues developed over the long term? How well is society prepared to meet the changing challenge of managing global environmental risks? What approaches have been tried and with what results? What have we learned about the effectiveness of such approaches and about prospects for enhancing that effectiveness?

We begin by synthesizing our main findings regarding the interplay among scientific research, policy analysis, and political action in the development of global environmental issues over the period 1957 to 1992 (section 22.1). Section 22.2 then summarizes the factors that we found to be most important in shaping this development. Finally, in section 22.3 we address the effectiveness of past efforts to manage global environmental risks, the prospects for doing better, and the implications of this study for the management of new issues of global environmental change that will surely emerge onto the scientific and policy agendas in years to come.

22.1 What is the Relationship between Issue Attention Cycles and the Performance of Management Functions in the Development of Society's Response to Global Environmental Risks?

This study has traced the historical development of society's efforts to manage the global environmental risks of acid rain, stratospheric ozone depletion, and climate change. It focused on the period from 1957 to 1992 and on events in—and interactions among—eleven arenas: Germany, the United Kingdom, the Netherlands, the former Soviet Union, Hungary, Japan, Mexico, Canada, the United States, the European Community, and the family of relevant international organizations. The story is a complex one, encompassing a fundamental transformation in the way that society dealt with global environmental issues over the generation preceding the Rio Conference in 1992.

Despite this complexity, however, several robust patterns have emerged in the relationship between changes in the attention society devoted to those issues and the performance of management functions directed toward them. Partial perspectives on those patterns have been developed through the analyses of chapters 14 and 21. Here we synthesize those partial perspectives to characterize the most significant patterns of overall issue development that have emerged from our history of global environmental risk management.

22.1.1 Long-Term Trends in Concern and Capacity

The broadest trend we observed consists of the slow, largely incremental growth in a number of relevant factors: public support for environmental action, the number and size of environmental nongovernmental organizations (NGOs), the engagement of industry, the amount and cost of environmental legislation, the number of international organizations involved in managing global environmental risk, and the number of international environmental treaties. The same incremental growth can be seen in the state of scientific understanding, in the development of management options and instruments to implement them, and even in the effort devoted to reflect critically on society's ways of managing global environmental risks. One view of the first generation of society's efforts to manage global environmental risks therefore highlights a pattern of gradual increase in the attention paid to global environmental risks, the complexity of how those risks were framed, and the capacity to manage them. This broad trend was a fairly general phenomenon in the arenas we studied, although individual arenas experienced episodes of decline as well as increase in their concern for and capacity to manage global environmental issues.

22.1.2 Partial Synchrony in Issue-Attention Cycles

These long-term trends, however, were punctuated over the thirty-five years of our study (1957 to 1992) by one or more episodes of greatly increased attention to each risk from the media, governments, and society in general. Such "issue-attention cycles" have been widely documented in other problem areas since the early work of Downs (1972). In our cases they generally consisted of a one- or two-year period of rapidly increasing attention, then a year or two with the issue in high profile, and finally a slower decline of public attention back to lower levels (see figures 14.1 to 14.3). Over sufficiently long periods, recurrent cycles of public attention are possible. They may well turn out to be the norm for global environmental issues, as suggested by our U.S. data on the ozone case (figure 11.2), by the return of the climate issue to a high-profile position on the international stage within five years of the close of our study period (figure 2.5), and by scholarship in other issue areas (e.g., Baumgartner and Jones 1993).

A more novel and potentially important finding of this study is that issue-attention cycles for our global environmental risks show considerable synchrony across arenas. In general, once attention to an issue had surged and it had reached the policy agenda in one arena, attention followed in other arenas within a year or two. Subsequent declines in issue attention were somewhat less synchronous but nonetheless similar. This general pattern of synchronous rise and fall of issue attention across arenas was not, however, universal. Sometimes an early surge of attention in one country would fail to spread to others, as in the attention accorded ozone depletion in the United States in the mid-1970s. And there were occasional late comers, such as Japan in the case of acid rain.

The similarity across most of the arenas in the timing of the rise and fall of issue attention, backed by evidence from the individual arena chapters of part II, suggests that there were powerful mechanisms of linkage, diffusion, or learning at work that helped to put global atmospheric issues onto the agenda in different national settings at roughly similar times. The existence of a few arenas where the timing of specific issue's take-off differed significantly from that in other arenas, however, suggests that national conditions, interacting with specific issue characteristics, may limit the effectiveness of such mechanisms.

22.1.3 Management Functions in the Issue-Attention Cycle

This study has revealed a systematic relationship between the patterns of issue attention and framing analyzed in chapter 14 and the patterns of linkage among management functions analyzed in chapter 21.

Recall that our research traced changes in the conception, performance, and interactions of a number of management functions. The management functions we studied were those that can be found in many models of the policy process: risk assessment (or problem identification), monitoring, option (policy) assessment, goal and strategy formulation, implementation, and evaluation.

Our research, however, did not presuppose a fixed relationship between the functions. Rather, we examined empirically how and to what extent efforts in one functional domain influenced those in another. In the summary of functional interactions presented in chapter 21 we found it helpful for some purposes to refer to three of our management functions (risk assessment, option assessment, and monitoring) as primarily knowledge-based and the remainder (goal and strategy formulation, implementation, and evaluation) as mainly action-based. We employ these descriptive groupings in the picture of issue development sketched below, before turning to possible explanations in section 22.2.

Synthesizing the results presented in chapters 14 and 21, our research suggests three phases of issue development. During the first phase, before the issue first achieved widespread public attention, the principle functional change was the gradual buildup of scientific and analytic capacity through research, monitoring, and assessment activities. Not uncommonly, this period was also characterized by the advocacy of a multitude of (mainly technological) management options only loosely connected to the scientific state of knowledge and often with little influence on the eventual course of action. Such goals as were articulated and achieved wide recognition generally did little more than support the need to nurture such capacity-building efforts. Few significant interactions between knowledge- and action-intensive functions occurred during this phase.

The subsequent period of rapid rise in public attention marks a second phase in issue development. As each global environmental issue began to move onto the policy agenda, linkages between knowledge- and action-intensive management functions began to emerge but were generally unidirectional.[1] Research, monitoring, and assessment activities at this point fed the development of goal statements on what to do about the risk in question. These goal statements were often widely shared across arenas and functioned as de facto common framings of the issue. They often continued to emphasize a need for capacity building. They did not, however, generally reflect elegant theories of causes, effects, and remedies. Instead, the management goals that became widely shared across arenas at this stage were generally simple, rigid calls for action on emissions—a 20 or 30 percent emission reduction, a production-capacity cap, and so on. Such simple goals often had a galvanizing impact on the subsequent issue debate.

A third phase of interactions among management functions was associated with the period following the peak in public attention and continuing through the subsequent decline in attention.[2] During this period, the linkages between knowledge- and action-intensive management functions increased in frequency and ran in both directions: knowledge influenced action and vice versa. As knowledge increased and participation in the management process broadened during this phase, goals became more complex, sophisticated, and differentiated at the local level. In the three cases studied here, international agreements in response to the global environmental risks began to be implemented during this third phase. They in turn set in motion regular evaluations of progress in dealing with the risks—a function not often found in the earlier phases. Subsequent evolution of the issue and its management followed the track of bureaucratic routinization, with an increasing emphasis on implementation structures and—at least in some cases—a decrease in basic scientific and political attention.

By virtue of its long-term perspective, however, our study also suggests that there is nothing final or irrevocable about such routinization. New developments in understanding of the problem, the available options, or the political environment can take an issue from relative bureaucratic oblivion back onto the front burner of global politics, with all the implications for changes in frames, advocates, and institutions that such reemergence entails. The "third coming" of the stratospheric ozone issue documented in our study for the mid-1980s provides one clear example of such "reemergence." The rebirth of high political concern for the climate issue in the late 1990s, though not described in the present study, provides another.

22.1.4 An Emergent International Rationality in Management Responses to Global Environmental Risks?

The pattern of issue development described above appears surprisingly rational: an extended period of low-profile capacity building in the scientific realm, followed by a brief period of high political attention when knowledge drives goal setting and other actions, followed by a more extended period of lower public attention but relatively intense interactions between the knowledge- and action-intensive functions of risk management.

Why is this surprising? Many models of the policy process in wide use indeed portray a linear sequence in which problem recognition triggers a search for solutions, goals are crafted that allow a choice among alternatives, and a policy is finally implemented (e.g., Jones 1984; Kates et al. 1985; von Prittwitz 1990). Most versions add a cyclical or iterative dimension to the model in which some version of the linear process is systematically repeated, presumably with adaptive improvements. Such linear, presumptively rationalistic models of how

the policy process works have, however, come in for a good deal of criticism from scholars of political science (e.g., Lindblom 1990), risk assessment (e.g., Beck 1992), social studies of science (e.g., Gieryn 1995), and policy studies (e.g., de Leon 1999). Sharing much of this skepticism, we were careful in this study not to *assume* any rational, linear relationship among management functions in our research. Instead, as described in chapter 1, we adopted a rather more contingent model of the policy process. This approach allows for the possibility that goal statements, agendas pushing particular options, or implementation of policies derived from neighboring issues might turn out to drive issue evolution, just as might scientific assessment or monitoring activities. We fully expected that our empirical work would justify our skepticism regarding the more rationalistic models, even if it did not lead to an alternative generalization.

For the empirical work on individual arenas reported in part II, our expectations were largely justified. The linear models were far too constraining to do justice to the complex, back-and-forth interactions of management functions we uncovered at the level of individual countries. This was partly due to the variety of reasons cited in the literature and was aggravated by the frequent intrusion of agendas and ideas from other countries into a particular national arena.[3]

Our surprise came when we shifted our perspective to the global level.[4] There, as suggested by the functional chapters of part III, the analysis of chapter 21, and the synthesis reported here, a significant degree of emergent rationality and linearity was in evidence. The net effect of interactions among complex, sometimes chaotic risk management processes at the local (arena) level was the unexpectedly orderly process of global issue development described here.[5]

To be sure, at some periods in the global story we have just told political goal statements drove scientific assessments. At others, the powerful advocacy of a particular option significantly affected issue development. Both of these are linkages with which classic rational models would presumably be uncomfortable. From the longer-term perspective of the overall issue-attention cycle, however, these apparently anomalous occurrences appear much more linear (if iterative) and rationalistic. For example, the goal statement on carbon dioxide (CO_2) reduction advanced at the Toronto Conference in 1988 did indeed drive the next round of technical assessments. But it in turn had been shaped by earlier technical assessments in classic rational tradition.

Our most general finding is that the kinds of interactions that occur among management functions differ greatly depending on where one is in this process of issue development. This is a systematic variation, characterized by a succession of periods in which

• Little interaction among management functions occurs before the issue enters the public-policy agenda;

• Brief episodes of event-driven but knowledge-shaped issue development are associated with initial bursts of political attention; and

• Intense interaction among management functions development in the wake of periods of peak attention and agenda setting.

These findings about the relation of management activities and issue attention over periods of a decade or more also suggest that the scholarly argument over the independence (or not) of Kingdon's problem, solution, and political "streams" (or Cohen and March's "garbage-can" contents) may be less a matter of fundamental disagreement and more a consequence of whether one is focusing on the pre- or postagenda ends of the history in question (Kingdon 1984; Cohen, March, and Olsen 1972).

22.2 How Did Ideas, Interests, and Institutions Shape the Development of Social Responses to Global Environmental Risks?

What were the factors that shaped the long-term evolution of societies' management of global environment risks summarized above? As noted in chapter 1, our principal goal in this initial historical reconnaissance was to document patterns in some potentially relevant variables rather than to test specific propositions. In our efforts to produce such documentation, we focused our attention on the usual suspects: ideas, interests, and institutions. This section summarizes our findings through reflections on possible answers to four sets of questions that we found to be particularly illuminated by our work:

• What were the processes by which certain environmental changes became viewed as global environmental risks? How did such risks become framed in terms of one set of causes and effects instead of others? (section 22.2.1)

• What were the actual roles played by the various interested "actors" in the issue histories we studied? (section 22.2.2)

• What drove the changing institutional capacity for management of global environmental risks? How did these processes interact with issue frames and actor coalitions to shape long-term issue development? (section 22.2.3)

• What were the major pathways and mechanisms for spreading new ideas and experience throughout the community of actors engaged in the management of global environmental risks? (section 22.2.4)

Section 22.2.5 draws conclusions on the development of global environmental risk management over time by linking the answers to the above questions to other published views of policy development.

22.2.1 Ideas and Issue Framing

What are the processes through which some interactions between human activities and the environment came to be characterized as risks worthy of international management attention while others did not? How did such risks become framed in terms of one set of causes and effects instead of others?

This study addressed such questions through its attention to *issue framing*—the processes by which certain features of a problem area are singled out for attention by particular actors or communities (Nelson 1984; Hajer 1995).

In the cases reviewed here, issue frames affected issue development in three related ways. First, to the extent that framing highlighted particular causes and effects of the issue at hand, it exerted a significant influence on which actors and institutions saw themselves as interested parties in the management process. Second, to the extent that issues were framed in ways that suggested a serious threat to important interests or values, this helped to raise the level of concern for the issue exhibited by media, public, and policy elites. Third, to the extent that common issue framings developed across groups of actors, institutions, or arenas, this enhanced the prospects for collective management responses.

Our work suggests several causal pathways through which new knowledge, interested actors, and available institutions may have shaped the framing of global environmental issues and thereby the development of society's response to their management. We summarize below several possible pathways that we found most interesting, important, and deserving of future research.

Old Local Frames for New Global Problems The initial framing of newly emerging global issues was shaped by existing framings of related local problems. This origin determined to a large extent which policy streams, actor coalitions, and institutions were involved in the initial shaping of the issue.

The "frames" of public debate on the global environmental issues studied here, though influenced by scientific and technical understanding, were shaped through a complex social process involving accident, interests, and advocacy as well. In all the cases examined here, natural scientists were instrumental in first raising the possibility that a particular set of human activities posed a potential risk to the environment and society. The initial framing of public and political debate, though inspired by this science, was nonetheless never more than partially shaped by it. Instead, the new knowledge was generally fit into frames derived from existing problems and policy programs. The causes of such early framing seem fairly straightforward in our studies. Advocates in existing policy debates found that the new scientific findings on, say, acid rain provided additional arguments to advance their own agendas. Such advocates became eager "adopters" of the new findings and showcased them within the context of existing problem frames. This pull of established ideas on the new issues brought with it established actor coalitions and institutions. As research programs or statutory frameworks these often became the initial "homes" for the emerging issues and often exerted a strong influence on their early adolescence.

These initial historically embedded frames seem to have had three quite different effects on subsequent issue development. First, by linking newly emerging issues with others already established on the policy agenda, early framing may have tended to accelerate growth in awareness of the new global issues and thereby to provide the institutional stability and resources those emerging issues otherwise would have lacked. Second, to the extent that such frames emphasized single sources of risk that were already under pressure to mitigate their supposed local pollution effects, they drew organized opposition into efforts to manage global environmental risks sooner than it might otherwise have taken form. Third, these early frames seem often to have cast a long shadow into the future. In particular, the institutions and actor coalitions associated with these initial frames probably tended to constrain the path of issue development for periods of several years to a decade or more. Nonetheless, the frames generally did grow broader in time, incorporating more causes, pathways, and impacts within their boundaries.

Stickiness of Early Issue Frames The observed "stickiness" of the early issue frames, together with their associated actor coalitions and institutions, often persisted until episodes of heightened public and political attention brought new perspectives and interests to bear on the problem.

The spates of increased attention that periodically punctuated the gradual maturation of the issues areas studied here reflected more than "windows of opportunity" in which existing knowledge and action agendas could come together within existing problem frames. In addition, by bringing new ideas, interests, and institutions into the debate, they provided a crucible of intense interactions from which new framings of the issue often emerged.

For both the ozone and acid rain cases major expansions of the frame of the debate beyond that inherited

from the issue's origin were made possible by underlying developments in science, technology, and politics. Catalyzed by powerful images of "the ozone hole" and forest dieback, a clear reframing occurred when the understanding of individuals (scientists and bureaucrats) or isolated institutions became a general understanding shared across nations. This reframing enabled wholly new actor coalitions to be formed and a wider range of sources and effects to be considered.

The evidence suggests that sudden reframing occurred at times when political and media attention was high but was not necessarily accompanied by a sudden change in the deeper understanding of complex cause-and-effect relationships. Sudden reframing coincided with efforts of actors from within or outside the bureaucracy to push issues onto the political agenda.

New Frames, New Interests New problem frames attracted new interests and actors to the issues. Their involvement in turn led to more complex issue frames to accommodate the concerns of those involved in the management process.

More expanded problem frames developed through a two-step process. First, new actors became involved because expanded frames enabled them to use the issue to advance their own agendas. Second, new actors thus entrained in issue development sought to further reshape—and often broaden—the issue frame to best accommodate their full range of concerns.

The evidence collected in this study strongly suggests that issue frames are the context within which the engagement (or not) of particular interests and actors within a given problem area is determined. Actors sorted out whether they cared about a particular problem based on whether its current framing of causes, pathways, or consequences coincided with their interests. They sorted out how much they cared based on whether, considered within this frame, the issue presented them with especially salient hazards or opportunities. Frames thus engaged the general public to the extent that they raised attention and promoted interpretation of the issue as being of personal interest. A frame engaged institutions when it allowed them to conclude that the issue merited a position on their agenda.

The actors studied here did not, however, merely react passively to issue frames. Instead, once the issue began moving up the action agenda, actors hoping to forge coalitions that would promote or block further development often did so by trying to change (or sustain) the prevailing issue frame. For example, industrial actors started to engage themselves in the ozone issue when they saw their direct interests threatened. They concentrated on questioning the risk without questioning the existing issue framing until some industrial actors started to see an opportunity in reducing the risk through specific measures from which they might benefit. The climate change issue was actively promoted by environmental NGOs only after they saw that the solution of the problem could be energy efficiency and new energy sources, not simply nuclear power.

Two-Level Frames Widely shared core frames of human-induced environmental change coexisted with a variety of actor and interest-specific frames dealing with the specific causes and effects of such changes.

The findings of this study strongly suggest that some shared framing of what "the issue" was all about emerged and was actively sought by coalitions seeking to promote international action. It was equally clear, however, that nothing like complete consensus or commonality of framing was either sought or achieved. Rather, as issues developed a two-tier framing structure emerged. The upper tier consisted of widely shared "global" framing elements such as "sulfur dioxide," "long-range transport," and "acidifying deposition." The lower tier consisted of specific "local" framing elements relevant to specific places and constituencies. The shared "global" frames (upper tier) served to facilitate communication between different actors, between countries at the international and national levels, and "perhaps" between science and policy. The specific "local" frames served to connect and make germane the broadly defined global issue to specific interests and constituencies.

Tentatively, we may conclude that there was a tension between frames that worked to move forward international agreements and frames that moved forward local support for action.[6] This tension seems to have been partially resolved by having a shared framing of the risk at the most general level (for example, "acid rain"), whereas at the same time local frames could differ according to local realities and concerns (such as forest death). We observe that the elements of the frames that were most broadly shared internationally generally related to "causes" (such as carbon dioxide or acidifying substances) rather than "effects" (forest death or sea-level rise). The status of initiating activities (forest clearing, point-source fossil fuel combustion, emissions from China versus from the United States) seems intermediate, with some sharing but a remarkable amount of local differentiation in emphasis. Local frames, in contrast, not only differed but were much more closely geared toward negative effects and possible ways of handling the issue.

One insight into the possible significance of these two-level frames is provided by our often repeated empirical observation that public and political attention to global environmental issues rose and fell in approximate synchrony for many (but not all) of the arenas we studied. A closer look at our data suggests that concern changed in synchrony with the "issue" that was framed in upper-tier, global terms—such as "transboundary acid rain from coal-fired electricity plants," "carbon dioxide–induced climate change," or "chlorofluorocarbon (CFC)-caused ozone depletion." This is the issue people "heard" when they "listened" to what was being discussed in other countries or in the international arena. Did the story warrant local (domestic) attention? This would happen only if it could also be framed in terms of local concerns— such as forest dieback, the national CFC industry, or a domestic debate on the future of nuclear power. Where such locally relevant frames did not emerge or were relatively uninteresting, the elite actors in the arena may have "heard" the international discussions about, say, acid rain but had no context in which to continue the discussion domestically. Thus in the ozone case we found that the original way of viewing ozone depletion as a cause of skin cancer did not resonate with Japanese publics and political elites, though Japanese scientists were certainly aware of the issue. Later, however, when the global frame broadened to encompass effects on ecosystems, agriculture, and fisheries, Japan had a "handle" on the issue and joined in both domestic and international discussions about it. Moreover, our acid rain data suggests that very different local frames were compatible with sustained cooperation on science and policy under a shared global frame at the international level.

22.2.2 Interests and Actor Coalitions

What were the actual roles played by the various interested "actors" in the issue histories we studied? The design of this study reflected our belief that explanations centered on nation-states alone are increasingly inadequate to capture the multiple players and multiple levels of interests and action involved in environmental affairs (Sabatier and Jenkins-Smith 1993 1999; Lipschutz and Conca 1993; Mathews 1997; Kates 1998). In particular, we assumed that the management of global environmental issues might be shaped not only by state actors and politicians but also by interested actors from the private sector, nongovernmental organizations (NGOs), the scientific community, and the media. And though we focused our attention at the national and international levels, we took account of influences on management at these levels stemming from subnational and local actors. This wider lens enabled us to examine where and when various interests and actors actually took a role in the development of management responses to global environmental risks and to assess their contributions.

The specific findings emerging from this approach have been related in the individual arena chapters found in part II of this book; cross-arena patterns for individual actor groups were analyzed in chapter 14. Beyond those details, however, four general findings stand out.

First, the numbers and kinds of actors involved in the management of global environmental risks rose appreciably over the period of our study. With this increase in participation came a growing challenge to legitimize proposed management actions over larger and more diverse constituencies, reflecting an increasingly wide range of interests and domestic issue framings. The emergence of two-level framings noted in the preceding section did much to accommodate the tension between increasingly diverse local interests and the quest for international coordination and cooperation in management response.

Second, individual national and regional arenas exhibited substantial variation in how they responded to the risks in question. For specific cases, it is possible to identify states that were distinctly out ahead of the issue in advocating international action and others that followed or resisted such action. Arenas that responded similarly—as, for example, in the U.S. and U.K. governments' responses to the acid rain issue—generally shared both political cultures and perceived vulnerabilities to the risk in question. In any event we found that no arena played the same distinctive role in relation to all three environmental risks. Rather, the roles played by arenas generally and nation-states in particular in the management of global environmental risks seem to be dependent on both their own sociopolitical characteristics and characteristics of the particular environmental issues that come before them.

Third, national governments were never the sole "mover" in the development of management responses to global environmental risks. The scientific community, NGOs, private sector, and media played sufficiently influential roles that a history or explanation that excludes them is likely to be seriously flawed. These nonstate actors operated not only at the national and local levels but also internationally. A general characteristic of all three issues was the seminal role of scientific research in defining the issues, in drawing attention to them, and in generating possible ways of viewing them. Nongovernmental organizations were largely absent from the early history of problem recognition and solution development but played—jointly with other actors—a crucial amplifier role in the agenda-setting stage and a predominantly monitoring role once implementation of serious management actions was underway. The media likewise were not agenda setters but played a pivotal role in amplifying concerns and in connecting international

and domestic issue frames. What we have called the "industry" sector[7] at first appears in the cases we studied in a reactive role, particular industrial actors actively promoted an open-ended search for reliable knowledge and effective actions.

Finally, for the most part it was not individual actors of any sort—local or international, state or nonstate—that figured as the most prominent agents of change in our histories. Rather, the motivating force for most of the changes we observed were coalitions of actors more or less loosely joined for the express purpose of affecting issue development. Many of the most influential coalitions were transnational in character. Some of them—such as the Scientific Committee on Problems of the Environment, the Climate Action Network, and some industry coalitions—were relatively homogeneous, with the shared beliefs and values that Haas (1990) and others have used to characterize "epistemic communities."[8] More commonly, however, the most influential groups were not obviously homogeneous in their beliefs and included individuals from various governmental, scientific, NGO, and even industrial communities. This makes them more like the domestic "advocacy coalitions" portrayed by Sabatier and Jenkins-Smith (1993) and the international "advocacy networks" discussed by Keck and Sikkink (1998). Our study was not designed as an exercise in developing or testing propositions about the growth and impacts of such actor coalitions or interest networks. We nonetheless draw from the histories we examined a deepened appreciation of how important multinational, multiactor advocacy communities are likely to be in the evolution of global environmental management and how inadequate approaches that focus on only traditional national actor or interest groups are likely to be.

22.2.3 Institutional Capacity and Public Attention

What drove the changing institutional capacity for management of global environmental risks? How did these factors interact with issue frames and actor coalitions to shape long-term issue development?

A striking aspect of the risk management histories explored in this study is the contrast between the "telescopic" perspective that shows relatively slow, smooth growth of knowledge and institutional activity during the entire period, and the rapid rise and fall of attention that dominates a more "microscopic" view of selected times and places. The long-term continuities in social response are especially striking when considering the growth of institutional capacity to address global environmental risks through performance of the management functions addressed in part III of this book. Our research leads us to suspect that part of the explanation for this secular growth in management capacity lies with the relationship between public attention and institutional agendas.

In particular, our findings suggest that over the long initial phase of issue development characterized by relatively low public attention, society's capacity to address them gradually accumulated within a relatively fixed group of institutions that historical circumstances and existing framings of the problem had generated. During these periods, the rate and pattern of capacity development was largely shaped by factors internal to the participating institutions and bound within constraints posed by those institutions' existing relationships with their political and organizational environments. Few new institutions became involved with the issue during these relatively quiescent periods.

In contrast, during the second phase of rapid increase in public and political attention, both issue framing and institutional participation often changed. These "windows of opportunity" enabled institutions not previously engaged to become involved in the issue and often precipitated renegotiation of leadership on the issue within already engaged institutions. They also provided an opening for new institutions to become engaged with the issue. The consequences of these perturbations to the institutional landscape worked themselves out rather slowly, however. This was in large part due to the relatively long lag times (that is, multiple years) involved in substantially altering institutions' patterns of resource allocation, networking, and influence. Over periods that in our studies ranged from several years to a decade after these occasional "spurts of attention," some institutions ultimately worked out stable or broadened roles in the management of the relevant risk, with consequences for what issue frames and actors would subsequently play major roles in issue development. Others gradually lost relative influence and prominence. Few, however, totally disappeared from the scene. The resulting pattern of capacity development was one of relatively slow and continuous growth in society's overall capacity to perform most of the management functions we studied.

This broadly based growth of management capacity in turn seems to have had two significant consequences. First, it provided a relatively stable and increasingly robust base for efforts to manage global environmental risks. This base served to sustain continued activity over long periods of relative political apathy or even hostility. Second, it fed subsequent growth in public and political concerns for the management of global environmental risks. These general relationships were not, of course, without exceptions. Some arenas failed to develop a sustained institutional capacity for management of the risks we studied. And the capacity to perform some management functions—particularly research and monitoring—not infrequently eroded after the initial adoption of policy measures. Nonetheless, the overall pattern seems clear: continually increasing—if qualitatively changing—management capacity, sporadically energized and redirected by the infusion of new institutions and actors during brief episodes of high political attention.

22.2.4 Pathways and Mechanisms for Spreading New Ideas and Experience

Which pathways and mechanisms were most important in spreading new ideas and experience throughout the community of actors engaged in the management of global environmental risks? This study found three main pathways to be particularly important: the dissemination of information through the mass media, exchanges among individuals, and the role of international institutions as forums for learning. These and other possible pathways are clearly related and not mutually exclusive.

Media The analysis of chapter 14 showed that the media indeed played a role in spreading internationally information about the issues of climate change, ozone depletion, and acid rain. This role seems to have been not so much one of bringing home the news on discoveries and experience in other places as it was of facilitating the use of high-profile foreign developments as occasions for exploring local ramifications of the issue at hand. The media-mediated European exposure to America's "greenhouse summer" of 1988 is an excellent case in point. But the tendency of the media to frame global issues in local terms meant that this mechanism may have done rather less than might be supposed to spread the idea that issues such as those we studied have truly global connections and are not just problems, like earthquakes, that happen everywhere.

Personal Contacts Personal contacts were highly important in spreading ideas and experience, especially in the early phase of issue evolution. The arena studies we report in part II of this book are replete with the seminal roles played by particular individuals—usually but not always scientists—in promoting initial interest and concern about the issues at hand. This is perhaps not surprising. Before climate change, ozone depletion, or acid rain were firmly established on scientific, public, or policy agendas and before there were international institutions committed to promoting their development, trust in particular individuals provided pathways for spreading ideas about global environmental risks internationally.

But if this role of personal contacts is not surprising, neither was it inevitable. Where the exchange among countries of individuals involved in particular global issues was common and regular, ideas flowed quickly and freely. By and large, in the early days of our issues' development, this was the case for a small group of elite Western scientists but for almost no one else. (Somewhat later, an even smaller cadre of international civil servants may have played a similar role, sharing experience and lessons on treaty negotiations.) But our studies suggest that countries lacking free and frequent exchanges of scientists and actor groups lacking the sorts of scientists involved in those exchanges were relatively unengaged in the early evolution of the issues we studied. As time passed and more and more people became involved in issue development, these "head starts"—and the personal contacts on which they were based—became less prominent among the available channels for spreading ideas internationally. It is nonetheless difficult to imagine the early history of globalizing environmental risk management without them.

International Fora The importance of personal contacts and the elite media notwithstanding, the most significant channels for spreading new ideas and experience internationally were provided by international fora and agreements. The role of internationalization for risk assessment and monitoring was particularly important in the early phases of issue development, when scientific capacity was building and the issue was not yet on the broader policy agenda. As attention to the issue began to increase, international fora played an important role in spreading ideas about goals and strategies beyond the core group of personally connected advocates initially involved. Ideas and experience about option assessment and evaluation were spread through international fora once the issue was firmly on the policy agenda. This third or "postemergence" phase of issue development also saw the most effective episodes of systematic "policy-oriented learning" of the sort explored by Sabatier (1993; Sabatier and Jenkins-Smith 1999). As suggested by scholars of U.S. science advisory processes (e.g., Jasanoff 1990; Gieryn 1995), one of the most effective vehicles for such learning is almost certainly the establishment of forums or occasions that encourage development of ongoing dialogues ("boundary negotiations") between those producing or assessing scientific information and those that might use that information to inform action. In our cases, international negotiations on emission reductions that are closely supported by finely targeted assessment activities—such as the Regional Acidification Information and Simulation (RAINS) model in Long

Range Transboundary Air Pollution (LRTAP) negotiations and the Technology Assessment Panels under the Montreal Protocol—have provided just such forums.

Differences in Receptivity at the National Level The overall conclusions about major pathways and mechanisms noted above must be modified somewhat to address how characteristics of specific arenas affected their receptivity to new ideas about global environmental risks.

For example, smaller countries were often particularly sensitive to the framing of issues and rise of attention in a large or economically important neighboring country. The attentiveness of the Netherlands to the acid rain debate in Germany or of Mexico to the climate debate in the United States illustrates the point. The dependence of the original Dutch position with respect to the ozone issue on the U.S. position shows that it is not necessarily the physical proximity that occasions such attentiveness but also the ready availability of messages that key actor groups in a particular arena want to hear.

Other national features also affected receptivity to "foreign" communications about global environmental risks. Depending on the national structures the press were either important in bringing global environmental risks to the fore (as in Mexico) or relatively unimportant (as in the former Soviet Union and Hungary). The role played by NGOs was in some cases limited (such as Mexico and Japan), but even where NGOs played a relatively larger role it was in monitoring progress rather than in putting the issues on the agenda. The way issues get on or off particular domestic agendas was often most clearly related to political shifts in that arena. Examples include the positions taken by political leaders (such as Thatcher before she "turned green" or the decline in importance of environmental issues in the Reagan administration in the United States) or the threat of issue capture by dynamic green parties in closely contested elections (Germany).

Notwithstanding the strong international ties of scientists, the disciplinary structure of science in particular arenas had a large influence on the questions addressed and on the problems that were recognized and accepted (a conclusion that is paralleled in the work of Hall 1989). The closeness of the relationship of scientists to policy and also the specific part of policy to which the science was tied strongly influenced the exchange rate of ideas and the part of policy affected.

22.2.5 Perspectives on Issue Development
Looking back on the issue histories assembled here, nothing leads us to conclude that the development of global environmental risk management has been conducted as a

self-conscious, rational process. There certainly are any number of diligent, well-intended, would-be international guardians at work out there who are identifying the important problems, discovering cause, effect, and options for actions, and then reporting on the optimal choice to the world. But the risk management process to which such individuals actually contribute—along with scientists, NGOs, political actors, negotiators, and other global environmental risk managers—is much more of a social and political enterprise than the science-driven rational-choice model would allow. The variety of alternative issue framings, political contingencies, and problem linkages we have found at the heart of our case histories would be sufficient to establish this social dimension of the risk management process, even if other scholarship on other issue areas had not (e.g., Beck 1992; Wynne 1995; NRC 1996).

Equally, however, we see no reason to adopt any of the more pessimistic perspectives that portray the development of policy as merely the product of chance encounters in a garbage can (Cohen, March, and Olsen 1972) or as totally refractory to meaningful scientific input (e.g., Collingridge and Reeve 1986). At small scales in place and time the process may indeed appear chaotic, and the "streams" of problem definition, solutions, and enabling politics may appear largely independent. But at larger scales, as we have shown, a certain orderliness does sets in—not least because of the continuing ability of scientific assessment to craft broadly convincing arguments about cause and effect. Our observations thus support the conclusion of van Eijndhoven and Groenewegen (1991) that scientific knowledge has an active role in policy development but not one in which action waits on scientific consensus. No rational actor may be directing the development of global environmental risk management. But that development, like development of economic markets, seems to take on a good deal of large-scale rationality— apparently as a result of individual actors and ideas struggling to find roles that work for them.

Our way of viewing the role of agency in the development of issues and the capacity to handle them has some parallels to the perspectives on issue development taken by Sabatier (1993) and Hajer (1995). Those authors stress the importance of actor coalitions in issue development, but with a focus on the level of domestic or national politics. Our study has shown that relationships among actors at the international level are indeed a major factor in management of global environmental risks. In underlining the importance of international fora and the groups operating within them, we find parallels with the epistemic communities discussed by Haas (1990) and the advocacy networks studied by Keck and Sikkink (1998). Those studies, however, focus on particular actor groups—experts for Haas, nongovernmental organizations for Keck and Sikkink. Our study has shown, however, the importance of multiactor, transnational coalitions in pushing issue development. We find that it is not a single closely connected epistemic community or advocacy network that furthers the management of a particular global environmental risk. Rather, the action lies in an assembly of loosely knit, multiactor coalitions held together by shared issue frames and international institutions. The issue frames connect the coalitions involved differently at different levels of social organization, allowing tacitly different meanings to be attached locally, nationally, internationally, and among different constituencies (such as scientists, international politics, and local implementation actors). The institutionalization at the international level—via organizations, treaties, and protocols—generates a common focus for management.

The perspective on issue development we have developed through this study leads us to see broad patterns of change as the thing to be explained, with the detailed timing of agenda setting as something in need of attention but in most cases relatively coincidental.[9] We view institutions as, on the one hand, accommodating to shifting perspectives but, on the other, as contributing to such change: institutions act as caretakers keeping issues alive but also as agents of changing issue framings when they see current frames as barriers to effectively addressing problems on their agendas. Each institution will be limited in the possibilities it has to shift frames and resources. If major changes are needed, it is likely that this can be achieved only by breaking up earlier institutionalization patterns. Actors may then attempt to generate new institutions (for example, by institutionalizing certain coalitions or by generating processes that may shift an agenda). Our emphasis has been on to the management process rather than the political process. We saw that the intense political debate did interact with the management processes we discussed and did change the terms of the debate by reestablishment of issue frames. But we also emphasized the importance of the new institutionalization emerging in the wake of such episodes.

Similarly, there are some interesting comparisons that can be made between the results of this study and research on networks stemming from the pioneering research of Heclo (1974). This study finds that issue networks do build up slowly at the domestic level, in line with Heclo's findings on social policies. However, this study concludes that international institutions also play a major role in establishing a strong transnational component in the management of risks. This addition of international institutions as linkage mechanisms provides a large qualitative difference between our description of the management process and descriptions that concentrate on individual national arenas.

Significantly, our study suggests that issue networks, though extending across borders, do not generally extend across issues. Few members of the transnational networks involved in the acid rain issue had significant contacts with, or learned much from, their opposite numbers in the networks addressing climate change or ozone depletion. Even the ozone and climate networks remained surprisingly isolated from one another, at least through the end of our study period. Rarely, if sometimes significantly, individuals overlapped issues. By and large, networks did not. Finally, as argued in chapter 20 (on evaluation), the very informality and fluidity of the transnational issue networks we observed in this study may well contribute to their failure to carry out systematic reflection on their work. In keeping with expectations from studies of risk management in general (e.g., Beck 1992; Wynne 1995), we found little organized, critical self-awareness of networks' successes and failures in contributing to the management of global environmental risks.

There is thus a considerable amount of empirical evidence in this study that supports modern discourse-oriented models of risk management (e.g., NRC 1996; Jasanoff 1990). Even more fundamentally, our findings resonate with Heclo's observation that politics is not merely about conflict and power but also finds its sources in uncertainty, with humans collectively wondering what to do. Society has "found out what to do" in response to issues like acidification and stratospheric ozone decline, so in that sense social (collective) learning—for better or worse—has occurred. Moreover, as shown in chapter 18 (on goal and strategy formulation), our study period also witnessed a modest amount of progressive change in the overarching goals, concepts, and beliefs that have guided social action. This suggests that "double-loop learning" of Argyris and Schon (1978), "third-order paradigm shifts" of Hall (1993), and reflexive changes in "core beliefs" of Sabatier and Jenkins-Smith (1993), though rare, have not been altogether absent in the history of efforts to manage global environmental change.

For social policy, Heclo (1974, p. 322) long ago concluded that "much of social policy has remained at the level of chance discovery and ad hoc invention, with little attention to accumulated evidence, experimentation, or questions of how the learning process itself might be improved." A bit more than a decade later, Beck claimed the same for social efforts to address the risks of science and technology (Beck 1986, 1992). Much of our study supports a similar depressing conclusion. Nonetheless,

the self-critical, reflexive processes of risk management established within the LRTAP Convention's Protocols for acid rain, the Montreal Protocol for ozone protection, and associated activities give some hope for the coming of a new era. We may indeed be learning to complement the seemingly haphazard development of risk management processes at small scales and times with more orderly processes managed internationally and encouraging reflexive learning through the inclusion of explicit evaluation processes.

22.3   What are the Implications of This Study for Improving Global Environmental Risk Management?

The first precondition for enhancing the management of global environmental risks is the existence of general management capacities in countries and, more generally, arenas and international institutions. The best way to further a better management of global environmental risks (and probably other issues) is to further possibilities for generally adequate management and to support the integration of countries and of actors in global management activities. This holds not only for international activities in general but also specifically with respect to the science involved in managing global environmental risks.

Our research has shown that the process of building capacity to address global environmental risks needs time. This seems to be irreducibly true, since it is not primarily the amount of resources (human resources and money) that is of importance but rather the generation of the coordination, cooperation, and trust needed to create an effective management process (Kay and Jacobson 1983; chapter 18 on goal and strategy formulation).

It appears that there was relatively little capacity transfer from one issue to another at least before 1992. Each of the issues studied here has its own assessment bodies, negotiating apparatus, and convention secretariat. Only in the cases of a few individuals was there a transfer of lessons learned from one issue to another. (For example, an important lesson learned within the LRTAP Convention regime was to secure long-term financing for the monitoring of emissions and effects through a separate protocol to the Convention. This example was not followed for climate change or stratospheric ozone depletion.) There is also evidence of incorrect lesson drawing between issues. For example, the use of "ozone-depletion potential" in the case of stratospheric ozone depletion was very useful in highlighting the order in which substances had to be dealt with. The attempt to use a similar concept, that of "global warming potential" in the case of climate change, led to considerable controversy because the global warming potential differed according to the time scale considered (see chapter 15 on risk assessment).

The results of our research suggest that there are at least two ways in which it is possible to strengthen and enhance learning and the prospect for progressive problem shifts in the management of global environmental risks: the first is through attention to issue framing; the second, through attention to the risk management functions. And it comes as no surprise that both routes interact.

22.3.1 Issue Framing

We concluded that tensions often exist between the framing or conceptualization of issues that work to move forward international agreements and the framing of issues that move forward local support for action. Frames that allowed progressive problems shifts were those that allowed differentiated local frames to co-exist with a shared frame of the risk at the most general level. However, one of the weaknesses we found in the management process, also reflected in its component management functions, was the existence of "stickiness": the lengthy survival of a certain operationalization long after the moment when it had its maximal effect. The broad framing in terms of "acid rain" allowed a progressive problem shift in addressing the negative effects of acidifying substances, as compared to the narrow framing in terms of sulfur dioxide (SO_2). The "ozone hole" framing allowed far-reaching shifts in the detailed understanding of the underlying causes and effects, without sticking to the original framing (spray cans). The framing in terms of climate change is in that respect "better" than the framing in terms of carbon dioxide because it allows other compounds to be included in the assessments. A reframing in terms of a lifestyle change reduces the possibilities for coalition building, whereas a reframing as the CO_2 problem enhances the risk of sticking to old solutions and coalitions, notwithstanding attempts to broaden CO_2 to "CO_2 equivalents."

We found that breaking old frames often led to new opportunities and enabled new coalitions to be built. Therefore, we suggest that it is of importance to further creative thinking and development of creative frames. However, it is not immediately clear how this could be achieved. Below, we address this question by going back to the functions we analyzed in our research and their performance.

Our conclusions suggest that the effectiveness of particular issue framings may differ depending on time, place, and use: cause-related framing works best internationally, effect- and remedy-related framing works locally, simple framing works early on, and more complex framing works when constituencies have been

built up. The framing also has to generate connotations that lead to concerted action instead of disagreement and to action instead of inaction. It is difficult for one framing of an issue to fulfill all those requirements. Therefore, effective frames are those that allow the interpretation to be tacitly different in different places and for different uses and that can act as anchoring devices for the overall issue-development process (van der Sluijs, van Eijndhoven, Shackley, and Wynne 1998). Those framings allow shifts in the underlying specific concepts, in the exact specifications of causes, effects, and remedies, and in the specific ways the implementation process is developing. Effective frames allow for a "thickening" of the underlying knowledge and action without openly rebuilding all of the building (see, e.g., van Eijndhoven and Groenewegen 1991). Effective framing allows for constant repairs as the result of new knowledge and actions carried out related to all of the above functions, without taking down the building.

A frame can be considered less effective when it fails to address concerns inherent in the way a risk is viewed: when new risks emerge from risk assessment or monitoring, when new causes need to be addressed, and when new actors need to be involved to address the issue. If the suboptimality of the framing cannot be taken away by tacitly changing some of the underlying building blocks of the management building, a clear shift of the framing is needed: a new design should be made. But building blocks of the old design can still be used as part of the new design. (When the framing of "the health threat from sulfur dioxide emissions" was replaced by the framing of "transboundary pollution of acid rain," much of the capacity to address the health issue could be brought to bear on the acid rain issue.)

22.3.2 Risk Assessment and Monitoring

Looking at major assessments of the magnitude and timing of the three global environmental risks over a period of twenty years or so gives clear indications of factors that can contribute to "successful" risk assessment. Major risk assessments that contributed to the further development of risk management have been multinational. The management of the climate change and ozone risks, in particular, were furthered by the use of international assessments, which were used to legitimize national assessments. However, it has been noted that problems arise if the representation of countries in the international groups, particularly the distribution of representatives of the developed and developing worlds, is not balanced.

For each of the issues studied here, monitoring played a significant role in reframing the debate and stimulating risk assessments, goal and strategy formulation, and implementation. The importance of monitoring in management of global environmental risks appears to be often underestimated. The motivation for the monitoring that did occur was often unrelated to the need to document environmental risks, which suggests that monitoring systems have not been designed to play a role in risk management, although in each case they have done so. Recognizing the importance of monitoring activities in the whole of the risk management process may help keep monitoring capacity in place.

Risk assessment and monitoring are action-oriented, knowledge-generating functions and in that respect are unlike more fundamental or disinterested forms of knowledge-generating activities. Risk assessment activities, in particular, have been discussed in the literature as examples of "postnormal science" in which the usual scientific peer-review procedures are no longer adequate to generate legitimate conclusions (Funtowicz and Ravetz 1993). Our findings show that indeed the validation and legitimization of the assessment becomes more critical in the sense of being more open to debate by a larger group of stakeholders. But we also see that the function of the openness is primarily to gauge the adequacy of the assessments (Have all relevant factors been taken into account? Have known pitfalls been avoided?), whereas the process and product hold most value if they end up as a thoroughly screened scientific product. In broad outline we can therefore conclude that approaches to risk assessment and monitoring have developed that are technically adequate and politically legitimate[10] international activities with a strongly scientific character and did play an important role in promoting the long-term development of management processes for global environmental risks.

22.3.3 Goals and Strategy Formulation and Implementation

Goals that were effective in promoting issue development were found in two categories:

• Simple, bold, fairly rigid goals, not tied to specific strategies, not necessarily consensually arrived at, and not necessarily scientifically sophisticated (for example, the Toronto 20 percent reduction target, cutting sulfur dioxide by 30 percent, or cutting CFCs by 50 percent) and

• Consensual and scientifically sophisticated goals (for example, critical-load targets in the acid rain case and the London and Copenhagen Amendments to the Montreal Protocol). Before acceptance of the seriousness of the problem is widespread, it appears that sophisticated, complex, flexible goals fall flat.

In the cases we studied, the implementation process became broader and more coherent through time in a number

of respects. In particular, implementation activities became more flexible and involved broader groups of actors.

Taken together these two functions show clear development from simple to more complex. It is, however, difficult to state that their operations have become more effective or more legitimate over time. It may be more true that over time a need arises to broaden the span of the action to optimize these functions (and therefore to stay efficient and legitimate). Or conversely, it may be even true that initially only very restricted action is feasible. For these functions to be adequately fulfilled, therefore, shifts should take place over time.

Chapter 18 (on goal and strategy formulation) and chapter 19 (on implementation) suggest that this is indeed what took place. In the early history of an issue, goals should be clear and simple to get action going. In the implementation track, easy targets are addressed first. Later on, the institutions involved become more experienced and can address more complex questions, but the inadequacies of the simple first goals and strategies also become more glaring. Also cooperation of more actors is needed to act on the issue, combined with stronger links between the functions. The management of the issue becomes more strongly embedded in society as a whole and may even be subsumed under other environmental issues (as happened in some arenas with the ozone issue) or even under more general policy concerns of a different kind.

The evidence suggests that it is not the complexity of goals, strategies, or implementation processes in itself that should be assessed but the adequacy of those processes in relationship to the development stage of the issue. "Old-fashioned" goal statements risk being too simple, too rigid, and too long-lasting, thereby hindering the development of an issue to a riper, better embedded stage.

22.3.4 Option Assessment and Evaluation

Option assessment and evaluation reside on the cusp of knowledge and action. Both are engaged in generating knowledge in relationship to action and might in that respect be seen as equivalent to risk assessment and monitoring. However, whereas we found that the legitimization of risk assessment and monitoring resided primarily in validation as "sound science" but with extended review processes, we did not see the same in the case of option assessment and evaluation.

As in risk assessment the validity of the data taken into account in option assessment or of perspectives from which an evaluation is carried out is often critically scrutinized, and this scrutiny is—parallel to what happens in risk assessment—coming from wider audiences than scientific peers. It is therefore critical that option assessment and evaluation activities are seen to be adequate (well analyzed) in the sense of using data that are acceptable to actors in the area analyzed. If no consensus exists between stakeholders, the existence of those multiple views should explicitly be taken into account. Without the sound data and information that takes due account of the perspectives of stakeholders playing a role in the decision processes, option assessments and evaluations will not become legitimate. Which actors need to be involved for legitimization differs over societies and over time.

The danger of being delegitimized may, however, not be the largest threat to the role of either option assessment or evaluation. A much greater danger is the threat of neglect. If the effectiveness of individual option assessments or evaluations were to be defined in terms of the tangible impact of an individual assessment on particular decisions, then most option assessments and evaluation efforts encountered in this study would be judged as extremely ineffective. Judging these activities by impact on specific decisions might, however, do injustice to what is happening because it is likely to underestimate the complexity of the decision processes. Therefore, our option-assessment chapter (chapter 17) suggested taking the impact of option assessment on the policy or option *agenda* as a measure. Even using that criterion for option assessment, the conclusion was still rather gloomy, whereas viewed in this perspective some of the evaluation efforts are evaluated rather more positively. The reason is probably not that evaluation is an easier task than option assessment. It seems rather that the intimate relationship of evaluation to the decision process has been better realized and therefore that "evaluators" learned earlier that it is no use to conduct evaluations as pure knowledge-based activities. Policy analysts involved in the option-assessment community still often tend to address option assessment as mainly a scientific enterprise. Our research showed that option assessment gets its value only after a decision maker has been identified that needs the assessment and can be engaged in negotiating its design. Such assessments must take into account the legitimate concerns of the decision maker to gain credibility. The difficulty, of course, is the relationship between the things that assessors can productively study and the things that users want to know from them.

The option-assessment chapter (chapter 17) suggested some directions for enhancing the effectiveness of option assessments. One was that assessment models encouraging "What if?" questions from potential users are a valuable tool for linking science and policy concerns. However, this is only an adaptation of a still primarily analytic activity. Slowly, policy analysts start to understand the importance of boundary negotiations between users and producers of assessment. Chapter 17 mentioned

the German Climate Enquete Commission and the International Institute for Applied Systems Analysis RAINS' support of the LRTAP Convention negotiators. These examples show that such boundary work (Jasanoff 1990) can be relatively ad hoc (Enquete) or permanently institutionalized (LRTAP). The crucial factor seems to be that option assessments produced through an intimate collaboration of analysts and specific users are more likely to prove worthwhile than those conducted for generic support of policy development. Such collaboration helps the actors involved understand the possibilities and needs of the collaborating parties and thereby presumably opens up new perspectives for viewing the issue and addressing the perceived problem.

Here we find a parallel with evaluation. Especially where evaluation is a standard integral part of a policy process, evaluation activities are structured in a way in which they are well designed for suggesting solutions for detected problems and for feeding them back into the process as well. The Montreal Protocol process is the outstanding example here.

We can conclude that for option assessment and evaluation to be of value to the management process of global environmental risks, they should be designed in such a way that they indeed work toward diminishing and bridging the gap between knowledge and action. When they are designed in such a way that they can reach these goals, it is likely that the details of the local framing of the issue change and get connected, thereby building new coalitions around the issue and actively generating feasible options and associated regimes: new or adapted agendas are created constituting progressive problem shifts.

22.3.5 Implications for Emerging Issues

On the basis of this study of the development of the management process for global environmental issues, what advice can be given with respect to issues that emerge onto the science and policy agendas in the future?

First, we noted at the beginning of this section that strengths and weaknesses in managing global environmental risks are very strongly related to strengths and weaknesses in management capacity more generally. To deal effectively with new issues as they emerge, the general management capacities of countries and more generally arenas—including international institutions—should be maintained and where necessary enhanced. That is not to say that management activities are the sole prerogative of governments or other official bodies. Indeed, what we have seen is that a variety of other bodies—for instance, industries or environmental NGOs—are increasingly important actors in codetermining the way in which (global) environmental risks are addressed and managed.

As new issues emerge, but before public and political attention to them has begun to increase dramatically, individual actor groups (scientists and some bureaucratic entrepreneurs) play an important, independent role in issue evolution. Based on our experience with other global environmental issues, it will not be NGOs, industry, or the media that will play a major role in pushing the issue forward at this early stage of issue evolution. Over a long period characterized by relatively low public attention, society's capacity to address new issues gradually accumulates within a relatively fixed group of institutions that is determined largely by historical circumstances and existing framings of the issue. It is unlikely that new institutions will become involved to a major extent with the issue over this relatively quiescent period.

During periods of rapid rise in public and political attention to a new issue, a renegotiation of leadership within already engaged institutions and a need for new institutions will emerge. At this stage of issue evolution it is important to recognize the need for coalitions of actors to push the issue forward. These coalitions provide the basis for a shared understanding of the problem and its possible solutions. Effective management of emerging issues will therefore encourage this coalition building rather than generally increased participation by individual, isolated groups of actors.

The process of building capacity to generate the coordination and cooperation needed to create an effective management process for global environmental issues needs time. As new issues emerge, the possibilities for lesson drawing from the management of other global environmental issues and transfer of institutional capacity from other issues should be consciously examined. Furthermore, as issues emerge it is important to recognize the important role played by issue framing. Frames most likely to promote progressive problem shifts are those where a shared frame of the risk at the most general level allows local frames to exist and to differ. At the same time, it is important to avoid the lengthy survival of a certain framing long after the moment when it had its maximal effect.

Effective management of emerging issues must recognize the importance of international fora and agreements for spreading ideas on knowledge and action. The role of internationalization in the science-based activities of risk assessment and monitoring is particularly important in the early phases of issue development, when the scientific capacity is building and the issue is not yet on the broader policy agenda. As attention to the issue begins to increase, international fora play an important role in spreading ideas about goals and strategies and ideas and experience about option assessment and evaluation are

spread through international fora once the issue is firmly on the policy agenda. International fora are important, however, not only for spreading ideas. Especially after the issue is firmly on the policy agenda, international fora should provide a locus for adaptive management.

Perhaps the greatest unmet need for promoting the long-term development of global environmental risk management is the institutionalization of periodic evaluations addressing how the overall management process is doing. To improve that process we must move from the ad hoc and accidental toward the organized and conscious pursuit of social learning.

Appendix 22A.   Acronyms

CFC	chlorofluorocarbon
CO$_2$	carbon dioxide
NGO	nongovernmental organization
NRC	National Research Council
RAINS	Regional Acidification Information and Simulation (model)
LRTAP	(Convention on) Long-Range Transboundary Air Pollution
SO$_2$	sulfur dioxide

Notes

1. Recall from chapter 21 that at the international level, the initial emergence of widely shared goals for managing acid rain occurred in the beginning of the 1980s, for climate change in the period 1985 to 1988, and for ozone depletion in 1974 and again in 1985. Comparing these dates with the figures illustrating newspaper attention to the global environmental risks in chapter 14, it is clear that for acid rain and climatic change the phase of "knowledge leading to action" coincided unambiguously with the time during which there was a steep rise in attention to the issue. In the case of stratospheric ozone depletion, the story is a little more complicated because there were really two distinctly separate times when attention to the issue rose steeply—in 1974 and in 1985. The rise in attention in 1974 was largely a U.S. phenomenon. Nevertheless, it is clear that the "knowledge leading to action" or "assessment leads to goal statements" phase occurred at the time of a sharp rise in public attention in 1974 and again in 1985.

2. The third phase began after 1983 in the case of acid rain, after 1987 in the case of stratospheric ozone depletion and after 1988 in the case of climate change.

3. In some countries initial strong interactions between knowledge and action can be driven by the action side, in contrast to the patterns described for the international level. This is particularly so for some of the smaller countries included in this study, such as the Netherlands and Hungary, where activities in other countries or at the international level led directly to action. For example, in the acid rain case, Hungary did not have a phase in which scientific capacity built up before scientific knowledge began to lead to goal statements and implementation. The results of assessment and monitoring in particular in Germany moved

the issue of acid rain onto the Hungarian policy agenda. Similarly, in the case of stratospheric ozone depletion, the Netherlands did not develop its own scientific capacity but acted on the basis of international monitoring and assessment efforts and international goals agreed to in the Montreal Protocol and subsequent amendments.

4. By *global* we mean, here as elsewhere in this book, "large scale and transnational" and not "the entire planet." The alternative phrase *international* doesn't work, both because the key relationships were not always among nation state actors and because we have reserved the term *international* to pair with *institutions* as the title of one of our arena studies.

5. On presenting these results in interdisciplinary seminars, we were not infrequently told of a (largely undocumented) conventional wisdom in fields such as economic policy making that order at the international level emerges from disorder at the national level. Explanations for this phenomenon were, however, ad hoc and contradictory.

6. This conclusion was also drawn by Robert Putnam (1988) when he argued that those engaged in international negotiations behaved as if they were simultaneously playing a game at two levels—first to reach agreement with their international opposite numbers and second to reach agreement with their domestic constituencies.

7. We have used the term *industry* here and throughout this study as a generic term meant to include actors who make things and provide services such as transportation as their principal function. In some countries, this could be referred to more directly as the *private sector.* But no or little private sector existed during most of the history of several of the countries we studied. The phrasing used here, we hope, is relatively clear in its intent without doing injustice to the diversity of social and economic systems we included in our effort.

8. We have not restricted ourselves to the term *epistemic communities* here because our results demand a broader term that does not require the sharing of both "common beliefs about cause-and-effect relationships" and "political values concerning ends to which policy should be addressed" that Haas (1990, xviii) uses to define that concept.

9. This is consistent with the biological analogy of "punctuated equilibria" brought to political issues by Baumgartner and Jones (1993), although it may be feasible in some cases to indicate that a certain shift could not have happened before a certain time because the knowledge was not yet available.

10. The use of the terms *adequacy, legitimacy, value,* and *effectiveness* is based on the work of Clark and Majone (1985). See also chapter 1, Managing Global Environmental Change: An Introduction to the Volume.

References

Argyris, C., and Donald Schon. 1978. *Organizational Learning: A Theory of Action Perspective.* Reading, Mass.: Addison Wesley.

Baumgartner, F.R., and B.D. Jones. 1993. *Agendas and Instability in American Politics.* Chicago: University of Chicago Press.

Beck, U. 1986. *Risikogeselsschaft: Auf dem Weg in eine andere Moderne.* Frankfurt: Suhrkampf.

———. 1992. *Risk Society: Towards a New Modernity.* London: Sage. (translation of Beck 1986).

Clark, W.C., and G. Majone. 1985. The critical appraisal of scientific inquiries with policy implications. *Science, Technology, and Human Values* 10(3): 6–19.

Cohen, M., J. March, and J. Olsen. 1972. A garbage can model of organizational choice. *Administrative Science Quarterly* 17: 1–25.

Collingridge, D., and C. Reeve. 1986. *Science Speaks to Power: The Role of Experts in Policy Making.* London: Frances Pinter.

de Leon, Peter. 1999. The stages approach to the policy process: What has it done? Where is it going? In Paul A. Sabatier, ed., *Theories of the Policy Process* (pp. 19–32). Boulder: Westview Press.

Downs, Anthony. 1972. Up and down with ecology: The "issue-attention cycle." *The Public Interest* 28: 38–50.

Funtowicz, S.O., and J.R. Ravetz. 1993. Science for the post normal age. *Futures* 25(7): 739–755.

Gieryn, T. 1995. Boundaries of science. In Jasanoff, Markle, Peterson, and Pinch (1995, 393–443).

Haas, Peter. 1990. *Saving the Mediterranean: The Politics of International Environmental Cooperation.* New York: Columbia University Press.

Hajer, M.A. 1995. *The Politics of Environmental Discourse, Ecological Modernization, and the Policy Process.* Oxford: Clarendon Press.

Hall, P., ed. 1989. *The Political Power of Economic Ideas: Keynesianism across Nations.* Princeton: Princeton University Press.

———. 1993. Policy paradigms, social learning, and the state. *Comparative Politics* 25(3): 275–296.

Heclo, H. 1974. *Modern Social Politics in Britain and Sweden.* New Haven: Yale University Press.

Jasanoff, S. 1990. *The Fifth Branch: Science Advisors as Policy Makers.* Cambridge: Harvard University Press.

Jasanoff, S., G.E. Markle, J.C. Petersen, and T. Pinch. 1995. *Handbook of Science and Technology Studies.* London: Sage.

Jones, Charles O. 1984. *An Introduction to the Study of Public Policy* (2nd ed.). North Scituate, Mass.: Duxbury Press.

Kates, Robert. 1998. Shifts in power (editorial). *Environment* 40(5): i.

Kates, Robert W., Christoph Hohenemser, and Jeanne X. Kasperson, eds. 1985. *Perilous Progress: Managing the Hazards of Technology.* Boulder: Westview Press.

Kay, D.A., and H.K. Jacobson. 1983. *Environmental Protection: The International Dimension.* Totowa, N.J.: Allanheld, Osmund.

Keck, Margaret E., and Kathryn Sikkink. 1998. *Activists Beyond Borders: Advocacy Networks in International Politics.* Ithaca: Cornell University Press.

Kingdon, J. 1984. *Agendas, Alternatives, and Public Policies.* Boston: Little Brown.

———. 1995. *Agendas, Alternatives, and Public Policies* (2nd ed.). New York: Harper Collins.

Lindblom, C.E. 1990. *Inquiry and Change: The Troubled Attempt to Understand and Shape Society.* New Haven: Yale University Press.

Lipschutz, Ronnie, and Ken Conca, eds. 1993. *The State and Social Power in Global Environmental Politics.* New York: Columbia University Press.

Mathews, Jessica T. 1997. Power shift: The changing role of central government. *Foreign Affairs* 76(1): 50–67.

National Research Council (NRC). 1996. *Understanding Risk: Informing Decisions in a Democratic Society.* Washington: National Academy Press.

Nelson, Barbara. 1984. *Making an Issue of Child Abuse: Political Agenda Setting for Social Problems.* Chicago: University of Chicago Press.

Putnam, Robert. 1988. Diplomacy and domestic politics: The logic of two-level games. *International Organization* 42(3): 427–460.

Sabatier, P.A. 1993. Policy change over a decade or more. In P.A. Sabatier and H.C. Jenkins-Smith, eds., *Policy Change and Learning: An Advocacy Coalition Approach.* Boulder: Westview Press.

———, ed. 1999. *Theories of the Policy Process.* Boulder: Westview Press.

Sabatier, P.A., and H.C. Jenkins-Smith, eds. 1993. *Policy Change and Learning: An Advocacy Coalition Approach.* Boulder: Westview Press.

———. 1999. The advocacy coalition framework: An assessment. In Paul A. Sabatier, ed., *Theories of the Policy Process* (pp. 117–166). Boulder: Westview Press.

van Eijndhoven, J.C.M., and P. Groenewegen. 1991. The construction of expert advice on health risks. *Social Studies of Science* 21(2): 257–278.

van der Sluijs, J., J. van Eijndhoven, S. Shackley, and B. Wynne. 1998. Anchoring devices in science for policy: The case of consensus around climate sensitivity. *Social Studies of Science* 28(2): 291–323.

von Prittwitz, V. 1990. *Das Katastrophenparadox. Elemente einer Theorie der Umweltpolitik.* Opladen, Germany: Leske und Budrich.

Wynne, B. 1995. Public understanding of science. In Jasanoff, Markle, Petersen, and Pinch (1995, 361–388).

APPENDIXES

Appendix A
Research Protocol for the Project:
Social Learning in the Management of
Global Environmental Risks

This appendix constitutes the Social Learning Project's common research protocol and is organized as follows. Following an overview and definitions of terms, sections 1 through 6 describe the individual risk management functions that constitute the heart of the protocol. Section 7 describes changes in how the risk is defined and framed in relation to, and connection with, other risks. Section 8 summarizes how the research treats background trends that provide the larger social, political, and environmental contexts for the study.

Overview and Definitions of Terms

The research focuses on efforts to understand and manage the risks in question from the mid-1950s through 1992. Significant events in earlier time periods will be taken into account. The intent of the project is to understand the patterns, nature, and origins of long-term change in risk management over the entire period addressed. We neither exclude nor become preoccupied with the more dramatic events of the last several years.

The unit of analysis for the project is the performance over time of a specific risk management *function* by a specific *actor* in a specific *arena* (such as a country) in one of three *case studies* of global environmental risks. These terms are defined in the following paragraphs.

Case Studies
The three case studies pursued in the research are

• Global climate change (including greenhouse gas and other possible causes),

• Global stratospheric ozone changes (including all possible causes), and

• Acid rain.

The following working definition is used to describe the global environmental risk being studied under the heading *acid rain:*

Air pollution that travels long distance and harms valued environmental assets as a result of direct acidic or corrosive effects or through mobilization of harmful chemical reactions.

This definition encompasses the evolution that the term *acid rain* has gone through since the late 1960s: sulfur that acidifies lakes and kills aquatic life, sulfur and nitrogen that acidify soil and kill terrestrial vegetation (and corrode materials), and nitrogen and other ozone precursors that harm vegetation, materials, human health, and visibility. In short, it includes acid rain and tropospheric ozone that is caused by long-range transport of pollutants.

This definition excludes acidification and ozone-creation problems that are caused by purely local transport mechanisms. In the 1870s acidification from sulfur emissions was conceived of as a purely local problem: we do not have to go back 120 years to cover this aspect of acidification. In the 1940s and 1950s, smog was conceived of as a purely local problem: we do not have to go back to the 1940s either. Today, these problems are considered both local and long-range, which complicates the analysis. The complication stems from the nature of the problem, however, and it would be a mistake to try to simplify our task through definitional sleight of hand.

The question of what to include or exclude is fairly straightforward. The rule of thumb for deciding whether to include something under this heading is whether the answer to both of the following questions is yes:

• *Form of harm that the pollutant engenders* Does the pollutant create acid deposition or tropospheric ozone?

• *Transport mechanism of the pollutant* Do the pollutants in question travel long distances (over 100 km)?

In short, it includes the *union* of acidification and ozone-creation as it *intersects* with long-range transport.

Arenas
The arenas studied by the core project consist of nine nation-states, the European Community, and the international arena. The nation-states studied are

• The Federal Republic of Germany (including some work on the Democratic Republic of Germany),

• The United Kingdom,

• Netherlands,

- The former Soviet Union (including some work on individual republics),
- Hungary,
- Japan,
- Mexico,
- Canada, and
- The United States.

The work on the European Community focuses on the activities of the Community and its constituent organizations as distinct from those of its member states. The work on the international arena addresses both the formal activities of international governmental organizations plus international scientific meetings, nongovernmental organization (NGO) activities, and legal arrangements that span multiple countries.

Actors

Data collection focuses on actors belonging to six different social groupings:

- The expert science communities,
- Business and industry groups,
- Other nongovernmental organizations (primarily environmental groups),
- Executive branches of government,
- Legislative or parliamentary branches of government, and
- The media.

Functions

Six functions are used in the study to characterize the process of risk management. These are grounded in a number of previous functional studies carried out in the fields of international organizations and hazard management. The functions provide a basic template that help us to organize data collection to produce the comparative material that is required to trace social learning in the management of global environmental risks. Definitions of the functions adopted for use in the project are summarized immediately below and discussed in more detail in sections 1 through 6 of this appendix:

- *Risk assessment* Research on this function traces changes in understanding the nature, causes, consequences, likelihood, and timing of the risk in question. Particular attention is paid to the subset of all causes and consequences addressed by particular actors.
- *Monitoring* Research on this function traces the evolution of efforts by any of the actors to document actual changes in aspects of the environment affected by the risk in question, relevant emissions, human responses, and results of management strategies and specific implementation measures.
- *Option assessment* Research on this function documents and explains changes in the assessment of possible response options. Response options are particular measures that an actor might undertake to help manage a risk. Assessments of response options are systematic examinations of the feasibility, costs, and benefits of particular options.
- *Goal and strategy formulation* Research on this function traces changes in management goals, the design of a package of responses appropriate for achieving them, and the selection of modes (such as command and control, incentives, and persuasion) for implementing those responses. Goals are statements of objectives or of conditions that an actor wishes to bring about. Strategies are plans for how—in what combination and at what time—particular response options will be combined to achieve a goal. Strategies thus organize particular means (response options) to achieve particular ends (goals). Note that strategy formulation includes not only making choices from among different kinds of strategies but also determining the resources allocated to different response options (reflecting the option assessment) and to the strategy as a whole (reflecting issue framing). We explicitly consider decisions made about the place of R&D in risk management under this function.
- *Implementation* Research on this function traces changes in the actions actually taken by various social actors with regard to management of the risk in question. Implementation may include persuasion through normative pronouncements, educational activities, the exchange or dissemination of information, rule making, provision of incentives, supervision or enforcement of compliance, and coordination of programs.
- *Evaluation* Research on this function documents the conscious efforts of actors to reflect on and evaluate their own and others' performance in contributing to management of the risk under consideration.

The ordering of the functions here and in subsequent lists should not be interpreted as an assumption that the functions are performed sequentially by first performing risk assessments, then monitoring, and so on. On the contrary, one of the goals of the research is to establish empirically which functions are performed when; and how the performance of one affects that of another.

For each of the six risk management functions described above, the generic questions to be addressed in the empirical work are defined by the learning-as-evolution model. In particular, we want to discover information on the following processes:

• *Innovation* What new ideas (facts, theories, values) emerged as relevant to performance of the specific function. When did they emerge? From what domestic institutions (the science community, NGOs, the private sector, and so on) or foreign sources did these emerge? In what institutions did they take hold?

• *Selection* Which choices were made in narrowing the range of possibilities presented by innovation? (For example, which of the control options suggested for limiting carbon dioxide release disappeared from broader debate? Which achieved wider acceptance among a larger group of domestic institutions?) What was the timing of these choices? What arguments, evidence, or other factors conditioned the choices? What institutions were important in influencing the choices?

• *Diffusion* Which ideas, policy proposals, or world views that survived the selection processes within a given state were successfully exported to other states? When did this diffusion occur? Which domestic institutions were involved as exporters? What foreign institutions were the most receptive targets for their exports?

The individual functions are described in greater detail in sections 1 through 6, below. Each of those sections is organized as follows:

• *Purpose and scope* A general statement about why we are gathering data on this function, our general approach to it, and notes on special difficulties or pitfalls expected in the work;

• *Indicators* A statement of the evidence that is required for characterizing performance of the function;

• *General questions* Broad questions that attempt to capture the pivotal issues that are to be understood through research on the functions, that elaborate on the definition of functions for the particular case of global environmental risks, and that help to provide a rationale for the selection of data; and

• *Research questions* Specific questions that each country team has agreed to examine and report on. The emphasis is on *changes* in functional performance rather than on the functions themselves. The research questions do not necessarily focus on the most important issues (importance is something the research needs to discover, not assume) but rather on issues that seem likely to illuminate the overall question of social learning. In most cases, the questions are presented in an order reflecting the priority assigned for allocation of research effort by

the project team. (This means that participants have agreed to ensure that everyone addresses, and thus can provide comparable data for, questions at the top of the list; they address those toward the bottom of the list as time and resources permit.)

Section 1: The Function "Risk Assessment"

Purpose and Scope
The purpose of research on the risk-assessment function is to trace changes in understanding the nature, causes, consequences, likelihood, and timing of the risk in question. Particular attention is paid to the subset of all causes and consequences addressed by particular actors.

Indicators

• Time at which a certain way of characterizing the risk becomes established;

• Period over which debates about alternative characterizations range, changes in the "dominance" of alternative views, and time at which such debates are resolved;

• Estimates of probability or timing of risk;

• Origin of innovations in characterizations and estimations (note that the origin could be in the risk assessment of a different issue or country or actor); and

• Confirmation of the risk assessment (for example, by studies with other models, at international meetings, in publications, and so on).

General Questions
General questions should be examined for each actor group based on publications and interviews. Some actor groups (especially the scientific experts) will have looked at most or perhaps even all aspects. For other actors we shall not find answers to all questions.

1. What does the actor consider to be the emissions and human activities responsible for the risk in question?

2. What does the actor believe the nature and size of the risk are?

3. What does the actor think the impacts of the risk are or will be?

Research Questions
Research questions should be examined for each actor group based on publications and interviews. Some actor groups (especially the scientific experts) will have looked at most or perhaps even all aspects. For other actors we shall not find answers to all questions. (Note that for this

function, an initial effort has been made both to link the general questions to the research questions and to provide research questions for each of the three risks addressed in the study. As a result, the research questions are listed in logical rather than priority order. An additional section therefore follows this one in which the research questions for the function "risk assessment" and the case of global climate change are listed in order of priority for attention by the research teams.)

1. What does the actor consider to be the emissions and human activities responsible for the risk in question?

1.1. Which emissions play a role in changing the atmosphere? What are the proportions of CO_2 and non-CO_2 greenhouse gases? Note the belated inclusion of non-CO_2 gases in global climate change assessments.

1.1.1. Which emissions are involved in stratospheric ozone depletion? What are thought to be the most important ozone-depleting gases (ODGs)?

1.1.2. Which emissions are involved in the acid and oxidant pollution debate? Is it mostly seen as a sulfur dioxide problem? What role do nitrogen oxides and volatile organic compounds play?

1.2. Which human activities lead to these emissions?

1.2.1. What are the relative contributions of energy use, industrial activity, deforestation, agriculture, population increase, economic growth, and so on? Note that all activities do not apply to each risk, though they may be perceived as relevant by some actors at some times.

1.2.2. Historically, what have been the absolute and relative contributions of energy use, industrial activity, deforestation, agriculture, population increase, economic growth, and so on. Note that all activities do not apply to each risk, though they may be perceived as relevant by some actors at some times.

1.2.3. In the future, what are expected to be the absolute and relative contributions of energy use, industrial activity, deforestation, agriculture, population increase, economic growth, and so on? Note that all activities do not apply to each risk, although they may be perceived as relevant by some actors at some times.

2. What does the actor believe the nature and size of the risk are?

2.1. What proportion of the relevant gases emitted by human activities remain in the atmosphere for what amount of time?

2.1.1. How much carbon dioxide emitted by human activities remains in the atmosphere for what time period? Attention here should focus on debates about "the airborne fraction," relative roles of terrestrial biosphere and oceans, "missing carbon sink," and so on.

2.1.2. How much SO_x and NO_x remains in the atmosphere and for how long? Wet and dry deposition, as well as transport distances, should be covered here.

2.1.3. How much of the various ozone-depleting substances remain in the atmosphere for what period of time?

2.2. How do relevant properties of the atmosphere (climate, stratospheric ozone, tropospheric acidity, and oxidant potential) respond to the emissions? What are the uncertainties and their sources?

2.2.1. How sensitive is the climate to increasing concentrations of the greenhouse gases? How large will the temperature change be for a given change in concentrations of greenhouse gases? Is this quantity estimated empirically or through models?

2.2.2. How much of the predicted global climate change is a result of emissions from each of the greenhouse gases identified in 1.1.1 above (such as carbon dioxide and chlorofluorocarbons). This question should track debates about indices of "global warming potential."

2.2.3. How is the stratospheric ozone layer expected to respond to increasing concentrations of the ozone-depleting gases identified in 1.1.2 above?

2.2.4. Which gases are thought to be responsible for what proportion of stratospheric ozone depletion? This question should track debates about the relative ozone-depleting potential (ODP) of various gases.

2.2.5. How are gases involved in changing the acid and oxidant properties of the atmosphere distributed through the atmosphere? What are their relative contributions?

2.3. Over which period of time do the atmospheric responses noted in 2.2 occur?

2.3.1. How soon is significant global climate change expected?

2.3.2. How long will stratospheric ozone depletion continue after ODG emissions are reduced?

2.4. What secondary changes in the global environment are thought to result from these primary atmospheric changes? What are their potential magnitudes? How long will they persist after emissions are reduced?

2.4.1. Sea-level rise;

2.4.2. Abrupt (nonlinear) responses of the earth's system to risk;

2.4.3. Increased ultraviolet radiation at the earth's surface;

2.4.4. Stratospheric cooling;

2.4.5. Forest dieback, lake and soil acidification;

2.4.6. Damage to biota (such as forest dieback).

3. What does the actor think the impacts of atmospheric change are or will be?

3.1. What are likely to be the most significant impacts of atmospheric change on society? How serious are they?

3.1.1. Human health (skin cancer, suppression of immune response, cataracts);

3.1.2. Unmanaged ecosystems (plankton);

3.1.3. Agriculture, forestry, and fisheries;

3.1.4. Material damage (such as flooding, corrosion, damage to cultural artifacts like monuments and books, degradation of paints and plastics).

3.2. Who are thought to be the winners and losers as a result of these impacts?

3.2.1. What serious references have been made to possible changes in the comparative advantages of nations as a result of the risks in question?

Section 2: The Function "Monitoring"

Purpose and Scope

The purpose of our research on the monitoring function is to trace the evolution of efforts by any of the actors to document actual changes in the following:

• Aspects of the environment affected by the risk in question (such as global temperature change, sea level, concentrations of gases in the atmosphere, natural sources, and sinks of greenhouse gases);

• Relevant emissions (such as CO_2 or CFCs);

• Human responses (such as adaptation to the risk or changes in behavior) to the risk in question; and

• The results (such as changes in public opinion or in consumption patterns) of management strategies and specific implementation measures (such as monitoring compliance). Note that in this case one actor (such as a nongovernmental organization) can monitor the results of measures taken by another actor (industry or the state).

Indicators

• Time at which particular types of monitoring systems are first put in place by an actor;

• Origin of innovations in monitoring; and

• Time at which particular types of monitoring data are first used (collected, reported, averaged, and so on) by an actor, whether or not the actor created it in the first place.

General Questions

For each question, attention should be given to the existence of monitoring programs and also to their technical adequacy and findings:

1. What provisions have been made for monitoring the emissions relevant to the risk in question?

2. What provisions have been made for monitoring the human activities leading to the emissions?

3. What provisions have been made for monitoring changes in the natural system?

4. What provisions have been made for monitoring the impacts of the risk in question?

5. What provisions have been made for monitoring the human responses to the risk in question?

Research Questions

These expand on the general questions given above.

1. What provisions have been made for monitoring:

1.1. The emissions of greenhouse gases, sulphur dioxide, NO_x, CFCs, particles, and so on?

1.2. The emissions of ODGs in the atmosphere?

2. What provisions have been made for monitoring the following human activities:

2.1. Energy use;

2.2. Changes in land use through deforestation, agriculture, urbanization, agricultural practices (such as cattle or pig rearing);

2.3. Use of particular products (such as spray cans, air conditioning, or solvents);

2.4. Fossil fuel combustion;

2.5. Intensive agriculture; and

2.6. Transportation?

3. What provisions have been made for monitoring changes in the following:

3.1. Temperature;

3.2. The concentrations of greenhouse gases, SO_2, NO_x, CFCs, particles, and stratospheric ozone;

3.3. The extent of snow and ice;

3.4. The sea level;

3.5. The cloud distribution;

3.6. Natural sources and sinks of relevant gases (such as volcanoes and biomass);

3.7. Ecosystems and biodiversity;

3.8. Ultraviolet radiation reaching the earth's surface;

3.9. Concentrations of ODGs in the atmosphere;

3.10. The pH of precipitation, water, and soils;

3.11. The extent of forest damage?

4. What provisions have been made for monitoring the impacts of stratospheric ozone depletion? Research could look at the impacts on:

4.1. Human mortality (malignant melanoma skin cancer);

4.2. Human morbidity (nonmelanoma skin cancer, eye disorders, suppression of immune responses);

4.3. Ecosystem effects (decreases in fecundity and plankton growth);

4.4. Reduced crop productivity and accelerated degradation of plastics and paints.

5. What provisions have been made for monitoring the following human responses:

5.1. Changes of public opinion;

5.2. Changes of consumer behavior;

5.3. Other adjustments (such as migration, wearing of sunglasses, hats or sunscreen);

5.4. Changes in agriculture, fisheries, and forestry practices;

5.5. Compliance with specific implementation measures (such as emission controls, standards, speed limits, banning of spray cans, recycling, or regulation of refrigerators)?

Section 3: The Function "Option Assessment"

Purpose and Scope

The purpose of research on the option assessment function is to document and explain changes in the assessment of possible response options, including the likely effectiveness and cost of such options.

Response options are particular measures that an actor might undertake to help manage a risk. Generally, response options may be technological, organizational, or behavioral in nature. Examples include switching from coal to nuclear fuels for electricity generation, taxing carbon emissions, issuing tradable permits allocating a given total emission budget, and launching a media campaign to make consumers more aware of the consequences of their decisions.

Assessments of response options are systematic examinations of the feasibility, costs, and benefits of particular measures. Assessments do not necessarily involve conclusions or recommendations regarding "best" options. For example, the U.S. Congressional Office of Technology Assessment (OTA) completed an assessment of major response options for dealing with the risk of climate change, focusing on the building, transport, and energy sectors. The OTA report makes no recommendations but tries to compare the costs and benefits of alternative response options in as unbiased a manner as possible.

Reminders:

• Responses should be considered for all actors, avoiding a bias toward the state. Research should show how each actor has assessed the likely effectiveness and costs of response options for managing the risk (for example, "We seek to document the emergence of proposals from the expert community that planting forests would allow the recovery of x tons of atmospheric carbon dioxide per hectare of area planted.").

• Changes through time in response functions, rather than responses at any specific time, are the focus of the study (for example, "We seek to document and explain the change from emission-reduction discussions focused on carbon dioxide and fossil fuels alone in 1979 to discussions focused on all the greenhouse gases and deforestation in 1985.").

• Options should include the full range of possible responses, including technical, institutional, and behavioral measures to alter emissions, environmental changes, and impacts. However, we should focus on specific options only when some actor has explicitly identified as possible means of reducing the risk at hand. (Thus, for example, we would not be interested in documenting the cost-effectiveness of nuclear energy until nuclear has been presented by some actor as an option for reducing the risk of climate warming.)

• Costs should be interpreted broadly to include financial, political, and other opportunity costs. We recognize, however, that formal cost assessments may occur very late, if at all, in the history of some assessment efforts. Much of the story to be told may therefore consist merely of documenting and explaining the appearance of different options on the agenda of response discussions.

Indicators

• Time at which specific response options first enter the debate;

• Time at which serious cost estimates are first applied to these options;

• Period over which debates about superiority of alternative options range, changes in the "dominance" of alternative views, and time at which such debates are resolved;

• Proportion of the "full range" of options given active consideration at any particular time; balance among technical, institutional, and behavioral options;

• Origin of innovations in options and their assessments; and

• Object of response options (whether domestic, bilateral, or international).

General Questions

1. What are the most important options considered by each actor for changing the emissions thought to be responsible for the risk in question? What are the expected costs and benefits? (Examples include banning the use of CFCs in spray cans, increasing energy efficiency, and fuel switching to reduce carbon dioxide emissions.)

2. What are the most important options considered by each actor for recovering gases thought to be responsible for the risk in question after they have been emitted? What are the expected costs and benefits? (Examples include planting trees to absorb carbon dioxide and scrubbing sulfur dioxide from stack gases.)

3. What are the most important options considered by each actor for changing the environment in ways that directly counter the effects of emissions? What are the expected costs and benefits? (Examples include liming acidified lakes and cooling the planet with solar reflectors.)

4. What are the most important options considered by each actor for adapting to the environmental changes associated with the risk in question? What are the expected costs and benefits? (Examples include shifting crop zones in response to climate change and wearing hats and sunglasses to protect human eyes from increased ultraviolet radiation.)

Research Questions

These questions focus on the climate case. They are presented in order of priority for research attention by the study teams.

1. What are the most important options considered by each actor that would

1.1. Change the carbon intensity of its energy system in ways relevant to the greenhouse risk? What are thought by each actor to be the likely costs, benefits, and constraints for these options? (These options would presumably include fuel-switching options, renewable resources, nuclear energy, and the like. Note that it may be appropriate to address options contemplated as solutions to other problems that would increase carbon intensity, such as synfuels.)

1.2. Change the emissions of ozone-depleting gases? (These options would presumably include switching from CFCs to propellants with less or no ozone-depleting potential in spray cans, recycling of refrigerants, and substituting solvents. Note that we could also look at options

contemplated as solutions to other problems that would increase emissions of ODGs: increasing insulation to reduce energy demand could increase CFC use and emissions, if CFCs are used in the production of insulating foams.)

2. What are the most important options considered by each actor that would change energy efficiency in ways relevant to the greenhouse risks? What are thought by each actor to be the likely costs, benefits, and constraints for those options? How do answers to this question change with time? (These options would presumably include potential improvements in the building, energy, transportation, and industrial sectors. Note that it may be appropriate to address options contemplated as solutions to other problems that would decrease energy efficiency.)

3. How quickly, how completely, and at what cost can research be expected to provide more usable knowledge for future decisions? (This research question would try to characterize changing assessments of the potential of research to illuminate central uncertainties in understanding the risk.)

4. What are the most important options considered by each actor that would change public or consumer behavior in ways relevant to the risks of climate change? What are thought by each actor to be the likely costs, benefits, and constraints for those options? How do answers to this question change with time? (This research question would include campaigns to make people more aware of the greenhouse consequences of their activities and to persuade consumers to use pump sprays instead of aerosol cans, to bring their refrigerators to the recycling center rather than throwing them into a dump, and so on.)

5. What options involving the planting of trees might play a role in this society's response to the greenhouse risk? What are the likely costs, benefits, and constraints for such options? (This research question would presumably address the wild claims and counterclaims about the potential for sequestering carbon in trees. The questions of disposal of the carbon in grown trees and the possible space and cost constraints might be central.) What options have been considered by each action that would remove ODGs from the atmosphere?

6. What are the most important options considered by each actor for adapting to climate change? Where does each actor expect adaptation to climate change to be relatively effective and inexpensive? Where, in contrast, is adaptation thought likely to be ineffective or prohibitively expensive? (This research question would try to assess the relative popularity of various measures, as well as their likely performance.)

7. What are the most important options considered by each actor that would change population growth in ways relevant to the risks of climate change? What are thought

by each actor to be the likely costs, benefits, and constraints for those options? How do answers to this question change with time? (This research question would presumably address the emergence of debates that name population control as an important means of reducing greenhouse risks. It could in principal deal with assessments of how much reduction of greenhouse emissions could be achieved by particular population control activities. Such assessments, however, seem to have been very rare to date.)

8. What are the most important options considered by each actor for controlling the emissions of greenhouse gases other than carbon dioxide? What are thought by each actor to be the likely costs, benefits, and constraints for these options? How do answers to this question change with time? (This research question would presumably address the emergence of debates on changing agricultural practices to cope with methane emissions and other greenhouse gases.)

9. What other "geoengineering" options have been considered? What are the likely costs, benefits and constraints for such options? (Presumably, this research question will discuss sun shields, fertilized oceans, and so on.)

Section 4: The Function "Goal and Strategy Formulation"

Purpose and Scope

The purpose of research on the goal and strategy formulation function is to trace changes in management goals, the design of a package of responses appropriate for achieving them, and the selection of modes (such as command and control, incentives, or persuasion) for implementing those responses.

Goals are statements of objectives or of conditions that an actor wishes to bring about (for example, "To reduce carbon dioxide emissions by x percent before year y"; "To limit ultimate temperature increase to 1 degree centigrade above 1985 levels."). *Strategies* are plans for how—in what combination and at what time—particular response options will be combined to achieve a goal. Strategies thus organize particular means (response options) to achieve particular ends (goals). (For example, one strategy for reaching the goal suggested above could be to require that all new electricity-generating capacity be nuclear fueled, that existing fossil-fueled electricity-generating facilities be required to improve their efficiency by 2 percent per year, that gasoline taxes increase by 2 percent per year, and that a tradable-permit scheme be implemented to allow efficient allocation of preventive efforts.)

Note that strategy formulation includes not only making choices from among different kinds of strategies but

also determining the resources allocated to different response options (reflecting the option assessment) and to the strategy as a whole (reflecting issue framing). We explicitly consider decisions made about the place of R&D in risk management under this function.

Reminders:

• Goals and strategies for managing global environmental risks may be formulated by all actors and not just by governments. We should adopt a balanced research approach that examines all actors.

• Goals tend to be of various types: high level and general or very specific. They can form hierarchies that are not necessarily found as hierarchical trees in published materials. Goals from various levels can relate to other activities.

• Goals are usually cast with an explicit time horizon (for example, emission stabilization at 1985 levels by the year 2010). We should be careful to document the presence or absence of such horizons.

• Strategies will depend on actors' frames and perspectives. Careful reference to "issue framing" should accompany research on this "goal and strategy formulation" function.

• Explanations should distinguish the extent to which goals or strategies are driven by direct concern for the risk, broader political or economic concerns, or other policies.

Indicators

• Time at which a certain goal or strategy is first established;

• Target of goal or strategy, such as a particular domestic actor, national policy, bilateral relationship, or international relationship;

• Period over which debates about alternatives range, changes in the "dominance" of alternative views, and time at which such debates are resolved; and

• Origin of innovations in goal or strategy formulation.

General Questions

1. What goals for managing the risk in question have been proposed? Research on this question might distinguish among the classic risk management goals of Kates et al. (1985)—that is:

1.1. Risk acceptance (deciding to let it happen);

1.2. Risk spreading (such as insurance schemes);

1.3. Risk reduction or prevention; or

1.4. Adjustment or adaptation to the consequences of risk.

2. What strategies regarding reduction or prevention of the emissions responsible for the risk in question have been proposed?

3. What strategies regarding adaptation or adjustment to the impacts of the risk in question have been proposed?

4. What strategies for gathering additional information to inform management choices have been proposed?

5. What has been the relationship among the potential elements of management strategy noted above?

Research Questions

1. What goals for managing the risk have been proposed? (For example, climate change research might look at the development of goals such as "stabilization," "no regrets," "Vorsorgeprinzip," "20 percent reductions," and the like. Ozone research might look at the developments of goals such as 30 percent CFC-production reduction. Acid rain research might look at 30 percent reduction of sulfur dioxide emissions and not exceeding defined critical loads.)

2. What strategies regarding emission reduction or prevention have been most strongly advocated by each actor? (This should distinguish between efficiency-enhancing and fuel-switching options.)

3. What, if any, strategies for avoiding action on the risk have been most strongly advocated by each actor? (This might note the use of research programs as excuses for doing nothing or the refusal to be drawn into international negotiations as an excuse for avoiding pressure to make commitments. For example, German delegations were told not to bring up the subject of acid rain at international meetings in the 1970s, and research to reduce uncertainties can delay decision making.)

4. What strategies for offsetting the risk have been advocated by each actor? (Examples include liming lakes and soils.)

5. What strategies regarding adaptation or adjustment to the effect of the risk have been most strongly advocated by each actor? (This strategy might note the difference between societies focusing on sea-level-oriented responses and societies focusing on agriculture.)

6. What (if any) balance between risk reduction and risk adaptation options has been explicitly struck by each actor in the debate over goals and strategies?

7. What has been the position of research and development options in the overall strategy? What resources were allocated to R&D tasks?

8. What use have actors made of the risk as an opportunity for social, economic, and technological innovation?

9. What strategies for increasing the capacity to directly manage the risk have been advocated by each actor? (Examples include carrying out public-opinion campaigns to get people worried about acidification and monitoring transport of acidifying substances.)

10. What strategies for increasing social capacity to manage the risk have been advocated by each actor? (Examples include establishing a public-outreach division in an environment ministry and establishing an international convention.)

Section 5: The Function "Implementation"

Purpose and Scope
The purpose of research on the implementation function is to trace changes in the actions actually taken by various social actors with regard to the risk in question. Implementation may include persuasion through normative pronouncements, educational activities and the exchange or dissemination of information, rule making, incentives, supervision or enforcement of compliance, and coordination of programs.

Reminders:

• Each actor will have a different position relative to implementation, and formal rule-making actions and authority may vary from country to country. For the moment, we will seek data on implementation for each actor, without rigidly structuring a view of *the* implementation process or the position of formal rule making within it.

• Research on implementation should not be restricted to preventive actions but should include adaptation or adjustment actions as well. (For example, the city of Charleston, South Carolina, has passed a new zoning plan to adjust coastal development back from the ocean front in order to avoid the need to build sea walls later on. This type of adjustment action should receive as much attention as does the passage of carbon tax laws.)

Indicators

• Time at which implementation actions are initiated by an actor;

• Estimates (by actor) of the time required to obtain a specified amount of adoption, or compliance with, an implemented action;

• Changes in the domain of the implemented action— that is, degree of adoption or recognition of implementation initiatives by other actors;

• Origins (internal or from other actors) of pressures to adopt those actions that are in fact implemented; and

• Object toward which implemented actions are directed (distinguishing among domestic actors, national policy, and bilateral or international relations) and extent to which this audience was in fact reached.

General Questions

1. What rules, incentives, or other instruments relevant to managing the risk in question have actually been undertaken? To what extent have such measures actually been complied with? What is the timing of first adoption, first compliance, and full compliance?

2. What efforts to organize or coordinate action on the risk in question have been undertaken by the various actors?

3. What major activities of education or information dissemination about the risk at hand have been undertaken by the various actors?

4. What important changes in technologies and production processes relevant to managing the risk in question have actually been adopted by various actors?

5. What normative pronouncements have various actors made regarding the risk at hand?

6. What major research and development programs were funded regarding the risk in question, and who funded them? What changes in funding have occurred?

Research Questions
These questions focus on the risk of climate change. They are listed in order of priority for research attention by the study teams.

1. What rules, incentives, or other instruments relevant to managing the risk of climate change have actually been undertaken? To what extent have such measures actually been complied with? What is the timing of first adoption, first compliance, and full compliance? (This research question could include zoning decisions affecting exposure to sea-level rise and government requirements for increased fuel efficiency. Note that we could include here either all measures that actually affect the risk or only measures explicitly implemented to address it. For example, fuel-efficiency standards have had a major impact on greenhouse emissions, though in many countries they have been adopted for other air-pollution reasons. If the latter, we should be sure that the "origin" indicator notes that the measure was imported from some "other" policy debate.)

2. What efforts to organize or coordinate activities on the risk of climate change have been undertaken by the various actors? (This research question could include internal reorganizations, such as setting up a Climate Office for the first time in a ministry or NGO, a new department in a university, or an Ozone Group; commissions and coordinating councils in the government or private sector, such as the Committee on Earth Sciences in the U.S. government and the Interministerial Working Group on Implementation of Carbon Dioxide Reductions in Germany; planning exercises in the science community; formal agreements between sectors; new programs in international organizations; and joint NGO campaigns.)

3. What other major activities of information dissemination about the risk of climate change have been undertaken by the various actors? (This research question might identify professional meetings, popular books, public hearings, television shows, press conferences, acid rain tours, school and university curricula, information packets, and petitions. Examples would include the publication of the German Enquete Commission Report in English.)

4. Which of the most important changes in technologies and production processes, relevant to managing the risk of climate change, have actually been adopted by various actors? (This research question could include recycling technologies, fuel shifts, efficiency measures, production of new semisafe fluorocarbons, flue-gas desulfurization, and catalytic converters. An example could be the production of relatively benign CFCs by DuPont.)

5. What normative pronouncements have various actors made regarding the risk at hand? (This research question might include norms for individual sectors or countries. It could include nonspecific advice to the world as a whole. This should include norms applied to preventive and adaptive options. Note that for many actors, such as international organizations or national presidents, normative pronouncements may be the primary action channel available. For example, a government minister's statement that emissions of carbon dioxide should be reduced by 25 percent by a particular year or that emissions of SO_2 should be reduced by 30 percent by a certain point in time.)

6. What major research and development programs were funded regarding the risk in question and by whom? What changes in funding have occurred? (Relevant indicators include time of initiation, amount of initial support, changes in budget, special constraints and termination of the program, participants, and policy or technology focus of program.)

Section 6: The Function "Evaluation"

Purpose and Scope

The purpose of research on the evaluation function is to document efforts of actors to reflect on and evaluate their own and others' performance in contributing to management of the risk under consideration.

We define *evaluation* broadly to include any form of conscious feedback between observations, actions, and objectives. Some degree of evaluation of performance is already embedded in other functions of the research protocol and provides a powerful social learning mechanism. It is useful to distinguish this function from the others and to pay explicit attention to it, especially since reflective activities may otherwise not receive systematic comparative attention. What we are after here is a determination of the degree to which actors both reflect on their performance of the other management functions over time and also learn better evaluation skills and practices.

Reminders:

• Care should be taken to distinguish between risk assessment and evaluation. The IPCC, for example, performed an assessment. A retrospective analysis of the IPCC process, with a view toward identifying its strengths and weaknesses as a mechanism for performing assessments, would be an evaluation.

• Evaluative effort may develop through the following phases: initiation, determination of terms of reference, conduct of analysis, and dissemination of results. Different actors can be involved in different phases.

Indicators

• Time at which a particular (kind of) evaluation was done;

• Time at which the results of an evaluation show up in relation to a particular function and particular actors;

• Time at which a particular type of evaluation becomes institutionalized;

• Timing and location (which actors at which place) of innovation in evaluation methods and procedures; and

• Timing, location, and nature of the evaluation of the human dimensions of the risk in question and the social science research needs.

General Questions

1. What have been the deliberate evaluations initiated in relation to acid and oxidant pollution, ozone depletion, and climate change? (Examples inlude evaluation by the IPCC, the ECE sulfur oxide targets, the Montreal Protocol, and the German Enquete Commission.)

2. Were the evaluations made by experts, corporate actors, journalists, or politicians or by combinations of these groups?

3. Which actors (scientific experts, other professionals, or political actors) dominate the successive parts of the evaluation (initiation, terms of reference, conduct of analysis, and dissemination)?

4. Who was the target of the evaluation (researchers, parliaments, and so on), and who was actually reached (the general public, scientists at the international level, and so on)?

5. At which level and with what scope were the evaluations attempted (in relation to goals, mechanisms, and instruments)?

6. To what extent were the evaluations ad hoc? To what extent did they create or utilize new institutional mechanisms?

7. Do the evaluations straddle more than one functional category or fall within one?

8. What innovations (elsewhere in the system) may facilitate (alter conditions of) evaluation? (For example, does the U.S. Freedom of Information Act assist public groups or the media in the United States?)

9. What is the arena of evaluation? (Examples include the Commission of Inquiry and Corporate Strategy Group.)

10. What purposes underlay, and what factors prompted, the evaluation?

11. What are the effects of the evaluations? Are they deliberate or accidental? Over what time scale do they develop? Do they have practical impacts (such as the adoption of new regulations or research programs)? Do they affect perceptions or assumptions underlying the risk management debate?

12. Are actors trying to avoid evaluation or to avoid acting on the results of evaluation? By what means?

13. Is there reflection on the roles actually played by various actors in the risk management process?

14. Have evaluative efforts themselves been evaluated? How?

Research Questions

It is not clear how widespread global environmental issues are evaluated. The general questions listed above explore relevant instances of evaluation. However, to do a comparative analysis of evaluation we need evidence of the types of evaluations, embedded or explicit, that have been used in the various cases and countries. Therefore, in the following research questions the meaning of *evaluation* is interpreted broadly:

1. What have been the most important evaluations of national or international efforts to manage the global environmental risks addressed in this study? When did they occur?

2. What actors were involved in initiating, conducting, and disseminating the results of these evaluations?

3. What audiences were targeted by the evaluation?

4. What resources were devoted to the evaluation?

5. What was the scope of the evaluations (actors' roles, level, issue)?

6. What evaluation mechanisms were used (consensus meetings or formal assessments of research returns)?

7. Was the evaluation accompanied by prescriptions or recommendations?

8. How did the evaluations feed back into the policy process?

9. What actions were derived from evaluations and by whom?

Section 7: Issue Framing

Purpose and Scope
In addition to collecting data on risk management functions, it is necessary to describe changes in how the risk is defined and framed in relation to other risks and issues. For each actor, this involves tracing the emergence of the risk onto that actor's agenda and its relative rank or position on the agenda through time. (We restrict ourselves to actors' environmental agendas and do not try to determine how the risk in question ranks relative to, say, defense or housing issues.) Research here also traces the images associated with the risk and its connection to other risks and issues as these connections and images change over time.

Reminders:

• Even if an actor believes scientific evidence for a risk to be weak, the risk could still occupy a high place on that actor's agenda if it pursues efforts to convince others not to take mistaken policy actions to combat the risk.

• In some cases it may be appropriate to distinguish an actor's personal framing of the issue from its perception of the way the issue is framed by institutions, nations, and the international community.

Indicators

• Dates at which issues, images, and risks emerge (note rank, connections to other issues, and rates of change);

• The permanency and saliency of the risk on actors' agendas as reflected in how often the risk is mentioned in plans, reports, and publicity; the level of commitment to the risk, as indicated in resource budgets, strength of language, and amount of space (note that these indicators may sometimes be misleading); and

• The origins of changes in issues, connections, and rankings in terms of self, other domestic actors, foreign actors, international institutions; journals, media, conferences, and personal contacts; important background events (such as drought).

General Questions

1. When and how does each actor recognize the risk for the first time?

2. How does the actor define the issue over time?

3. With what images do actors associate the issue, and how do these change over time?

4. How do actors rank and relate the risks addressed in this study (acid and oxidant pollution, ozone depletion, and climate change) to each other over time?

5. What is the position of the risk on each actor's overall (environmental) agenda?

Research Questions

1. When, and how, is the risk first mentioned (positively or negatively) as a research question, risk, or policy issue?

2. Did the risk emerge in its own right or as a subsidiary of another issue? If a subsidiary, when did it emerge independently? How do the connections between issues change over time? (For example, the risk of greenhouse-induced climate change first appears in the U.S. debate as a subsidiary issue, riding on higher-priority concerns for intentional weather modification or possible supersonic transport risks. We might examine when the greenhouse risk became the subject of government hearings and newspaper articles and was awarded its own budget. In the United States, the greenhouse risk did not have such independent status for most actors in the 1960s and did have it for most actors by the 1980s. In between, something happened, and we need to discover what it was. In Germany the acidification issue emerged as the health impacts of local air pollution were discussed.)

3. What images (pictures, cartoons, and slogans) of the risk are most powerful and widespread? Do they originate in the media or in other actors' public statements and reports? (This might, at different times, address Venus, the "great geophysical experiment," waves lapping on the Washington monument, the ozone hole, spray cans, and "ozone killers.")

4. How are our three risks ranked, interrelated, differentiated, combined, or confused with each other (or with related climate change and air-pollution issues) over time?

Section 8: Background Trends

In addition to collecting data on risk management functions, it is necessary to collect comparative data that allows us to place these functions in a broader context. These contextual trends may form part of the explanatory background of what we pick up under the risk management functions. Work in this area involves the documentation of major events, trends, and processes (economic, environmental, political, and institutional) that may be relevant to social learning about the risks. It includes setting important actors in their contexts (for example, describing the changing role of NGOs within society at large).

Examples of key events, trends, and processes are:

• Changes in the nature and quantity of energy and agricultural production;

• Major environmental changes (such as droughts and forest dieback);

• Economic recessions;

• Changing political economic relationships (such as the formation of the European Community and the dissolution of COMECON);

• Significant technological shifts (such as CFC alternatives, global telecommunications, and technological accidents);

• The rise of new social movements and coalitions (such as environmental movements and international scientific links); and

• Important cultural changes (such as television, alienation, religion, criticism, vegetarianism).

The initial research in this area will consist of the construction of a chronology of trends that seem to be important in the individual country studies.

Appendix B
About the Authors

Jeannine Cavender-Bares is a biologist at the Smithsonian Environmental Research Center in the United States and an Adjunct Faculty Member at Georgetown University. Her research focuses on plant physiology and evolution and the ecological impacts of global change.

William C. Clark is a professor at Harvard's Kennedy School of Government in the United States. His research focuses on how technical information mediates the interactions of environment, development, and security concerns in international affairs.

Ellis B. Cowling is a forest biologist at North Carolina State University in the United States. His research focuses on the response of terrestrial and aquatic ecosystems to human-induced changes in the chemical climate of North America and Europe.

Nancy M. Dickson is a Senior Research Associate at Harvard's Kennedy School of Government in the United States. Her research explores the role of assessment as a bridge between science and policy and focuses on large-scale environmental change.

Gerda Dinkelman is a political scientist. She works for the Advisory Council for Science and Technology Policy (AWT), an advisory body to the Dutch government.

Rodney Dobell is Professor of Public Policy at the University of Victoria in Canada. His current research focuses on citizen agency and institutions of government in shaping the flow of knowledge into public or individual action.

Renate Ell is a freelance science journalist working with German press and radio on environmental topics. She is also an adjunct research assistant at the Wuppertal Institute for Climate, Environment, and Energy.

Adam Fenech is an Associate Faculty Member at the Institute for Environmental Studies at the University of Toronto, and a Senior Science Advisor at the Meteorological Service of Canada.

Alexandre S. Ginzburg is a Senior Research Fellow at the Institute of Atmospheric Physics at the Russian Academy of Sciences. His research focuses on urbanization, urban climate, and environment. He is a professor of climate dynamics at the International University in Moscow.

Elena Goncharova is a journalist in Russia. She worked for many years for the Russian Press Agency "News" and now writes for several Russian magazines and newspapers on social and economic issues.

Peter M. Haas is a professor of political science at University of Massachusetts at Amherst in the United States. He studies international environmental governance and the interplay of international institutions and scientific knowledge in shaping how states address transboundary environmental threats.

Éva Hizsnyik is an economist at the Budapest Institute for Environmental Studies in Hungary. Her work is related to methodologies of integrated environmental assessments and their applications in climate change and other global environmental issues.

Michael Huber is employed at the Institute of Sociology at the University of Hamburg in Germany. His research focuses on European regulatory policy making in the field of the environment and problems of organizational learning.

Peter Hughes is Senior Lecturer and Programme Leader for Environmental Studies at the University of Sunderland in the United Kingdom. His research and teaching focus on the relationship between environment and development processes from local to global scales.

Jill Jäger is Executive Director of the International Human Dimensions Programme in Germany. Her research focuses on the design of assessment processes. She has a strong interest in developing effective interdisciplinary and international research projects dealing with global environmental issues.

Marc A. Levy is Associate Director for Science Applications at the Center for International Earth Science Information Network (CIESIN) at the Columbia University Earth Institute in the United States. His research is on institutional effectiveness and environment-security connections.

Angela Liberatore works in the Research Directorate General of the European Commission in Belgium. She focuses on the research-policy interface, environmental policy, European governance and citizenship, and new models of development, and is the author of a book on Chernobyl.

Diana Liverman is Professor of Geography and Director of the Latin American Studies Program at the University of Arizona in the United States. Her research focuses on the social causes, consequences, and policy responses to environmental change in Latin America.

Justin Longo is a Ph.D. student at the University of British Columbia in Canada. His current research focuses on the validity and relevance of policies as perceived by the individuals whose compliance with the intentions raises important questions for implementation.

David McCabe is a Ph.D. candidate in the Department of Political Science at the University of Massachusetts at Amherst in the United States. His dissertation focuses on the role of new information on state interest formation and international collaboration to limit chemical warfare.

Donald Munton is professor and chair of the International Studies Program at the University of Northern British Columbia in Canada. His research focuses on international environmental issues and the relationship between science and policy making.

Elena Nikitina is a senior researcher at the Institute of World Economy and International Relations at the Russian Academy of Sciences. Her research is on domestic implementation and effectiveness of international environmental agreements, particularly in Russia.

Karen L. O'Brien is a senior research fellow at the Center for International Climate and Environmental Research–Oslo (CICERO) at the University of Oslo in Norway. Her research focuses on climate change impacts and adaptations, with an emphasis on vulnerability to climate change.

Edward A. Parson is Associate Professor of Public Policy at Harvard's Kennedy School of Government in the United States. His research examines international environmental policy, institutions, and negotiations.

Vladimir Pisarev is a Senior Research Fellow at the Institute of USA and Canada Studies at the Russian Academy of Sciences. He is interested in sustainable development, environmental security in U.S. policy, and Russian-American relations.

Ruud Pleune is a staff member of Waterpakt, the Dutch water protection organization. His main interest is the relation between values and action in strategies of environmental actors. He was affiliated with Utrecht University.

Miranda A. Schreurs is an Assistant Professor in the Department of Government and Politics at the University of Maryland at College Park in the United States. Her research addresses national policy making in response to global environmental issues, particularly in Japan and Germany.

Peter Simmons is Lecturer in Environmental Risk in the School of Environmental Sciences, University of East Anglia. His research focuses on the sociocultural dimensions of public and institutional responses to technological and environmental risk.

Simon Shackley is a lecturer in environmental management at the School of Management at the University of Manchester Institute of Science and Technology in the United Kingdom. He is interested in the science-policy interface and regional climate change impacts and adaptation.

Heather A. Smith is an Assistant Professor in International Studies at the University of Northern British Columbia in Canada. Her work focuses on Canadian foreign policy and the issue of climate change.

Vassily Sokolov chairs the Department of Canada at the Institute of USA and Canada Studies at the Russian Academy of Sciences and is CIS Academic Director for International Program Leadership for Environment and Development. He is interested in environmental management and economics.

Ferenc L. Tóth is Associate Professor at the Department of Economic Geography at the Budapest University of Economic Sciences in Hungary. His research activities cover economic and policy dimensions of global environmental change, especially climate change.

Jeroen van der Sluijs is a research associate and lecturer at the Department of Science, Technology, and Society of Utrecht University in The Netherlands. His main research area is the management of uncertainties and values in integrated assessment of climate change.

Josee van Eijndhoven is director of the Rathenau Institute, the Dutch national organization for technology assessment. The institute focuses its activities on assessment of information and communication technology, biotechnology, and environmental issues.

Claire Waterton is a lecturer at the Centre for the Study of Environmental Change at Lancaster University in the United Kingdom. Her research interests center around science and the relationship between scientific knowledge and decision making in environmental policy.

Cor W. Worrell is a chemist at the Department of Science, Technology, and Society at the University of Utrecht in The Netherlands. His research focuses on risk assessment and risk communications, particularly on the implementation of Dutch governmental risk policy.

Brian Wynne is Professor and Director of the Centre for the Study of Environmental Change at Lancaster University in the United Kingdom. He is interested in the ways in which scientific knowledge about environmental issues is shaped and understood by scientists, policymakers, and the public.

Author Index

Subject Index